Antimicrobial Resistance in Agriculture and its Consequences

This book offers comprehensive coverage of all manifestations of resistance in combating infectious diseases and explores advances in antimicrobial resistance in agriculture and their applications in the fight against microbes. It discusses and compares biological, biochemical, and structural aspects of resistance and its evolution. This is a comprehensive tool covering all manifestations of antimicrobial resistance and microbial resistance genes. In addition, it also provides a variety of photographs, diagrams, and tables to help illustrate the material. Novel strategies to combat antimicrobial resistance are also described, emphasizing collaborative measures of control. The underlining molecular mechanisms, which depend not only on the microbe but on the specific drug (target) molecule, are highly diverse and are covered in great detail.

Students, researchers, scientists, practitioners, academics, computational biologists, stakeholders, and policymakers can benefit from using *Antimicrobial Resistance in Agriculture and its Consequences* as a resource that addresses microbial biotechnology, microbiology, ethnopharmacology, toxicology, medicinal plant products, and all disciplines related to antimicrobial research.

Features of the book:

- Covers antimicrobial resistance in agriculture with up-to-date research.
- Includes recent references on each plausible antimicrobial resistance in agriculture.
- Details the possible spread of antibiotic resistance bacteria from animals to humans.
- Provides several perspectives in the resistance flux with modern agricultural practices.
- Describes the public health impact of the use of antibiotics in agriculture.
- Presents cutting-edge research on epigenetics, nanotechnology, and emergent antimicrobial technologies.
- Outlines recent laws and regulatory guidelines in the federal agency, responsibility, and authority.

Current Trends in Antimicrobial Research

Series Editor: Arti Gupta

Antimicrobials in Pharmaceutical and Medicinal Research
Edited by Arti Gupta and Ram Prasad

Antimicrobials in Food Science and Technology
Edited by Arti Gupta and Ram Prasad

Antimicrobial Resistance in Agriculture and its Consequences
Edited by Arti Gupta and Ram Prasad

For more info: https://www.routledge.com/Current-Trends-in-Antimicrobial-Research/book-series/CTAR

Antimicrobial Resistance in Agriculture and its Consequences

Edited by Arti Gupta and Ram Prasad

CRC Press
Taylor & Francis Group
Boca Raton London

CRC Press is an imprint of the
Taylor & Francis Group, an **informa** business

Cover image credit: Shutterstock

First edition published 2024
by CRC Press
2385 NW Executive Center Drive, Suite 320, Boca Raton FL 33431

and by CRC Press
4 Park Square, Milton Park, Abingdon, Oxon, OX14 4RN

CRC Press is an imprint of Taylor & Francis Group, LLC

ISBN: 9781032216447 (hbk)
ISBN: 9781032216461 (pbk)
ISBN: 9781003269380 (ebk)

DOI: 10.1201/9781003269380

Typeset in Times
by Apex CoVantage, LLC

Contents

Acknowledgements

First and foremost, we would like to thank God for his never-ending grace, mercy and provision during what ended up being one of the toughest time of our life.

We are indebted to the many people who helped bring this book to light. I wish to thanks series editors Dr. Gustavo Molina and Dr. Vijai Kumar Gupta; Dr. Marc Gutierrez and Dr. Bonita Glanville-Morris, senior production editor Dr. Jyotsna Jangra, and commissioning editor Dr. Renu Upadhyaya, Taylor & Francis/CRC Press, for their generous assistance, constant support, and patience in initializing the series volume. We are very thankful to Vice Chancellor Faizabad University, Ayodhaya, and Mahatma Gandhi Central University, Motihari, for the kind support and constant encouragement. Special thanks are due to well-wishers, family members, and friends.

Dr. Arti Gupta gives special thanks to her exquisite husband, Mr. Saket Agrawal, and her son, Reyansh, who encouraged her to start the work, persevere, and put everything together. Dr. Ram Prasad gives his special regards to his family for their constant, invaluable support and motivation to publish. We both also give special thanks to all faculty colleagues and senior faculties of our team.

Arti Gupta
Shri Avadh Raj Singh Smarak Degree College, Gonda, UP, India

Ram Prasad
Mahatma Gandhi Central University, Motihari, Bihar, India

Preface

Antimicrobial resistance (AMR) is considered the most significant trial to human efforts in controlling communicable diseases and recently caused 10 million deaths annually and is considered to distribute off a massive cumulative US$100 trillion from the economy by 2050 as a consequence of action to contain AMR. Antimicrobial resistance in agriculture is a valuable industrial resource that involves multifaceted topics regarding farm, wild, and friendly animals and how the environment has as an active role in amplification and transmission of resistant genes into the human food chain. Information of phenotypic and genotypic assets of microbiomes associated with antimicrobial resistance, transmission dynamics from different reservoirs, and control measures with diverse therapy, such as phytobiotics and nanomaterials, are provided. The transmission of resistant microbes and genes from animals to humans is a complex process influenced by numerous aspects.

This book explores the evidence-based information on various aspects of AMR in agriculture and their perceived impact on human health, including updated information about all the genera of livestock and pathogenic microbes associated with AMR. The valuable information may be exploited by policymakers and various groups of experts including academic and research personnel engaged in clinical and medical microbiology and industrial microbiology, along with all veterinary and medical practitioners. Researchers, scientists, and healthcare and management practitioners will find this an essential resource on the legal use of antibiotics in animals and humans.

The first chapter by Singh et al. reviews antimicrobial resistance and its causes, effects, and mitigation. Chapter 2 by Al Fattah Amara and his collaborators highlights antibiotic resistance and its facts. In Chapter 3, Patel et al. describe features of antimicrobial resistance in agriculture: mechanistic approaches, types, and future prospects. Extensive studies and application of nanobiotechnology in detection and diagnosis of phytopathogens are discussed by Joji et al. in Chapter 4, and Chapter 5, by Chaudhary and his collaborators, highlights antimicrobial resistance in agrarian distress. In Chapter 6, Dwivedi and Pandit emphasize a host plant resistance approach against insect pest and their sustainable management. In Chapter 7, Dwivedi and Sonawane describe sustainable management practices in pod borer, *Helicoverpa armigera* (Hubner), in chickpea, and in Chapter 8, Sagheer et al. highlight the recent trends in antimicrobial drug resistance and consequences for microbial toxicology research. In Chapter 9, Kumar et al. highlight emerging tools for approaches of artificial intelligence in antimicrobial research. Finally, in Chapter 10, tracking of antimicrobial resistance through metagenomics approaches is addressed by Dangar and Ramani.

Arti Gupta and Ram Prasad

Preface

Editor Biographies

Arti Gupta, Ph.D., obtained her doctorate from Mahatma Jyotiba Phule Rohilkhand University, Bareilly, India, in 2010 in animal science. Dr. Gupta's current research interests include animal biotechnology, molecular plant biotechnology, molecular animal biotechnology, bioprocess technology, and microbiology. Since July 2014, Dr. Gupta has been employed at Shri Avadh Raj Singh Smarak Degree College, Gonda, Uttar Pradesh, India, and has been engaged in editorial work with Springer Nature. Dr. Gupta has 10 years of teaching and research experience in plant and animal biotechnology and microbial biotechnology; in 2023–24 she served as external examiner for central evaluation at MJP Rohilkhand University, Bareilly.

Dr. Gupta has been awarded University Topper (Gold Medal), M.Sc. (Biotech.) by Ch.C.S. University, Meerut; Young Scientist Award (Gold Medal) by the Zoological Society of India, Lucknow; Best Poster Presenter by *Asian Journal of Experimental Science*, Jaipur; Best Poster Presenter by International Consortium of Contemporary Biologists (ICCB) and Madhawai-Shyam Educational Trust, Ranchi; Fellowship Award by International Consortium of Contemporary Biologists (FICCB) and Madhawi Shayam Educational Trust (FMSET); and Dr. V.P. Agarwal (Gold Medal) by D.A.V. (P.G.) College, Muzaffarnagar. She has published 1 monograph, 7 book chapters, around 50 National and International research papers, and 9 edited books; she has presented 28 abstracts in national and international symposia/seminars/conferences/workshops. Dr. Gupta has lifetime membership of the Indian Science Congress Association, Biotech Research Society of India, Zoological Society of India, and International Consortium of Contemporary Biologists. She has served as a member of several editorial boards. Previously, Dr. Gupta was employed as Teaching Personnel, G.B. Pantnagar University, Pantnagar, Uttarakhand, India, and Visiting Academic Researcher, Biorefining and Advanced Material Research Center (BAMRC), Scotland's Rural College (SRUC), University of Edinburgh, Scotland, United Kingdom.

Ram Prasad, Ph.D., is associated with the Department of Botany, Mahatma Gandhi Central University, Motihari, Bihar, India. His research interests include applied and environmental microbiology, plant-microbe interactions, sustainable agriculture, and nanobiotechnology. Dr. Prasad has more than 275 publications (total citations 13,840 with an h-index 59, i10-index 189) to his credit, including research papers, review articles, and book chapters, as well as seven patents issued or pending, and he has edited or authored several books. Dr. Prasad has 14 years of teaching experience and has been awarded the Young Scientist Award and Prof. J. S. Datta Munshi Gold Medal by the International Society for Ecological Communications; Fellowship of Biotechnology Research Society of India; Fellow of Agricultural Technology Development Society, India; Fellow of the Society for Applied Biotechnology; the American Cancer Society UICC International Fellowship for Beginning Investigators, United States; Outstanding Scientist Award in the field of Microbiology; and BRICPL Science Investigator Award and Research Excellence Award. He served as a member of several editorial boards. Previously, Dr. Prasad served as Assistant Professor at Amity University Uttar Pradesh, India; Visiting Assistant Professor, Whiting School of Engineering, Department of Mechanical Engineering at Johns Hopkins University, Baltimore, United States; and Research Associate Professor at the School of Environmental Science and Engineering, Sun Yat-Sen University, Guangzhou, China.

Editor Biographies

Contributors

Mohamed Morsi Mohamed Ahmed
Nucleic Acids Research Department
Genetic Engineering and Biotechnology Research Institute (GEBRI)
City of Scientific Research and Technological Applications (SRTA-City), Alexandria, Egypt

Amro Abd Al Fattah Amara
Protein Research Department
Genetic Engineering and Biotechnology Research Institute (GEBRI)
City of Scientific Research and Technological Applications (SRTA-City), Universities and Research Center District, New Borg El-Arab, P.O. Box 21934, Alexandria, Egypt

Neha Batra
Discipline of Biotechnology
IIS (Deemed to be University)
Jaipur, Rajasthan, India

Nikhilesh Bhadauriya
Bioinformatics Laboratory
Biotech Park, Jankipurum, Sector G, Kursiroad, Lucknow, Uttar Pradesh-226021, India

Neelam Chaudhary
Lab of Microbial Technology and Plant Pathology
Department of Botany, Dr. Harisingh Gour Vishwavidyalaya
Sagar, Madhya Pradesh, 470003, India

Kruti Dangar
Department of Biochemistry
Saurashtra University
Rajkot-360005, Gujarat, India

Sunilkumar Anilkumar Dwivedi
Department of Entomology
School of Agriculture, Lovely Professional University
Phagwara Punjab, India

Nawal Abd El-Baky
Protein Research Department
Genetic Engineering and Biotechnology Research Institute (GEBRI)
City of Scientific Research and Technological Applications (SRTA-City)
Universities and Research Center District, New Borg El-Arab, P.O. Box 21934, Alexandria, Egypt

Nazia Iqbal
Department of Applied Chemistry
University Polytechnic, Integral University
Lucknow-226026, India

Sohani Singh Jain
Program Centre, Corporate Services, Department of Jobs, Precincts and Regions
Melbourne, Australia

Riya Rachel Joji
Department of Plant Pathology
School of Agriculture, Lovely Professional University
Phagwara-144411, India

Ravindra Kumar
Department of Developmental Biology
Washington University School of Medicine
St. Louis, Missouri

Sakshi Malik
Department of Chemistry
Maharishi Markandeshwar (Deemed to be University)
Mullana, Haryana-133207, India

Malik Nasibullah
Department of Chemistry
Integral University
Lucknow-226026, India

Trupti Rajesh Pandit
Department of Entomology
School of Agriculture, Lovely Professional University
Phagwara Punjab, India
Sakshi Patel
Department of Zoology
University of Rajasthan
Jaipur, Rajasthan, India

Malavika Ram AK
Department of Plant Pathology
School of Agriculture, Lovely Professional University
Phagwara-144411, India

Ketaki Ramani
Department of Microbiology
School of Science, RK University
Rajkot-360020, Gujarat, India

Meenakshi Rana
Department of Plant Pathology
School of Agriculture, Lovely Professional University
Phagwara-144411, India

Rajrani Ruhel
Plant and Microbial Biology
NC State University
Raleigh, North Carolina

Rida Sagheer
Department of Chemistry
Integral University, Lucknow-226026, India
AND
Department of Applied Chemistry
University Polytechnic, Integral University
Lucknow-226026, India

Mounir Mohamed Salem-Bekhit
Department of Pharmaceutics
College of Pharmacy, King Saud University, Riyadh 11451, Saudi Arabia
AND
Kayyali Chair for Pharmaceutical Industry, Department of Pharmaceutics
College of Pharmacy, King Saud University, Riyadh 11451, Saudi Arabia
AND
Department of Microbiology and Immunology, Faculty of Pharmacy
Al-Azhar University, Cairo 11884, Egypt

Ameeta Sharma
Department of Zoology
University of Rajasthan
Jaipur, Rajasthan, India

Chandan Singh
Lab of Microbial Technology and Plant Pathology, Department of Botany
Dr. Harisingh Gour Vishwavidyalaya
Sagar, Madhya Pradesh, 470003, India.

Kuldeep Singh
Department of Applied Chemistry
Amity University Madhya Pradesh, Gwalior
Madhya Pradesh-474005, India

Raman Singh
WTL Clean and Renewable Energy Private Limited, New Delhi-110003, India
AND
Department of Chemistry
Maharishi Markandeshwar (Deemed to be University)
Mullana, Haryana-133207, India

Vishal Kailash Sonawane
Department of Entomology
School of Agriculture, Lovely Professional University
Phagwara Punjab, India

Ajay Tomar
Amity Institute of Organic Agriculture
Noida-201313, India

Pravin Chandra Trivedi
Jai Narayan Vyas University
Jodhpur, Rajasthan, India

Deepak Vyas
Lab of Microbial Technology and Plant Pathology, Department of Botany
Dr. Harisingh Gour Vishwavidyalaya
Sagar, Madhya Pradesh, 470003, India

Khairy Mohamed Abdallah Zoheir
Cell Biology Department
Biotechnology Research Institute, National Research Center
33 Bohouth St., 12622, Dokki, Cairo, Egypt

1 Antimicrobial Resistance
Causes, Effects, and Mitigation

Raman Singh, Sakshi Malik, Nikhilesh Bhadauriya, and Kuldeep Singh

1.1 INTRODUCTION

Antimicrobials were designed to limit or prevent the growth of microbes that reside in a living body. Microbes could be bacteria, viruses, fungi, algae, or protozoa. Antimicrobials acted as a boon to living beings in protecting themselves from a wide range of infections caused by microbes (Kaur, Singh, Ahlawat, et al. 2020). Antimicrobial resistance (AMR) is a global crisis created by humans that cannot be treated with any known antimicrobial agent. The widespread and intense applications of antimicrobials by humans make microbes more adaptive. The continuous use of antimicrobials places selective pressure on microbes, resulting in resistance against these antimicrobials.

The introduction of penicillin during World War II was effective against a wide range of bacteria. Due to its excessive use, half of the *Streptococcus aureus* strains became resistant, and in the 1990s, 90% of the strains had become resistant (Jones 1996).

Several pathogens are resistant to penicillin and its derivatives. Furthermore, in the case of malaria, *Plasmodium* sp., the causative organism of malaria, has now become resistant to all known antimalarial medicines (Michael, Dominey-Howes, and Labbate 2014).

1.1.1 History

Salvarsan was the world's first antimicrobial agent, developed as a treatment for Syphilis (Saga and Yamaguchi 2009). Salvarsan was synthesized by Ehrlich in 1910. In 1928, Fleming accidentally discovered penicillin. The observation that the fungus of the *Penicillium* genus in culture dishes inhibits the growth of *Staphylococcus aureus* led to the conclusion that a microorganism would produce some substances that could inhibit the growth of other microorganisms. It saved the lives of many wounded soldiers during World War II (Saga and Yamaguchi 2009).

1.1.2 Natural and Acquired Resistance

Natural resistance is the one that the new generation possesses from its parents through heredity. It can be of two types; one is intrinsic, and the second is induced (C Reygaert 2018). Intrinsic natural resistance is common for a particular bacterial species and can always be expressed (Cox and Wright 2013). Natural resistance can be expressed after exposure to antimicrobials. The most common example is the reduced permeability of the outer membrane, especially in gram negative bacteria, and the natural working of the efflux pump (Cox and Wright 2013; Fajardo et al. 2008). A multidrug efflux pump is an example of an induced natural mechanism.

1.1.3 Resistant and Persistent Bacteria

If a bacterium can survive even after the action of an antimicrobial agent, it is called resistant bacteria. Daughter cells of resistant bacteria will also be resistant. However, persistent bacterial cells are in their dormant growth phase during the action of antimicrobials, which do not affect cells that

DOI: 10.1201/9781003269380-1

are not actively growing or multiplying. Antimicrobial resistant bacteria have the ability to produce offspring resistant to antimicrobials. Persistent cells, on the other hand, produce normal daughter cells (C Reygaert 2018; Wood, Knabel, and Kwan 2013; Keren et al. 2004).

1.1.4 Access vs Excess

The number of deaths is increasing worldwide, not only due to the excessive use of antimicrobials but also due to the lack of access to antimicrobials for the public. Untreated pneumonia and sepsis kill one million children each year. Therefore, a balance is needed between the use of antimicrobials and access to antimicrobials. Improving global access to antimicrobials again creates AMR crises, and reducing the use of antimicrobials is necessary; however, it can create hurdles in the development of new antimicrobial drugs (Roope et al. 2019).

1.2 THE COST OF ANTIMICROBIAL RESISTANCE

Antimicrobial resistance has become the most serious threat to the public health system worldwide, but most people are still unaware of it. Antimicrobials already discovered can no longer be effective for most infectious diseases due to their overuse and misuse. Different levels of cost associated with antimicrobial resistance have been shown in Figure 1.1. According to the CDC (Centers for Diseases Control and Prevention), nearly two million people in the United States suffer from antibacterial resistant diseases, resulting in 23,000 deaths every year (Dadgostar 2019).

1.2.1 Patient Perspective

Infection with resistant bacteria is more dangerous than infection with nonresistant bacteria because it increases the likelihood of a serious condition and triples the risk of death. The number of cases of HIV and malaria is on the rise in developing resistance to the medicine. Due to drug-resistant diseases, approximately 700,000 patients have lost their lives around the world (Dadgostar 2019).

1.2.2 Healthcare Perspective

AMR is also the most significant challenge faced by hospitals. Patients with resistant bacterial infections may need hospitalization longer than patients with normal bacterial infections. The hospital system may also need to arrange more intensive care units for patients. Surgeries cannot be performed if the infection spreads in the hospital because such patients are more exposed to infections. It can lead to serious health problems, which can even become the reason for the closure of hospitals (Dadgostar 2019).

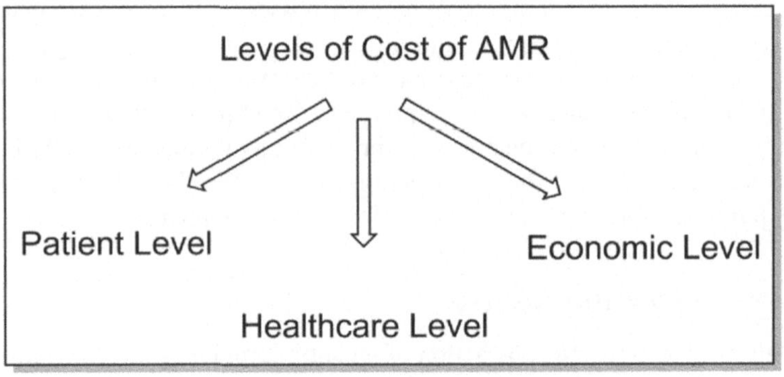

FIGURE 1.1 Levels of cost of AMR.

1.2.3 Economic Perspective

If AMR is not controlled in the coming years, we would see a significant difference between developing and developed countries. Developing countries might go below the poverty line. The economic status of a country depends on its labor. The spread of infection among workers can cause high levels of illness and death, reducing the workforce and ultimately affecting the economy of the country (Dadgostar 2019).

1.3 CAUSES OF ANTIMICROBIAL RESISTANCE

The development of antimicrobial resistance is a complex phenomenon driven by numerous factors (Figure 1.2). Bacteria can naturally develop antimicrobial resistance through mutations and by acquiring resistance genes. However, human activities like overusing antibiotics in healthcare and livestock production have accelerated the pace at which bacteria become resistant.

1.3.1 Microbial Population

Microbes have the potential to multiply and expand in a short period of time. In addition, they can easily and effectively adapt to environmental change. Heredity plays a vital role in microbial life, by which they can quickly transfer their antimicrobial resistant genes to the next generation. Thus, the microbial community adapted to environmental change both quickly and effectively. Another source for the microbial community to become genetically strong is mutation. For an organism, a mutation is a change in the genetic sequence that results in the formation of new genes that improve its ability to survive in its environment (Michael, Dominey-Howes, and Labbate 2014).

The new antimicrobial treatment will be dominated by microbes that can survive and reproduce under antimicrobial treatment. Because of this, they develop resistance to antimicrobial treatment and pass this resistance on to their offspring through their genes (Michael, Dominey-Howes, and Labbate 2014).

1.3.2 Human Population

Every country has seen a steady increase in the number of people over the years. Humans are by far the most important factor in the spread of disease. Even more alarming is that most people are

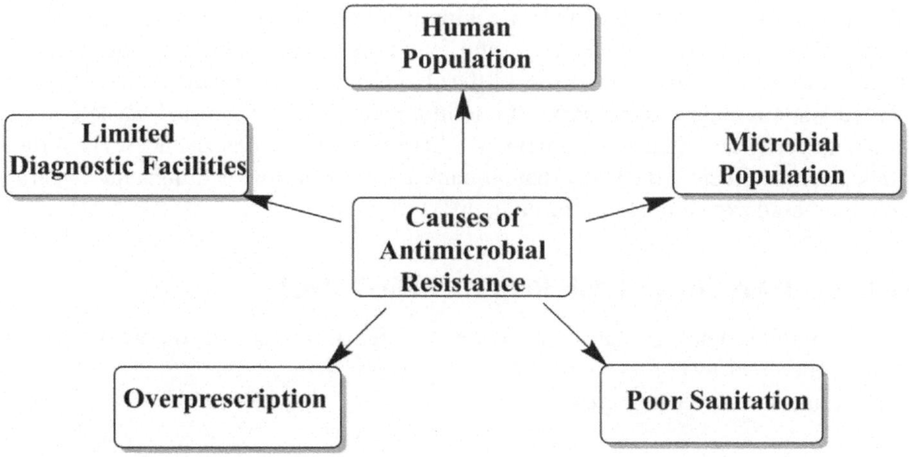

FIGURE 1.2 Causes of antimicrobial resistance.

moving to urban areas, where they are more likely to be infected by germs. In 2007, more than half of the world's population lived in urban areas. People can travel from one end of the earth to another in one or two days, so they can easily transmit infections to other countries' populations (Michael, Dominey-Howes, and Labbate 2014).

1.3.3 Limited Diagnostic Facilities

Regarding diagnostic facilities, developed countries do not lack high-tech facilities. They can quickly identify the types of microbes and provide appropriate treatment to patients, thus restricting the resistance of the microbes. But developing countries suffer from a lack of advanced technology facilities and cannot identify most infections (Bebell and Muiru 2014).

1.3.4 Over-the-Counter Drugs

When antimicrobials were introduced, people thought they could be treated quickly with no side effects. These beliefs of people lead to the overuse of antimicrobials, and consequently, intense selective pressure on microbes makes them resistant to antimicrobials (Michael, Dominey-Howes, and Labbate 2014).

Today, clinicians do not feel the need to identify the type of pathogen, which often takes days or weeks. They prescribe a list of antimicrobials, stating that one of these would work against that pathogen. But this action can exert strong and polarized selective pressure on the microbial community and make it resistant to all prescribed antimicrobials (Michael, Dominey-Howes, and Labbate 2014).

Second, people visit general practitioners to save money on hospitals. Based on past experiences, a general practitioner tests any antimicrobials related to the patient's symptoms. In this way, repeated visits to physicians would benefit the patients. It is also a belief that a high dose over a long period will completely ruin the bacterial community in the body of the organism, which consequently will show resistance to further treatment (Michael, Dominey-Howes, and Labbate 2014).

1.3.5 Poor Sanitation

Urban lakes can be called an ideal environment for the aggregation of antimicrobial resistance genes (ARGs) affected by human activities and the surrounding environment (Czekalski et al. 2015). A minimum number of investigations are carried out for the occurrence, composition, and diversities of ARGs in lakes (Chen et al. 2019; Yang et al. 2018). Experiments have been carried out in two urban lakes (Lake Tai and Lake Baiyang) in China that resulted in the extraction of some ARGs, such as types of MCR and carbapenemases, and found resistome, mobilome, and virulome (Chen et al. 2021). Antimicrobial resistance is one of the six growing environmental challenges identified by the United Nations Environment Programme in a recent study. If animal waste and municipal sewage are not appropriately treated, humans will have to face the consequences. Bacteria containing resistance genes will enter the water that humans use for drinking, bathing, and irrigating crops, resulting in increased exposure (Marathe et al. 2017).

1.4 MECHANISMS OF ANTIMICROBIAL RESISTANCE

To combat the antimicrobial resistance crisis, we must reach the roots of the working of drugs on microbes and microbial responses against drugs to ensure their survival. So, we can classify the overall mechanism into two categories:

1. Mechanism of antimicrobial action on microbes.
2. Mechanism of the response of microbes to applied antimicrobials.

1.4.1 Mechanism of Antimicrobial Action on Microbes

1 Inhibit cell wall synthesis.
2 Depolarize the cell membrane.
3 Inhibit protein synthesis.
4 Inhibit nucleic acid synthesis.
5 Inhibit metabolic pathways.

Antimicrobials that inhibit the synthesis of bacteria cell walls are β-lactams and glycopeptides. Lipopeptides depolarize the cell membrane of bacteria. In case of inhibition of protein synthesis, a few antimicrobials bind to 30S ribosomal units such as aminoglycosides and tetracyclines. Antimicrobials like chloramphenicol, lincosamides, macrolides, etc. bind to the 50S ribosomal unit to inhibit protein synthesis in bacteria. Quinolones inhibit nucleic acid synthesis. Sulfonamides and trimethoprim are useful in inhibiting the metabolic pathways of bacteria (C Reygaert 2018).

A large number of antimicrobials have been discovered to kill a variety of microbes, and, although initially effective against microbes, they became ineffective because of their excessive use. Such antimicrobials can be seen in Table 1.1, along with their year of discovery, the year of first observed resistance, and the mechanism.

TABLE 1.1
Antimicrobials with Their Year of Discovery, Year of First Observed Resistance, and Mechanism

Entry	Antimicrobial	Year of discovery	Resistance mechanism
1	Penicillin	1928 (Tan and Tatsumura 2015)	β-lactamases (C Reygaert 2018)
2	Streptomycin	1944 (Woodruff 2014)	Mutation in the rpsL gene (Springer et al. 2001)
3	Bacitracin	1943 (Bernard et al. 2007)	The bac A gene encodes for undecaprenyl pyrophosphate phosphatase (UppP) (Charlebois et al. 2012)
4	Chloramphenicol	1947 (Wiest, Cochran, and Tecklenburg 2012)	Enzymatic inactivation (chloramphenicol phosphotransferases) by acetylation, target site mutation, decreased outer membrane permeability, efflux pump (Fernández et dl. 2012)
5	Polymyxin	1947 (Biswas et al. 2012)	Alteration of LPS (lipopolysaccharide), efflux pump, capsule formation (Olaitan, Morand, and Rolain 2014)
6	Chortetracycline	1948 (Martens and Demain 2017)	Efflux and ribosomal protection (Speer, Shoemaker, and Salyers 1992)
7	Cephalosporin	1948 (Sykes 2000)	Altered PBPs (penicillin-binding proteins) (Moosdeen 1997)
8	Nystatin	1950 (Gupte, Kulkarni, and Ganguli 2002)	Transport alterations, target alterations, utilization of compensatory pathways, and presence of complex multicellular structures (Sanglard, Coste, and Ferrari 2009)
9	Erythromycin	1952 (Farzam, Nessel, and Quick 2022)	Ribosomal methylation (Leclercq and Courvalin 2002)
10	Spiramycin	1954 (Sutherland 1962)	Inducible ribosomal resistance, target modification (Pernodet et al. 1993)
11	Kanamycin	1957 (Umezawa 2006)	Polyphenylalanine synthesis in cell free system (Hotta et al. 1981)
12	Vancomycin	1956 (Levine 2006)	Altered peptidoglycan terminus (Alby and Miller 2018)
13	Gentamicin	1963 (Coutinho et al. 2018)	1 Aminoglycoside 6-N-acetyl transferase (CDC 2019) 2 Gentamicin transferase (Dowding 1977)
14	Lincomycin	1962 (Bal 2021; Papich 2012)	Methylation of 23S rRNA (Wang et al. 2017)
15	Fusidic acid	1960 (Fernandes 2016)	Mutation in fusA gene (Chen et al. 2011)

(Continued)

TABLE 1.1 ***(Continued)***
Antimicrobials with Their Year of Discovery, Year of First Observed Resistance, and Mechanism

Entry	Antimicrobial	Year of discovery	Resistance mechanism
16	Griseofulvin	1939 (Tsunematsu et al. 2020; Waugh 2007)	Absence of a prolonged energy-dependent transport system (Cowen et al. 2015)
17	Abacavir	1998 (Martin and Kroetz 2013)	M184V mutation (Melroy and Nair 2005)
18	Zidovudine	1987 (Cochrane 2000)	Multiple mutation in HIV reverse transcriptase (Arion and Parniak 1999)
19	Indinavir	1994 (Lyle 2006)	Alteration in enzyme catalysis and reshaping of the active site (Tredwell and Gouverneur 2012)
20	Zanamivir	1999 (Ison 2010)	Mutation in viral neuraminidase or hemagglutinin or both (Ison 2010)
21	Caspofungin	2001 (CDC 2019)	Caspofungin-resistant *Candida*
22	Daptomycin	2003 (CDC 2019)	Daptomycin-resistant methicillin-resistant *Staphylococcus aureus*
23	Ceftaroline	2010 (CDC 2019)	Mutation within PBP2A, PBP4, thus lowering the affinity of PBP with antibiotic
24	Ceftazidime-avibactam	2015 (CDC 2019)	Ceftazidime-avibactam-resistant Klebsiella pneumoniae carbapenemase (KPC)-producing *Klebsiella pneumoniae*

1.4.2 Mechanism of Responses of Microbes against Applied Antimicrobials

1.4.2.1 Limiting Drug Uptake

Different microbes have different properties to protect themselves from the effect of the drug. Depending on the structural and functional properties, a drug may be limited to one or more microbes. Grammeme-negative bacteria are resistant to many antimicrobial agents due to the different functions of the LPS (lipopolysaccharide) layer (Blair, Richmond, and Piddock 2014). The size of the cell membrane porin can also be responsible for the limitations of drug uptake. The size of the porin of Gramme-negative bacteria allows access to hydrophilic drugs (Gill et al. 1998; Blair, Richmond, and Piddock 2014). Hydrophilic drugs do not affect mycobacteria due to the presence of a high content of lipids in the cell wall. However, these bacteria can be easily treated with hydrophobic drugs such as fluoroquinolones (Kumar and Schweizer 2005; Lambert 2002). Drugs like β-lactam and glycopeptides attack the cell wall of bacteria. Bacteria such as *Mycoplasma* that do not possess cell walls do not affect such drugs (Bebear and Pereyre 2005). Some organisms, such as *Enterobacteriaceae*, reduce the number of porins to become resistant to carbapenems (Cornaglia et al. 1996; Chow and Shlaes 1991). Another way bacteria protect themselves is through the production of biofilms. The biofilm consists of polysaccharides, proteins, and DNA from resistant bacteria, making it difficult for the antimicrobial agent to attack (Mah 2012).

1.4.2.2 Altered Target Enzyme or Receptor

When there is a change in the order of the amino acids that make up the active site of the target enzyme, the effectiveness of the antimicrobial drug's ability to bind to the enzyme decreases. The enzyme with the changed sequence of amino acids displays characteristics different from those of the normal enzyme. An enzyme called dihydrofolate reductase that has been altered by adding a single amino acid is resistant to the antibiotic trimethoprim (Spratt 1994; Then and Hermann 1981). Another example is Gramme-positive bacteria that become resistant against β-lactam by altering the structure/number of PBP (penicillin-binding protein) (Beceiro, Tomás, and Bou 2013). *Staphylococcus aureus* (MRSA, or methicillin–resistant *Staphylococcus aureus*) and enterococci have become resistant to vancomycin. Due to the acquisition of *van* genes, the structure of the peptidoglycan precursor has changed, reducing the binding ability of vancomycin6,69. Drugs targeting ribosomal units become ineffective because of ribosomal mutation or methylation of ribosomal subunits in bacteria. Such changes in ribosomal subunits do not allow antibiotics to attack (Kumar, Mukherjee, and Varela 2013; Roberts 2003; 2004; C Reygaert 2018). Enzymes gyrase and topoisomerase IV play an important role in DNA twisting and unwinding in nucleic acid synthesis (Levine, Hiasa, and Marians 1998). Modifying these two drugs makes bacteria resistant to fluoroquinolones that target nucleic acid synthesis due to their ability to bond to the modified structure of enzymes (Hawkey 2003; Sharma et al. 2017; Redgrave et al. 2014).

1.4.2.3 Drug Inactivation

A drug can be inactivated by hydrolyzing, or transfer, of a chemical group to the drug (Blair et al. 2015). The first antibiotic discovered, penicillin, became ineffective due to hydrolysis of the β-lactam ring by β-lactamase enzyme (Wilke, Lovering, and Strynadka 2005). Aminoglycoside antibiotics inhibit protein synthesis by binding to the ribosomal RNA of the organism (e.g., kanamycins and neomycins). Bacteria acquire such enzymes, which perform the functions of acetylation, adenylation, and phosphorylation leading to the modification of the hydroxyl and amino group of these antibiotics and hence blocking the binding of antibiotics to ribosomal RNA (Mingeot-Leclercq, Glupczynski, and Tulkens 1999). The DNA-cleaving antitumor drug bleomycin gets inactivated by bleomycin hydrolase, which hydrolyses the carboxamide group of antibiotics (Galm et al. 2005).

1.4.2.4 Overproduction of Target Enzyme

Inducing additional copies of the gene that encodes the enzyme leads to overproduction of the enzyme. For example, the increased number of genes encoding the enzyme dihydrofolate reductase makes cells resistant to methotrexate (Alt et al. 1978).

1.4.2.5 Drug Efflux

Efflux is a bacteria phenomenon that excludes toxic substances such as drugs or chemicals outside their body to keep themselves regulated (Ryan et al. 2001). The efflux mechanism can be of two types: multidrug and drug specific. These mechanisms are different in origin and coding. Chromosomes encode a multidrug efflux mechanism, and due to mutation in regulatory genes, they possess intrinsic origin, while the drug-specific efflux mechanism is encoded by plasmids and bacteriophage elements and possesses acquired origin. These mechanisms can also determine the resistance (intrinsic or acquired) to the antimicrobials (Poole 2007). Tetracycline resistance is an example of efflux-mediated resistance because tetracycline is exported by the Tet efflux pump using proton exchange as a source of energy (Munita and Arias 2016). The following Table 1.2 shows five major families of efflux pumps, their source of energy, functions, and examples.

1.5 OBSTACLES TO THE DEVELOPMENT OF NEW ANTIMICROBIALS

As antimicrobial resistance is growing alarmingly, developing new antimicrobials is highly demanding. But this process is expensive and risky too, and it is not predictable that the drug will be successful on the market. If it becomes successful, it is challenging to conserve this effective antimicrobial because microbes can also become resistant. All these things can push pharmaceutical companies to a loss (Roope et al. 2019).

1.6 ANTIMICROBIAL RESISTANCE MITIGATING TECHNOLOGIES/RESPONSES TO ANTIMICROBIAL RESISTANCE OR CONTROL STRATEGIES

Developing countries have poor sanitation conditions. Poor hygiene conditions worsen the situation, and the threat of antimicrobial resistance in these countries is extremely high. Researchers have

TABLE 1.2
Different Types of Drug Efflux

Entry	Efflux type	Full form	Source of energy	Function	Example
1	ABC	ATP-binding cassette (Blair, Richmond, and Piddock 2014)	ATP hydrolysis (Kumar and Schweizer 2005)	Transport sugars, amino acids, ions, drugs, polysaccharides, and proteins (Lubelski, Konings, and Driessen 2007)	*Lactococcus lactis*, *Haloferaxvolcaniietc* (Putman, van Veen, and Konings 2000)
2	MATE	Multidrug and toxic compound extrusion (Blair, Richmond, and Piddock 2014)	Na+ gradient (Kuroda and Tsuchiya 2009)	Cationic dyes and fluoroquinolone drugs (Kuroda and Tsuchiya 2009)	*Vibrio parahaemolyticus*, *Haemophilus influenza*, etc. (Piddock 2006)
3	SMR	Small multidrug resistance (Blair, Richmond, and Piddock 2014)	Proton motive force (Bay, Rommens, and Turner 2008)	Lipophilic cations (Bay, Rommens, and Turner 2008)	*S. aureus*, *Halobacterium salinarium*, etc. (Bay, Rommens, and Turner 2008)
4	MFS	Major facilitator superfamily (Blair, Richmond, and Piddock 2014)	Solute\cation symport or antiport (Kumar, Mukherjee, and Varela 2013)	Transport of anions, drugs, metabolites, and sugars (Kumar and Schweizer 2005)	*Acinetobacter baumannii*, *Bacillus subtilis*, etc. (Kumar, Mukherjee, and Varela 2013)
5	RND	Resistance-nodulation-cell division (Blair, Richmond, and Piddock 2014)	Substrate/H+ antiport (Kumar and Schweizer 2005)	Bile salts, fatty acids, dyes, and steroid hormones (Blair and Piddock 2009)	*Pseudomonas aeruginosa*, *Stenotrophomonas maltophiliaetc* (Alvarez-Ortega, Olivares, and Martínez 2013)

shown their concern for antimicrobial resistance since the introduction of penicillin and reported resistance in 1940 (Michael, Dominey-Howes, and Labbate 2014). In May 2015, the World Health Organization endorsed a global plan to combat antimicrobial resistance, including antibiotic resistance (Kaur, Singh, Ahlawat, et al. 2020). New antimicrobials are always needed, either new molecules with unique mechanisms or new molecules with resistance to AMR (Kaur et al. 2019; Kaur, Singh, Kumar, et al. 2020; Singh, Kaur, and Singh 2020; Southgate 1994).

1.6.1 Role of Vaccines in AMR

Vaccines can reduce the need for antimicrobial medication because they alleviate the symptoms that are caused by the antimicrobial resistance gene. Vaccines that have already been developed can protect against infections caused by antimicrobial resistant pathogens such as *Streptococcus pneumoniae* and *Haemophilus influenzae*. The development of cutting-edge technology has made it feasible to manufacture vaccines in significantly higher volumes (Baker et al. 2018).

1. **Recombinant DNA technology** enabled the development of the hepatitis B vaccine in the late 1970s. It provides an innovative idea and platform to develop and produce vaccines (Baker et al. 2018).
2. **Glycoconjugate vaccine** consists of bacterial polysaccharides covalently linked to a protein carrier. Vaccines against *H. influnzae* type B, *Pneumococcus*, and *Meningococcus* were developed through this technology. These vaccines have protected millions of children's lives during the past three decades (Baker et al. 2018).

3 **Reverse vaccinology** is a technology to rearrange the genome of a pathogen using a computer setup and an experimental screening program. The four component *Meningococcus* B vaccine (4CMenB) has also been developed using reverse vaccinology, proving that it is efficient in the treatment of infections, including antimicrobial resistance (Baker et al. 2018).
4 **Structural vaccine development** is the method of vaccine development that determines the structure of antigen and antibody-antigen complexes. This is being used to improve the treatment of HIV and to invent a vaccine against influenza (Baker et al. 2018).
5 The **vaccine adjuvant** is an essential substance added to a vaccine to enhance patient immunity, particularly stimulating innate immune system cells (Agrawal, Owais, and Singh 2013). A vaccine named 4C-Staph formed with a new adjuvant dependent on TLR7 (T7-alum) against *Staphylococcus aureus* has the ability to protect mice from death (Mancini et al. 2016). A well-known example is the Adjuvant AS01, which is recently licensed for vaccines against malaria and new herpes zoster (Otieno et al. 2016; Cunningham et al. 2016). It can be seen that vaccines added with adjuvants will contribute to escaping the trap of AMR in the future.

1.6.2 Antibiotic Resistance Breakers

Antibacterial resistance is continuously rising, and the discovery of new antibacterials is highly needed, but it could take decades. Antibiotic resistance breakers (ARBs) came as a co-administered solution with failed antibiotics and restored antibacterial activity (Figure 1.3) (Bakkali et al. 2008). ARBs are also known to increase the host defense mechanism and are broadly classified into three categories: modifying enzyme inhibitors, efflux pump inhibitors, and membrane permeabilizers (Laws, Shaaban, and Rahman 2019).

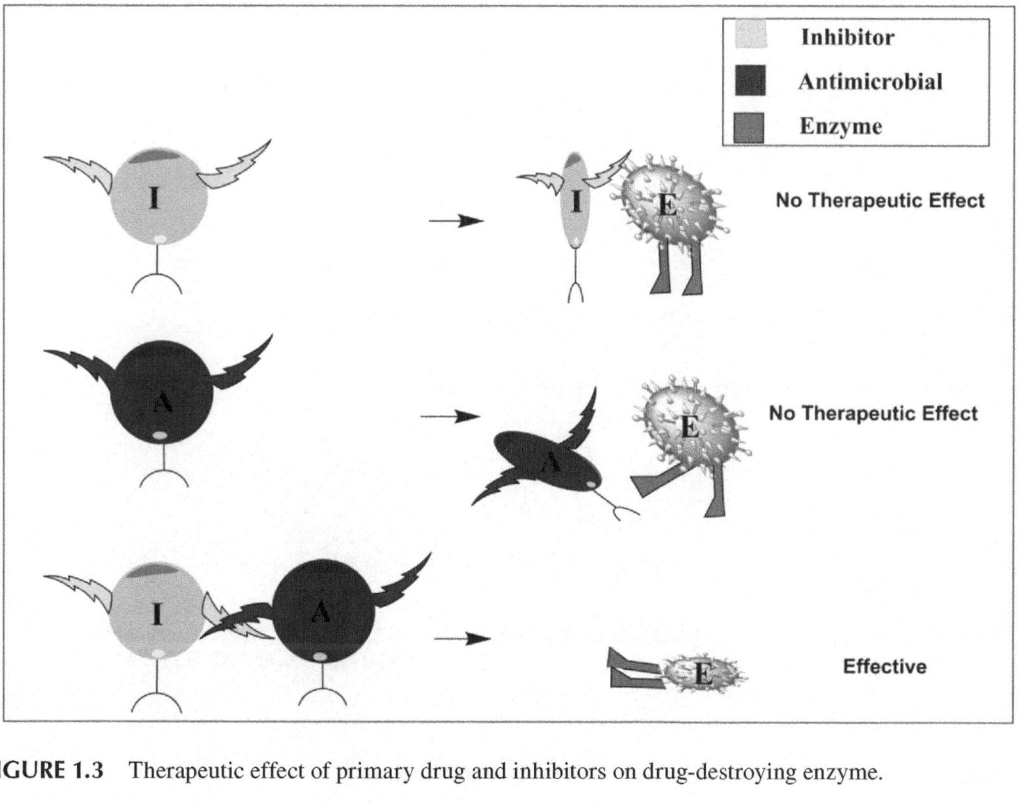

FIGURE 1.3 Therapeutic effect of primary drug and inhibitors on drug-destroying enzyme.

1.6.2.1 Modifying Enzyme Inhibitors

Antibiotic resistance mediated by bacterial enzymes is the biggest challenge, as an enzyme can destroy a whole class of antibiotics with a similar mechanism. There are two possible approaches to combating antimicrobial resistance. The first approach is to alter the structure of the β-lactam, i.e., β-lactamase-stable antibiotics stable to β-lactamase or insensitive to hydrolysis by the β-lactamase, maintaining its potency as an antibiotic. The other approach uses a β-lactamase inhibitor with classical antibiotics such as penicillin. In this way, β-lactamase inhibitors would mimic the β-lactam ring structure to fit the catalytic center of the enzyme. When it binds to the enzyme, it works either way. It may react to the enzyme and form a catalytically inactive derivative, or it may only occupy the catalytic site and block access to the substrate molecule. Therefore, no effective inhibitor has been developed, so there is a need to develop new β-lactamase inhibitors with different modes of action. Modifying enzyme inhibitors has no therapeutic effect if administered alone. It only obstructs the drug-destroying enzyme, but it shows the therapeutic effect when administered with an antimicrobial drug, as shown in Figure 1.3. β-lactamases and aminoglycoside-modifying enzymes are responsible for the enzyme-mediated antibacterial resistance (Bush and Miller 1998). Therefore, we can broadly classify enzyme inhibitors modifying enzymes into two classes: β-lactamase inhibitors (Figure 1.7) and aminoglycoside-modifying enzyme inhibitors (Figures 1.4–1.6).

1.6.2.2 Efflux Pump Inhibitors

The efflux pump of bacteria throws out toxic substances, including all classes of antibiotics, from the bacterial cell to the external environment. Thus, the efflux pump makes bacteria resistant to

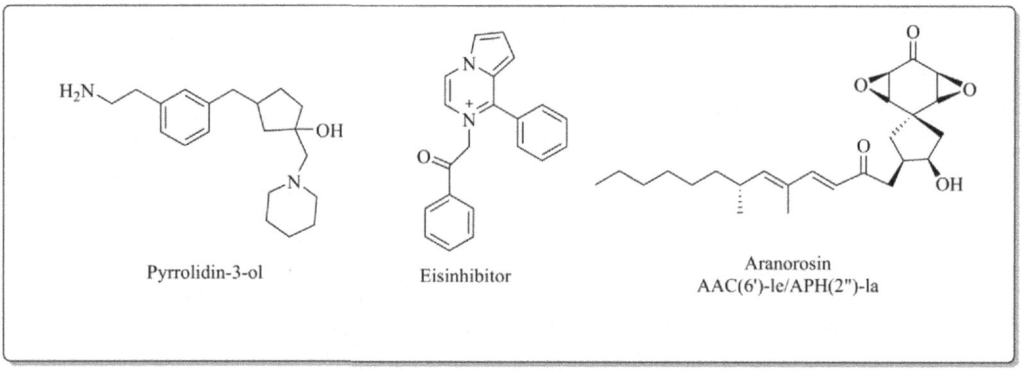

FIGURE 1.4 Inhibitors of aminoglycoside acetyltransferases (AACs).

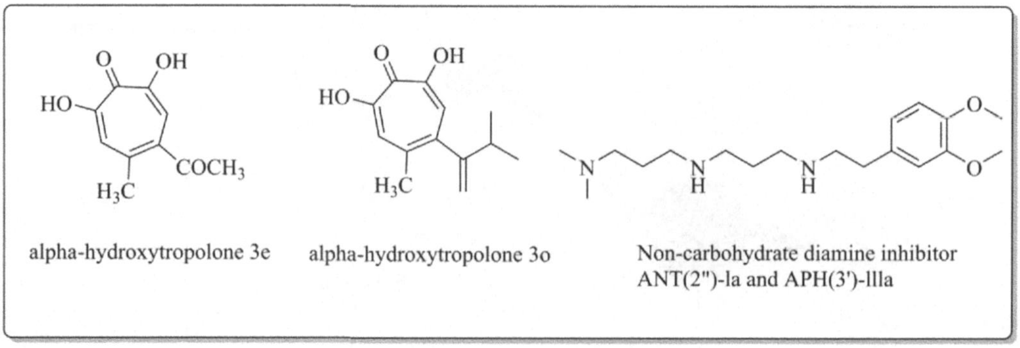

FIGURE 1.5 Inhibitors of aminoglycoside nucleotidyltransferases (ANTs).

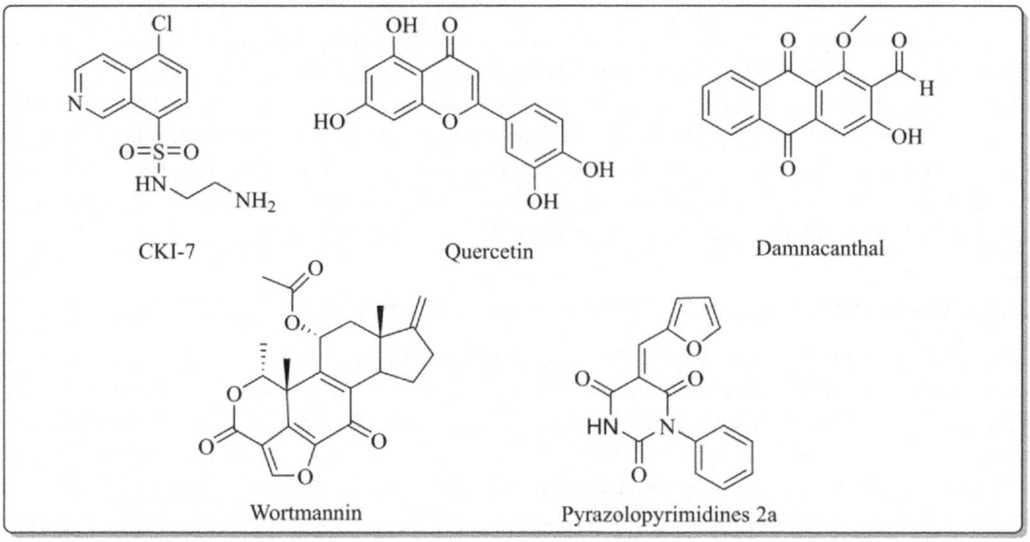

FIGURE 1.6 Inhibitors of aminoglycoside phosphotransferases (APHs).

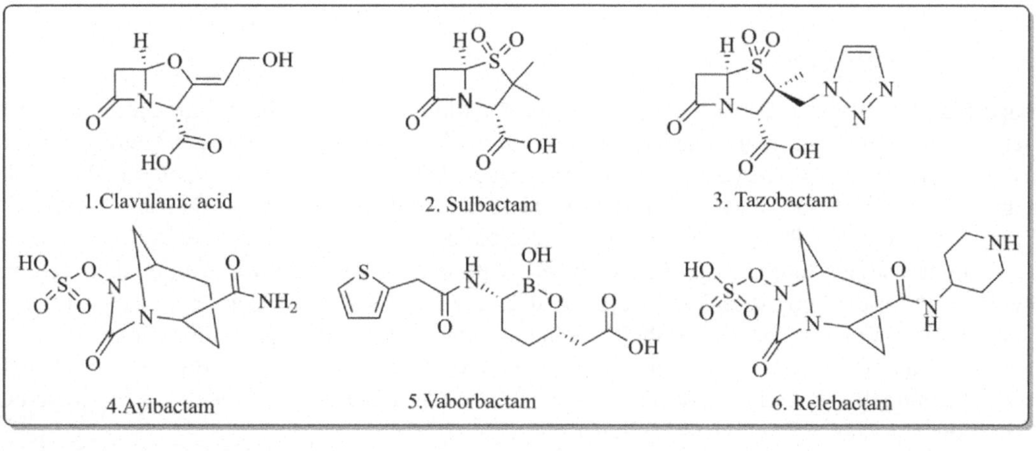

FIGURE 1.7 Inhibitors of β-lactamase.

antibacterials (Webber 2003). Efflux pump inhibitors (EPIs) are the drugs that resist the bacteria to perform this activity of antibiotic extrusion. When EPIs are co-administered with antibacterials, they lower the rate of antibiotic extrusion, leading to an increased concentration of antimicrobials within the bacterial cell. Hence, a disruption of the bacterial cell takes place (Lamut et al. 2019).

1.6.2.3 Membrane Permeabilizers

The membrane permeabilizer is used to increase the permeability of the bacterial cell membrane in order to increase the influx of antibacterials within the bacterial cell. Mutation in bacteria can lead to reduced porin size and blocks the activity of antibacterials (Delcour 2009). Polymyxin, colistin, and their derivatives are a few examples of membrane permeabilizers. Besides this, peptides and peptidomimetics are also used to increase cell permeability. Antimicrobial peptides are the short chain of amino acids that act as the first-line defense against foreign attacks; these may be natural or synthetic (Lei et al. 2019). Alexander Fleming discovered the first antimicrobial peptide (AMP), lysozyme, in 1922 (Zhang and Gallo 2016). Due to the net positive charge and hydrophobicity of AMP, they can easily interact with the negatively charged cell membrane of microbes, resulting in the destruction of microbial cells (J. Wang et al. 2019). These AMP have certain disadvantages, such as their complicated synthesis leading to low yield and high production cost, instability, short half-life, and attacks of proteases, resulting in proteolytic digestion of such AMPs (Sierra and Viñas 2021). Certain peptidomimetics have the ability to mimic the bactericidal mechanism of AMPs. By utilizing these peptidomimetics, the disadvantages of AMPs can be overcome (Mendez-Samperio 2014). Peptidomimetics are more stable to enzymatic degradation, cell specificity, and better tolerability (Mendez-Samperio 2014).

1.6.3 Use of Nanoparticles to Mitigate Antibiotic Resistance

Nanoparticles are fine particles whose size ranges between 10 and 1,000 nm. The drug to be administered can be dissolved, encapsulated, trapped, or attached to the nanoparticle matrix (Mohanraj and Chen 2007). Nanoparticles are used to shield drugs from degradation. Their reduced size and increased surface area increase their solubility (making them available to all target tissues), allowing them to cross the blood-brain barrier (BBB) and be absorbed through the skin's endothelial cells (Moineau 2013). Nanoparticles are known as delivery vehicles whose efficacy is highly dependent on size and shape (Farokhzad and Langer 2009). The nanostructures can be metallic, organic, inorganic, or polymeric (Patra et al. 2018). Liposomal-encapsulated doxorubicin was the first liposome-based (organic nanoparticle) drug that was used in the treatment of Kaposi's sarcoma (related to HIV) in 1995 and approved by the US Food and Drug Administration (Petros and DeSimone 2010).

1.6.4 Phage Therapy

Phage (bacteriophage) is a bacteria-infecting virus (Moineau 2013). Phages are first attached to bacteria and then inject their genetic material into the bacteria, which replicate. Finally, the bacteria explode, releasing a large number of new phages that can attach to new bacteria (Brives and Pourraz 2020). In 1919, de Herelle had used phages to kill *Salmonella gallinarum* in chickens (Kortright et al. 2019). Phage treatment offers several benefits over antibiotic therapy, including the fact that phages focus on the target molecule and begin to multiply there, while antibiotics disseminate throughout the body (Sulakvelidze, Alavidze, and Morris 2001). Phage treatment is the best alternative to antibiotics because antibiotics can take a long time to become available on the market, along with the expenditure of millions in clinical trials (Weber-Dąbrowska, Mulczyk, and Górski 2001). Additionally, bacteria can develop resistance to phages, but phages have the ability to mutate to overcome resistance from bacteria (Matsuzaki et al. 2005).

1.6.5 Essential Oils against Antimicrobial Resistance

Essential oils are aromatic and volatile liquids that are obtained from plant parts such as leaves, stems, roots, seeds, flowers, bark, fruits, etc. through distillation and mechanical methods such as cold pressing (Ríos 2016). Essential oils protect plants from microbes due to their antifungal, antibacterial, antiviral, and insecticidal properties (Bakkali et al. 2008). In addition, these properties aid in the protection of living organisms against microbes. The benefits of essential oils include their low toxicity, minimal environmental impact, widespread acceptance, and lack of antimicrobial resistance after prolonged use (Helal et al. 2019). Essential oil micro- and nano-emulsions exhibit significant antimicrobial activity when prepared with oil, surfactant, and water (Franklyne, Mukherjee, and Chandrasekaran 2016). Several factors influence the antimicrobial activity of essential oils, including concentration, solubility, time, pH, oxygen availability, temperature, type of microbe, bacterial cell count, and dispersing and emulsifying agent (Seow et al. 2014).

1.7 CONCLUSIONS AND FUTURE PROSPECTS

Antimicrobials are a vital resource that should not be wasted. The widespread use of antimicrobial drugs in both human and animal medicine has led to the evolution of resistant microorganisms. This has created a serious public health problem that leads to increased morbidity and mortality due to the inability of existing antibiotics to effectively treat bacterial infections. AMR can be combated through the use of inhibitors, nanoparticles, vaccines, essential oils, peptides, and peptide-mimicking chemicals, among others. In certain circumstances, these strategies can offer some measure of protection to living creatures. Antimicrobials that require prescription should be used with caution and in accordance with the recommendations on the label. Research and development of new antimicrobial and antimicrobial resistance techniques are necessary to develop a more holistic approach to truly prevent AMR. As antimicrobial resistance continues to increase, it is important that healthcare professionals and patients practise appropriate stewardship of existing resources.

1.8 ABBREVIATIONS

AACs – aminoglycoside acetyltransferases, AMPs – antimicrobial peptides, AMR – antimicrobial resistance, ANTs – aminoglycoside nucleotidyltransferases, APHs – aminoglycoside phosphotransferases, ARBs – antibiotic resistance breakers, BBB – blood-brain barrier, CDC – Centers for Diseases Control and Prevention, EPIs – efflux pump inhibitors, LPS – lipopolysaccharide, MRSA – methicillin–resistant *Staphylococcus aureus*, PBP – penicillin-binding protein.

1.9 ACKNOWLEDGMENTS

The authors acknowledge infrastructural support from their respective institutes.

REFERENCES

Agrawal, Anshu, Mohammad Owais, and Udai P. Singh. 2013. "Editorial Novel Vaccine Adjuvants." *BioMed Research International*. https://doi.org/10.1155/2013/835105.

Alby, Kevin, and Melissa B. Miller. 2018. "Mechanisms and Detection of Antimicrobial Resistance." In *Principles and Practice of Pediatric Infectious Diseases*, 1467–1478.e4. Elsevier. https://doi.org/10.1016/B978-0-323-40181-4.00290-5.

Alt, F. W., R. E. Kellems, J. R. Bertino, and R. T. Schimke. 1978. "Selective Multiplication of Dihydrofolate Reductase Genes in Methotrexate-Resistant Variants of Cultured Murine Cells." *Journal of Biological Chemistry* 253 (5): 1357–1370. https://doi.org/10.1016/s0021-9258(17)34875-5.

Alvarez-Ortega, Carolina, Jorge Olivares, and José L. Martínez. 2013. "RND Multidrug Efflux Pumps: What Are They Good For?" *Frontiers in Microbiology* 4 (February): 7. https://doi.org/10.3389/fmicb.2013.00007.

Arion, Dominique, and Michael A. Parniak. 1999. "HIV Resistance to Zidovudine: The Role of Pyrophosphorolysis." *Drug Resistance Updates* 2 (2): 91–95. https://doi.org/10.1054/drup.1999.0076.

Baker, Stephen J., David J. Payne, Rino Rappuoli, and Ennio De Gregorio. 2018. "Technologies to Address Antimicrobial Resistance." *Proceedings of the National Academy of Sciences* 115 (51): 12887–12895. https://doi.org/10.1073/pnas.1717160115.

Bakkali, F., S. Averbeck, D. Averbeck, and M. Idaomar. 2008. "Biological Effects of Essential Oils – A Review." *Food and Chemical Toxicology* 46 (2): 446–475. https://doi.org/10.1016/j.fct.2007.09.106.

Bal, Abhijit M. 2021. "Lincosamide Antibiotics." *Reference Module in Biomedical Sciences* 130 (26): 715–722. https://doi.org/10.1016/B978-0-12-820472-6.00111-0.

Bay, Denice C., Kenton L. Rommens, and Raymond J. Turner. 2008. "Small Multidrug Resistance Proteins: A Multidrug Transporter Family That Continues to Grow." *Biochimica et Biophysica Acta (BBA) – Biomembranes* 1778 (9): 1814–1838. https://doi.org/10.1016/j.bbamem.2007.08.015.

Bebear, C., and S. Pereyre. 2005. "Mechanisms of Drug Resistance in Mycoplasma Pneumoniae." *Current Drug Target-Infectious Disorders* 5 (3): 263–271. https://doi.org/10.2174/1568005054880109.

Bebell, Lisa M., and Anthony N. Muiru. 2014. "Antibiotic Use and Emerging Resistance: How Can Resource-Limited Countries Turn the Tide?" *Global Heart* 9 (3): 347. https://doi.org/10.1016/j.gheart.2014.08.009.

Beceiro, Alejandro, María Tomás, and Germán Bou. 2013. "Antimicrobial Resistance and Virulence: A Successful or Deleterious Association in the Bacterial World?" *Clinical Microbiology Reviews* 26 (2): 185–230. https://doi.org/10.1128/CMR.00059-12.

Bernard, Remi, Annick Guiseppi, Marc Chippaux, Maryline Foglino, and François Denizot. 2007. "Resistance to Bacitracin in Bacillus Subtilis : Unexpected Requirement of the BceAB ABC Transporter in the Control of Expression of Its Own Structural Genes." *Journal of Bacteriology* 189 (23): 8636–8642. https://doi.org/10.1128/JB.01132-07.

Biswas, Silpak, Jean-Michel Brunel, Jean-Christophe Dubus, Martine Reynaud-Gaubert, and Jean-Marc Rolain. 2012. "Colistin: An Update on the Antibiotic of the 21st Century." *Expert Review of Anti-Infective Therapy* 10 (8): 917–934. https://doi.org/10.1586/eri.12.78.

Blair, Jessica M. A., and Laura J. V. Piddock. 2009. "Structure, Function and Inhibition of RND Efflux Pumps in Gram-Negative Bacteria: An Update." *Current Opinion in Microbiology* 12 (5): 512–519. https://doi.org/10.1016/j.mib.2009.07.003.

Blair, Jessica M. A., Grace E. Richmond, and Laura J. V. Piddock. 2014. "Multidrug Efflux Pumps in Gram-Negative Bacteria and Their Role in Antibiotic Resistance." *Future Microbiology* 9 (10): 1165–1177. https://doi.org/10.2217/fmb.14.66.

Blair, Jessica M. A., Mark A. Webber, Alison J. Baylay, David O. Ogbolu, and Laura J. V. Piddock. 2015. "Molecular Mechanisms of Antibiotic Resistance." *Nature Reviews Microbiology* 13 (1): 42–51. https://doi.org/10.1038/nrmicro3380.

Brives, Charlotte, and Jessica Pourraz. 2020. "Phage Therapy as a Potential Solution in the Fight against AMR: Obstacles and Possible Futures." *Palgrave Communications* 6 (1): 1–11. https://doi.org/10.1057/s41599-020-0478-4.

Bush, Karen, and George H. Miller. 1998. "Bacterial Enzymatic Resistance: β-Lactamases and Aminoglycoside-Modifying Enzymes." *Current Opinion in Microbiology* 1 (5): 509–515. https://doi.org/10.1016/S1369-5274(98)80082-9.

C. Reygaert, Wanda. 2018. "An Overview of the Antimicrobial Resistance Mechanisms of Bacteria." *AIMS Microbiology* 4 (3): 482–501. https://doi.org/10.3934/microbiol.2018.3.482.

CDC. 2019. "Biggest Threats and Data." *Prevention, Center for Disease Control and Prevention*. www.cdc.gov/drugresistance/biggest-threats.html.

Charlebois, Audrey, Louis-Alexandre Jalbert, Josée Harel, Luke Masson, and Marie Archambault. 2012. "Characterization of Genes Encoding for Acquired Bacitracin Resistance in Clostridium Perfringens." Edited by Herman Tse. *PLoS One* 7 (9): e44449. https://doi.org/10.1371/journal.pone.0044449.

Chen, Chih Ming, Mei Huang, Huei Fen Chen, Se Chin Ke, Chia Ru Li, Jen Hsien Wang, and Lii Tzu Wu. 2011. "Fusidic Acid Resistance among Clinical Isolates of Methicillin-Resistant Staphylococcus Aureus in a Taiwanese Hospital." *BMC Microbiology* 11: 98. https://doi.org/10.1186/1471-2180-11-98.

Chen, Haiyang, Lijun Jing, Zhipeng Yao, Fansheng Meng, and Yanguo Teng. 2019. "Prevalence, Source and Risk of Antibiotic Resistance Genes in the Sediments of Lake Tai (China) Deciphered by Metagenomic Assembly: A Comparison with Other Global Lakes." *Environment International* 127 (June): 267–275. https://doi.org/10.1016/j.envint.2019.03.048.

Chen, Haiyang, Chang Liu, Yanguo Teng, Zulin Zhang, Yihan Chen, and Yuyi Yang. 2021. "Environmental Risk Characterization and Ecological Process Determination of Bacterial Antibiotic Resistome in Lake Sediments." *Environment International* 147 (February): 106345. https://doi.org/10.1016/j.envint.2020.106345.

Chow, Joseph W., and David M. Shlaes. 1991. "Imipenem Resistance Associated with the Loss of a 40 KDa Outer Membrane Protein in Enterobacter Aerogenes." *Journal of Antimicrobial Chemotherapy* 28 (4): 499–504. https://doi.org/10.1093/jac/28.4.499.

Cochrane, James M. T. 2000. "Zidovudine's Patent History." *The Lancet* 356 (9241): 1611–1612. https://doi.org/10.1016/S0140-6736(05)74463-9.

Cornaglia, Giuseppe, Annarita Mazzariol, Roberta Fontana, and Giuseppe Satta. 1996. "Diffusion of Carbapenems Through the Outer Membrane of Enterobacteriaceae and Correlation of Their Activities with Their Periplasmic Concentrations." *Microbial Drug Resistance* 2 (2): 273–276. https://doi.org/10.1089/mdr.1996.2.273.

Coutinho, A. G. G., S. M. P. Biscaia, R. Fernandez, and A. L. Tararthuch. 2018. "The Aminoglycoside Antibiotic Gentamicin Is Able to Alter Metabolic Activity and Morphology of MDCK-C11 Cells: A Cell Model of Intercalated Cells." *Brazilian Journal of Medical and Biological Research* 51 (10). https://doi.org/10.1590/1414-431x20187417.

Cowen, Leah E., Dominique Sanglard, Susan J. Howard, P. David Rogers, and David S. Perlin. 2015. "Mechanisms of Antifungal Drug Resistance." *Cold Spring Harbor Perspectives in Medicine* 5 (7): a019752. https://doi.org/10.1101/cshperspect.a019752.

Cox, Georgina, and Gerard D. Wright. 2013. "Intrinsic Antibiotic Resistance: Mechanisms, Origins, Challenges and Solutions." *International Journal of Medical Microbiology* 303 (6–7): 287–292. https://doi.org/10.1016/j.ijmm.2013.02.009.

Cunningham, Anthony L., Himal Lal, Martina Kovac, Roman Chlibek, Shinn-Jang Hwang, Javier Díez-Domingo, Olivier Godeaux, et al. 2016. "Efficacy of the Herpes Zoster Subunit Vaccine in Adults 70 Years of Age or Older." *New England Journal of Medicine* 375 (11): 1019–1032. https://doi.org/10.1056/NEJMoa1603800.

Czekalski, Nadine, Radhika Sigdel, Julia Birtel, Blake Matthews, and Helmut Bürgmann. 2015. "Does Human Activity Impact the Natural Antibiotic Resistance Background? Abundance of Antibiotic Resistance Genes in 21 Swiss Lakes." *Environment International* 81 (August): 45–55. https://doi.org/10.1016/j.envint.2015.04.005.

Dadgostar, Porooshat. 2019. "Antimicrobial Resistance: Implications and Costs." *Infection and Drug Resistance* 12 (December): 3903–3910. https://doi.org/10.2147/IDR.S234610.

Delcour, Anne H. 2009. "Outer Membrane Permeability and Antibiotic Resistance." *Biochimica et Biophysica Acta (BBA) – Proteins and Proteomics* 1794 (5): 808–816. https://doi.org/10.1016/j.bbapap.2008.11.005.

Dowding, J. E. 1977. "Mechanisms of Gentamicin Resistance in Staphylococcus Aureus." *Antimicrobial Agents and Chemotherapy* 11 (1): 47–50. https://doi.org/10.1128/AAC.11.1.47.

Fajardo, Alicia, Nadia Martínez-Martín, María Mercadillo, Juan C. Galán, Bart Ghysels, Sandra Matthijs, Pierre Cornelis, et al. 2008. "The Neglected Intrinsic Resistome of Bacterial Pathogens." Edited by Matthew Falagas. *PLoS One* 3 (2): e1619. https://doi.org/10.1371/journal.pone.0001619.

Farokhzad, Omid C., and Robert Langer. 2009. "Impact of Nanotechnology on Drug Delivery." *ACS Nano* 3 (1): 16–20. https://doi.org/10.1021/nn900002m.

Farzam, Khashayar, Trevor A. Nessel, and Judy Quick. 2022. *Erythromycin. StatPearls*. StatPearls Publishing.

Fernandes, Prabhavathi. 2016. "Fusidic Acid: A Bacterial Elongation Factor Inhibitor for the Oral Treatment of Acute and Chronic Staphylococcal Infections." *Cold Spring Harbor Perspectives in Medicine* 6 (1). https://doi.org/10.1101/cshperspect.a025437.

Fernández, Matilde, Susana Conde, Jesús De La Torre, Carlos Molina-Santiago, Juan Luis Ramos, and Estrella Duque. 2012. "Mechanisms of Resistance to Chloramphenicol in Pseudomonas Putida KT2440." *Antimicrobial Agents and Chemotherapy* 56 (2): 1001–1009. https://doi.org/10.1128/AAC.05398-11.

Franklyne, J. S., A. Mukherjee, and N. Chandrasekaran. 2016. "Essential Oil Micro- and Nanoemulsions: Promising Roles in Antimicrobial Therapy Targeting Human Pathogens." *Letters in Applied Microbiology* 63 (5): 322–334. https://doi.org/10.1111/lam.12631.

Galm, Ute, Martin H. Hager, Steven G. Van Lanen, Jianhua Ju, Jon S. Thorson, and Ben Shen. 2005. "Antitumor Antibiotics: Bleomycin, Enediynes, and Mitomycin." *Chemical Reviews* 105 (2): 739–758. https://doi.org/10.1021/cr030117g.

Gill, M. J., S. Simjee, K. Al-Hattawi, B. D. Robertson, C. S. F. Easmon, and C. A. Ison. 1998. "Gonococcal Resistance to β-Lactams and Tetracycline Involves Mutation in Loop 3 of the Porin Encoded at the PenB Locus." *Antimicrobial Agents and Chemotherapy* 42 (11): 2799–2803. https://doi.org/10.1128/aac.42.11.2799.

Gupte, P., B. Kulkarni, and M. Ganguli. 2002. "Antifungal Antibiotics." *Applied Microbiology and Biotechnology* 58 (1): 46–57. https://doi.org/10.1007/s002530100822.

Hawkey, P. M. 2003. "Mechanisms of Quinolone Action and Microbial Response." *Journal of Antimicrobial Chemotherapy* 51 (90001): 29–35. https://doi.org/10.1093/jac/dkg207.

Helal, Ismail M., Ashraf El-Bessoumy, Erwa Al-Bataineh, Martin R. P. Joseph, Prasanna Rajagopalan, Harish C. Chandramoorthy, and Sami Ben Hadj Ahmed. 2019. "Antimicrobial Efficiency of Essential Oils from Traditional Medicinal Plants of Asir Region, Saudi Arabia, Over Drug Resistant Isolates." *BioMed Research International* 2019. https://doi.org/10.1155/2019/8928306.

Hotta, Kunimoto, Hirokazu Yamamoto, Yoshiro Okami, and Hamao Umezawa. 1981. "Resistance Mechanisms of Kanamycin-, Neomycin-, and Streptomycin-Producing Streptomycetes to Aminoglycoside Antibiotics." *Journal of Antibiotics* 34 (9): 1175–1182. https://doi.org/10.7164/antibiotics.34.1175.

Ison, Michael G. 2010. "Antiviral Agents against Respiratory Viruses." In *Infectious Diseases*, edited by Jonathan Cohen, 2: 1464–1470. Elsevier. https://doi.org/10.1016/B978-0-323-04579-7.00147-7.

Jones, Ronald N. 1996. "Impact of Changing Pathogens and Antimicrobial Susceptibility Patterns in the Treatment of Serious Infections in Hospitalized Patients." In *American Journal of Medicine*, 100: 3S–12S. Elsevier Inc. https://doi.org/10.1016/s0002-9343(96)00102-7.

Kaur, Rajneesh, Raman Singh, Priyanka Ahlawat, Parul Kaushik, and Kuldeep Singh. 2020. "Contemporary Advances in Therapeutic Portfolio of 2-Azetidinones." *Chemical Biology Letters* 7 (1): 13–26.

Kaur, Rajneesh, Raman Singh, Antresh Kumar, Satvinder Kaur, Nitesh Priyadarshi, Nitin Kumar Singhal, and Kuldeep Singh. 2020. "1,2,3-Triazole β-Lactam Conjugates as Antimicrobial Agents." *Heliyon* 6 (6): e04241. https://doi.org/10.1016/j.heliyon.2020.e04241.

Kaur, Rajneesh, Raman Singh, Antresh Kumar, and Kuldeep Singh. 2019. "Click Chemistry Approach to Isoindole-1,3-Dione Tethered 1,2,3-Triazole Derivatives." *SynOpen* 3 (2): 67–70. https://doi.org/10.1055/s-0039-1690326.

Keren, Iris, Niilo Kaldalu, Amy Spoering, Yipeng Wang, and Kim Lewis. 2004. "Persister Cells and Tolerance to Antimicrobials." *FEMS Microbiology Letters* 230 (1): 13–18. https://doi.org/10.1016/S0378-1097(03)00856-5.

Kortright, Kaitlyn E., Benjamin K. Chan, Jonathan L. Koff, and Paul E. Turner. 2019. "Phage Therapy: A Renewed Approach to Combat Antibiotic-Resistant Bacteria." *Cell Host & Microbe* 25 (2): 219–232. https://doi.org/10.1016/j.chom.2019.01.014.

Kumar, A., and H. Schweizer. 2005. "Bacterial Resistance to Antibiotics: Active Efflux and Reduced Uptake." *Advanced Drug Delivery Reviews* 57 (10): 1486–1513. https://doi.org/10.1016/j.addr.2005.04.004.

Kumar, Sanath, Mun Mun Mukherjee, and Manuel F. Varela. 2013. "Modulation of Bacterial Multidrug Resistance Efflux Pumps of the Major Facilitator Superfamily." *International Journal of Bacteriology* 2013 (December): 1–15. https://doi.org/10.1155/2013/204141.

Kuroda, Teruo, and Tomofusa Tsuchiya. 2009. "Multidrug Efflux Transporters in the Mate Family." *Biochimica et Biophysica Acta (BBA) – Proteins and Proteomics* 1794 (5): 763–768. https://doi.org/10.1016/j.bbapap.2008.11.012.

Lambert, P. A. 2002. "Cellular Impermeability and Uptake of Biocides and Antibiotics in Gram-Positive Bacteria and Mycobacteria." *Journal of Applied Microbiology* 92 (1): 46S–54S. https://doi.org/10.1046/j.1365-2672.92.5s1.7.x.

Lamut, Andraž, Lucija Peterlin Mašič, Danijel Kikelj, and Tihomir Tomašič. 2019. "Efflux Pump Inhibitors of Clinically Relevant Multidrug Resistant Bacteria." *Medicinal Research Reviews* 39 (6): 2460–2504. https://doi.org/10.1002/med.21591.

Laws, Mark, Ali Shaaban, and Khondaker Miraz Rahman. 2019. "Antibiotic Resistance Breakers: Current Approaches and Future Directions." *FEMS Microbiology Reviews* 43 (5): 490–516. https://doi.org/10.1093/femsre/fuz014.

Leclercq, Roland, and Patrice Courvalin. 2002. "Resistance to Macrolides and Related Antibiotics in Streptococcus Pneumoniae." *Antimicrobial Agents and Chemotherapy* 46 (9): 2727–2734. https://doi.org/10.1128/AAC.46.9.2727-2734.2002.

Lei, Jun, Li Chun Sun, Siyu Huang, Chenhong Zhu, Ping Li, Jun He, Vienna Mackey, David H. Coy, and Quan Yong He. 2019. "The Antimicrobial Peptides and Their Potential Clinical Applications." *American Journal of Translational Research* 11 (7): 3919–3931.

Levine, Cindy, Hiroshi Hiasa, and Kenneth J. Marians. 1998. "DNA Gyrase and Topoisomerase IV: Biochemical Activities, Physiological Roles during Chromosome Replication, and Drug Sensitivities." *Biochimica et Biophysica Acta (BBA) – Gene Structure and Expression* 1400 (1–3): 29–43. https://doi.org/10.1016/S0167-4781(98)00126-2.

Levine, Donald P. 2006. "Vancomycin: A History." *Clinical Infectious Diseases* 42 (Supplement_1): S5–S12. https://doi.org/10.1086/491709.

Lubelski, Jacek, Wil N. Konings, and Arnold J. M. Driessen. 2007. "Distribution and Physiology of ABC-Type Transporters Contributing to Multidrug Resistance in Bacteria." *Microbiology and Molecular Biology Reviews* 71 (3): 463–476. https://doi.org/10.1128/mmbr.00001-07.

Lyle, T. A. 2006. "Ribonucleic Acid Viruses: Antivirals for Human Immunodeficiency Virus." In *Comprehensive Medicinal Chemistry II*, edited by John B. Taylor and David J. Triggle, 7: 329–371. Elsevier. https://doi.org/10.1016/b0-08-045044-x/00213-3.

Mah, Thien-Fah. 2012. "Biofilm-Specific Antibiotic Resistance." *Future Microbiology* 7 (9): 1061–1072. https://doi.org/10.2217/fmb.12.76.

Mancini, Francesca, Elisabetta Monaci, Giuseppe Lofano, Antonina Torre, Marta Bacconi, Simona Tavarini, Chiara Sammicheli, et al. 2016. "One Dose of Staphylococcus Aureus 4C-Staph Vaccine Formulated with a Novel TLR7-Dependent Adjuvant Rapidly Protects Mice through Antibodies, Effector CD4+ T Cells, and IL-17A." *PLoS One* 11 (1). https://doi.org/10.1371/journal.pone.0147767.

Marathe, Nachiket P., Chandan Pal, Swapnil S. Gaikwad, Viktor Jonsson, Erik Kristiansson, and D. G. Joakim Larsson. 2017. "Untreated Urban Waste Contaminates Indian River Sediments with Resistance Genes to Last Resort Antibiotics." *Water Research* 124 (November): 388–397. https://doi.org/10.1016/j.watres.2017.07.060.

Martens, Evan, and Arnold L. Demain. 2017. "An Overview of the Industrial Aspects of Antibiotic Discovery." In *Microbial Resources: From Functional Existence in Nature to Applications*, 149–168. Elsevier Inc. https://doi.org/10.1016/B978-0-12-804765-1.00007-2.

Martin, Michael A., and Deanna L. Kroetz. 2013. “Abacavir Pharmacogenetics – From Initial Reports to Standard of Care.” *Pharmacotherapy: The Journal of Human Pharmacology and Drug Therapy* 33 (7): 765–775. https://doi.org/10.1002/phar.1278.

Matsuzaki, Shigenobu, Mohammad Rashel, Jumpei Uchiyama, Shingo Sakurai, Takako Ujihara, Masayuki Kuroda, Shosuke Imai, et al. 2005. “Bacteriophage Therapy: A Revitalized Therapy against Bacterial Infectious Diseases.” *Journal of Infection and Chemotherapy* 11 (5): 211–219. https://doi.org/10.1007/s10156-005-0408-9.

Melroy, J., and V. Nair. 2005. “The Antiviral Activity, Mechanism of Action, Clinical Significance and Resistance of Abacavir in the Treatment of Pediatric AIDS.” *Current Pharmaceutical Design* 11 (29): 3847–3852. https://doi.org/10.2174/138161205774580642.

Mendez-Samperio, Patricia. 2014. “Peptidomimetics as a New Generation of Antimicrobial Agents: Current Progress.” *Infection and Drug Resistance* 7 (August): 229. https://doi.org/10.2147/IDR.S49229.

Michael, Carolyn Anne, Dale Dominey-Howes, and Maurizio Labbate. 2014. “The Antimicrobial Resistance Crisis: Causes, Consequences, and Management.” *Frontiers in Public Health* 2 (September). https://doi.org/10.3389/fpubh.2014.00145.

Mingeot-Leclercq, Marie-Paule, Youri Glupczynski, and Paul M. Tulkens. 1999. “Aminoglycosides: Activity and Resistance.” *Antimicrobial Agents and Chemotherapy* 43 (4): 727–737. https://doi.org/10.1128/AAC.43.4.727.

Mohanraj, V. J., and Y. Chen. 2007. “Nanoparticles – A Review.” *Tropical Journal of Pharmaceutical Research* 5 (1): 561–573. https://doi.org/10.4314/tjpr.v5i1.14634.

Moineau, S. 2013. “Bacteriophage.” In *Brenner's Encyclopedia of Genetics*, edited by Stanley Maloy and Kelly Hughes, 2nd ed., 26: 280–283. Elsevier. https://doi.org/10.1016/B978-0-12-374984-0.00131-5.

Moosdeen, Faridah. 1997. “The Evolution of Resistance to Cephalosporins.” *Clinical Infectious Diseases* 24: 487–493.

Munita, Jose M., and Cesar A. Arias. 2016. “Mechanisms of Antibiotic Resistance.” *Microbiology Spectrum* 4 (2): 464–472. https://doi.org/10.1128/microbiolspec.VMBF-0016-2015.

Olaitan, Abiola O., Serge Morand, and Jean-Marc Rolain. 2014. “Mechanisms of Polymyxin Resistance: Acquired and Intrinsic Resistance in Bacteria.” *Frontiers in Microbiology* 5 (November). https://doi.org/10.3389/fmicb.2014.00643.

Otieno, Lucas, Martina Oneko, Walter Otieno, Joseph Abuodha, Emmanuel Owino, Chris Odero, Yolanda Guerra Mendoza, et al. 2016. “Safety and Immunogenicity of RTS,S/AS01 Malaria Vaccine in Infants and Children with WHO Stage 1 or 2 HIV Disease: A Randomised, Double-Blind, Controlled Trial.” *The Lancet Infectious Diseases* 16 (10): 1134–1144. https://doi.org/10.1016/S1473-3099(16)30161-X.

Papich, Mark G. 2012. “Antimicrobial Drugs.” In *Canine and Feline Gastroenterology*, 471–476. Elsevier. https://doi.org/10.1016/B978-1-4160-3661-6.00039-0.

Patra, Jayanta Kumar, Gitishree Das, Leonardo Fernandes Fraceto, Estefania Vangelie Ramos Campos, Maria Del Pilar Rodriguez-Torres, Laura Susana Acosta-Torres, Luis Armando Diaz-Torres, et al. 2018. “Nano Based Drug Delivery Systems: Recent Developments and Future Prospects.” *Journal of Nanobiotechnology* 16 (1): 71. https://doi.org/10.1186/s12951-018-0392-8.

Pernodet, J. L., M. T. Alegre, M. H. Blondelet-Rouault, and M. Guerineau. 1993. “Resistance to Spiramycin in Streptomyces Ambofaciens, the Producer Organism, Involves at Least Two Different Mechanisms.” *Journal of General Microbiology* 139 (5): 1003–1011. https://doi.org/10.1099/00221287-139-5-1003.

Petros, Robby A., and Joseph M. DeSimone. 2010. “Strategies in the Design of Nanoparticles for Therapeutic Applications.” *Nature Reviews Drug Discovery* 9 (8): 615–627. https://doi.org/10.1038/nrd2591.

Piddock, Laura J. V. 2006. “Clinically Relevant Chromosomally Encoded Multidrug Resistance Efflux Pumps in Bacteria.” *Clinical Microbiology Reviews* 19 (2): 382–402. https://doi.org/10.1128/CMR.19.2.382-402.2006.

Poole, Keith. 2007. “Efflux Pumps as Antimicrobial Resistance Mechanisms.” *Annals of Medicine* 39 (3): 162–176. https://doi.org/10.1080/07853890701195262.

Putman, Monique, Hendrik W. van Veen, and Wil N. Konings. 2000. “Molecular Properties of Bacterial Multidrug Transporters.” *Microbiology and Molecular Biology Reviews* 64 (4): 672–693. https://doi.org/10.1128/mmbr.64.4.672-693.2000.

Redgrave, Liam S., Sam B. Sutton, Mark A. Webber, and Laura J. V. Piddock. 2014. “Fluoroquinolone Resistance: Mechanisms, Impact on Bacteria, and Role in Evolutionary Success.” *Trends in Microbiology* 22 (8): 438–445. https://doi.org/10.1016/j.tim.2014.04.007.

Ríos, José-Luis. 2016. “Essential Oils.” In *Essential Oils in Food Preservation, Flavor and Safety*, 3–10. Elsevier. https://doi.org/10.1016/B978-0-12-416641-7.00001-8.

Roberts, Marilyn C. 2003. "Tetracycline Therapy: Update." *Clinical Infectious Diseases* 36 (4): 462–467. https://doi.org/10.1086/367622.

Roberts, Marilyn C. 2004. "Resistance to Macrolide, Lincosamide, Streptogramin, Ketolide, and Oxazolidinone Antibiotics." *Molecular Biotechnology* 28 (1): 47–62. https://doi.org/10.1385/MB:28:1:47.

Roope, Laurence S. J. J., Richard D. Smith, Koen B. Pouwels, James Buchanan, Lucy Abel, Peter Eibich, Christopher C. Butler, et al. 2019. "The Challenge of Antimicrobial Resistance: What Economics Can Contribute." *Science* 364 (6435): eaau4679. https://doi.org/10.1126/science.aau4679.

Ryan, Brenda M., Thomas J. Dougherty, Danielle Beaulieu, Julia Chuang, Brian A. Dougherty, and John F. Barrett. 2001. "Efflux in Bacteria: What Do We Really Know about It?" *Expert Opinion on Investigational Drugs* 10 (8): 1409–1422. https://doi.org/10.1517/13543784.10.8.1409.

Saga, Tomoo, and Keizo Yamaguchi. 2009. "History of Antimicrobial Agents and Resistant Bacteria." *Japan Medical Association Journal* 52 (2): 103–108.

Sanglard, Dominique, Alix Coste, and Sélène Ferrari. 2009. "Antifungal Drug Resistance Mechanisms in Fungal Pathogens from the Perspective of Transcriptional Gene Regulation." *FEMS Yeast Research* 9 (7): 1029–1050. https://doi.org/10.1111/j.1567-1364.2009.00578.x.

Seow, Yi Xin, Chia Rou Yeo, Hui Ling Chung, and Hyun-Gyun Yuk. 2014. "Plant Essential Oils as Active Antimicrobial Agents." *Critical Reviews in Food Science and Nutrition* 54 (5): 625–644. https://doi.org/10.1080/10408398.2011.599504.

Sharma, Barkha, Parul, A. K. Verma, Udit Jain, Janaradan K. Yadav, Ravneet Singh, and Raghvendra Mishra. 2017. "Occurrence of Multidrug Resistant Escherichia Coli in Groundwater of Brij Region (Uttar Pradesh) and Its Public Health Implications." *Veterinary World* 10 (3): 293–301. https://doi.org/10.14202/vetworld.2017.293-301.

Sierra, Josep M., and Miguel Viñas. 2021. "Future Prospects for Antimicrobial Peptide Development: Peptidomimetics and Antimicrobial Combinations." *Expert Opinion on Drug Discovery* 16 (6): 601–604. https://doi.org/10.1080/17460441.2021.1892072.

Singh, Kuldeep, Rajneesh Kaur, and Raman Singh. 2020. Synthesis of 1,2,3-Triazole Carboxamides and Derivatives Thereof. 201811028429 A, issued 2020.

Southgate, R. 1994. "The Synthesis of Natural Beta-Lactam Antibiotics." *Contemporary Organic Synthesis* 1: 417–431.

Speer, B. S., N. B. Shoemaker, and A. A. Salyers. 1992. "Bacterial Resistance to Tetracycline: Mechanisms, Transfer, and Clinical Significance." *Clinical Microbiology Reviews* 5 (4): 387–399. https://doi.org/10.1128/CMR.5.4.387.

Spratt, Brian G. 1994. "Resistance to Antibiotics Mediated by Target Alterations." *Science* 264 (5157): 388–393. https://doi.org/10.1126/science.8153626.

Springer, B., Y. G. Kidan, T. Prammananan, K. Ellrott, E. C. Böttger, and P. Sander. 2001. "Mechanisms of Streptomycin Resistance: Selection of Mutations in the 16S RRNA Gene Conferring Resistance." *Antimicrobial Agents and Chemotherapy* 45 (10): 2877–2884. https://doi.org/10.1128/AAC.45.10.2877-2884.2001.

Sulakvelidze, Alexander, Zemphira Alavidze, and J. Glenn Morris. 2001. "Bacteriophage Therapy." *Antimicrobial Agents and Chemotherapy* 45 (3): 649–659. https://doi.org/10.1128/AAC.45.3.649-659.2001.

Sutherland, R. 1962. "Spiramycin: A Reappraisal of Its Antibacterial Activity." *British Journal of Pharmacology and Chemotherapy* 19 (1): 99–110. https://doi.org/10.1111/j.1476-5381.1962.tb01430.x.

Sykes, R. B. 2000. "From Moulds to Drugs." *Clinical Microbiology and Infection* 6 (S3): 10–12. https://doi.org/10.1111/j.1469-0691.2000.tb02034.x.

Tan, S. Y., and Yvonne Tatsumura. 2015. "Alexander Fleming (1881–1955): Discoverer of Penicillin." *Singapore Medical Journal* 56 (7): 366–367. https://doi.org/10.11622/smedj.2015105.

Then, R. L., and F. Hermann. 1981. "Mechanisms of Trimethoprim Resistance in Enterobacteria Isolated in Finland." *Chemotherapy* 27 (3): 192–199. https://doi.org/10.1159/000237977.

Tredwell, M., and V. Gouverneur. 2012. "1.5 Fluorine in Medicinal Chemistry: Importance of Chirality." In *Comprehensive Chirality*, edited by Erick M. Carreira and Hisashi Yamamoto, 1: 70–85. Elsevier. https://doi.org/10.1016/B978-0-08-095167-6.00106-3.

Tsunematsu, Yuta, Yuichiro Hirayama, Takahiro Masuya, and Kenji Watanabe. 2020. "Oxidative Modification Enzymes in Polyketide Biosynthetic Pathways." In *Comprehensive Natural Products III*, 479–505. Elsevier. https://doi.org/10.1016/b978-0-12-409547-2.14637-2.

Umezawa, Hamao. 2006. "Kanamycin: Its Discovery." *Annals of the New York Academy of Sciences* 76 (2): 20–26. https://doi.org/10.1111/j.1749-6632.1958.tb54688.x.

Wang, Guojun, Masumi Izawa, Xiaoge Yang, Dongbo Xu, Yukinori Tanaka, and Kozo Ochi. 2017. "Identification of a Novel Lincomycin Resistance Mutation Associated with Activation of Antibiotic Production in Streptomyces Coelicolor A3(2)." *Antimicrobial Agents and Chemotherapy* 61 (2). https://doi.org/10.1128/AAC.02247-16.

Wang, Jiajun, Xiujing Dou, Jing Song, Yinfeng Lyu, Xin Zhu, Lin Xu, Weizhong Li, and Anshan Shan. 2019. "Antimicrobial Peptides: Promising Alternatives in the Post Feeding Antibiotic Era." *Medicinal Research Reviews* 39 (3): 831–859. https://doi.org/10.1002/med.21542.

Waugh, Christine D. 2007. "Griseofulvin." In *XPharm: The Comprehensive Pharmacology Reference*, 1–4. Elsevier. https://doi.org/10.1016/B978-008055232-3.61849-5.

Webber, M. A. 2003. "The Importance of Efflux Pumps in Bacterial Antibiotic Resistance." *Journal of Antimicrobial Chemotherapy* 51 (1): 9–11. https://doi.org/10.1093/jac/dkg050.

Weber-Dąbrowska, Beata, Marian Mulczyk, and Andrzej Górski. 2001. "Bacteriophage Therapy of Bacterial Infections: An Update of Our Institute's Experience." In *Inflammation*, edited by Andrzej Górski, Hubert Krotkiewski, and Michał Zimecki, 48: 201–209. Dordrecht: Springer Netherlands. https://doi.org/10.1007/978-94-015-9702-9_15.

Wiest, Donald B., Joel B. Cochran, and Fred W. Tecklenburg. 2012. "Chloramphenicol Toxicity Revisited: A 12-Year-Old Patient with a Brain Abscess." *The Journal of Pediatric Pharmacology and Therapeutics* 17 (2): 182–188. https://doi.org/10.5863/1551-6776-17.2.182.

Wilke, Mark S., Andrew L. Lovering, and Natalie C. J. Strynadka. 2005. "β-Lactam Antibiotic Resistance: A Current Structural Perspective." *Current Opinion in Microbiology* 8 (5): 525–533. https://doi.org/10.1016/j.mib.2005.08.016.

Wood, Thomas K., Stephen J. Knabel, and Brian W. Kwan. 2013. "Bacterial Persister Cell Formation and Dormancy." *Applied and Environmental Microbiology* 79 (23): 7116–7121. https://doi.org/10.1128/AEM.02636-13.

Woodruff, H. Boyd. 2014. "Selman A. Waksman, Winner of the 1952 Nobel Prize for Physiology or Medicine." *Applied and Environmental Microbiology* 80 (1): 2–8. https://doi.org/10.1128/AEM.01143-13.

Yang, Yuyi, Wenjuan Song, Hui Lin, Weibo Wang, Linna Du, and Wei Xing. 2018. "Antibiotics and Antibiotic Resistance Genes in Global Lakes: A Review and Meta-Analysis." *Environment International* 116 (July): 60–73. https://doi.org/10.1016/j.envint.2018.04.011.

Zhang, Ling-juan, and Richard L. Gallo. 2016. "Antimicrobial Peptides." *Current Biology* 26 (1): R14–R19. https://doi.org/10.1016/j.cub.2015.11.017.

2 Antibiotic Resistance from Theory to Facts

Amro Abd Al Fattah Amara, Nawal Abd El-Baky, Mounir Mohamed Salem-Bekhit, Khairy Mohamed Abdallah Zoheir, and Mohamed Morsi Mohamed Ahmed

Table of Abbreviations

ABC	ATP-binding cassette
ARGs	antibiotic resistance genes
GIM	gastrointestinal microflora
GIS	gastrointestinal system
GTP	Guanosine triphosphate
MATE	multi-antibiotic and toxic compound extrusion family
MFS	major facilitator subfamily
MGC	minimum growth concentration
MIC	minimum inhibitory concentrations
MRSA	methicillin-resistant *Staphylococcus aureus*
PABA	para aminobenzoic acid
PCR	polymerase chain reaction
RF	resistance factor
RND	resistance-nodulation-cell division
SMR	small multi-antibiotic resistance family
Tet(M)	tetracycline resistance gene M
Tet(O)	tetracycline resistance gene O

2.1 INTRODUCTION

Microbes are variants concerning their cells' structures. Therefore, they respond differently against antibiotics. For that, within the genus level, every species (or even each strain) should be treated on case-by-case basis. They should be handled with high specificity and selectivity. Many lessons can be extracted from antibiotics, antibiotic resistance history, mode of actions, structures, and the like. There is still a chance to reduce and control the spreading of different antibiotic resistance factors among microbes. Drugs, chemicals, phages, plant extracts, peptides, vaccines, and the like are alternatives that could be used to decrease our dependence on antibiotics.

Pathogenic microbes (for us) are not necessarily harmful for other microbes (around). Different types of microbes (including moderate pathogens) in our body adapted themselves to live friendly (e.g., as bodies' microflora). They adapt themselves to live friendly in different places, such as in the gastrointestinal system and on our skin. They reduce their virulence factors to calm down our immune system. Meanwhile, if they are subjected to different kinds of stresses, such as a low dosage of antibiotics, or sense the body's weakness, they will start to attack us. In case of applying antibiotic treatment, they will search around to gain new resistance factors (e.g., resistance genes carried on plasmids) or build their own resistance strategy. If they start to re-attack us and have antibiotic

 DOI: 10.1201/9781003269380-2

resistance factor or developed/activated one or more than one resistance mechanism, they are now ready to depress and incapacitate our weapons, antibiotics.

Our history of using antimicrobial agents is deep in the past; Imhotep, around 2600 BCE, has repeatedly been considered as the first-known scientist. He wrote many formulas to treat diseases. The different types of knowledge from old civilizations have been transferred everywhere. For that, many human groups have similar knowledge and practices for treating different kinds of illness. In ancient Egypt and in different parts of the world nowadays, similar practices are still used, particularly by lay people. For example, they have been using moldy bread to treat skin and infected injuries (Alnaimat et al. 2015). Another example is castor oil, which has been used as a purgative, and poppy, which have been used as juice to relieve pain. The thousands who built the pyramids and temples ate radishes, garlic, and onions. Egyptian priests believed that these vegetables prevented epidemic diseases. There has been also a tendency to mix medicines with wine or beer. The mummies are another example. In the old Arabic books such as EL-Bokhari, a text contains a discussion about an infection, and a question has been proposed, "Who infected the first one?" One could understand from this text that there is a sign that the infection happened only in particular cases and not every time! Before the Pasteurian era, a number of scientists have established the first modern scientific steps toward our ability to control microbes. For example, Abu Bakr al-Razi (945 CE) chose the correct place to build a hospital in Baghdad by designing an experiment based on the best place that keeps meat subjected to the air for a longer time without spoiling. There are many other examples from the old civilizations, particularly in Egypt, China, India, Greek and Roman empires, the Arab world, and the like (Hutchings, Truman, and Wilkinson 2019).

In the 16th century, Paracelsus applied strong chemicals to cure disease. Mercury compounds are examples. Louis Pasteur has benefited from the previous research of many scientists who worked on bacterial attenuation using different chemical, physical, and biological methods. Louis Pasteur and Jules François Joubert observed (in 1877) that bacteria from the air can kill anthrax bacilli. They also knew that soil does not harbor most disease-causing bacteria. Substances in soil, probably produced by common soil bacteria or fungi, seem to kill other bacteria. Pasteur and Joubert (1887) defined antagonism of microorganisms. This is the relation between opposing principles (forces or factors). Pasteur suggested that microbes could secrete compounds to kill different microorganisms. The term "antibiosis" was coined in 1889 by the French mycologist and bacteriologist Jean-Paul Vuillemin. Its primary scientific use was during the 1890s. By the end of the 19th century, medical researchers had identified some pathogens, such as syphilis, anthrax, tuberculosis, and cholera. Garré, in 1887, defined opposed consequences of *Pseudomonas fluorescens*, which at that time was named *Bacillus fluorescens putidus*. Primary antibiotics had already been isolated in the 1890s by Gosio (1896). From moldy maize, Gosio isolated a fungus, which he named *Penicillium glaucum*. He also isolated an antimicrobial substance (most likely a strain of *P. brevi compactum*). Gosio found that this substance had a phenolic resin that inhibited expansion of *B. anthracis*. Emmerich and Löw used an extract of *P. aeruginosa*, at the time named *Bacillus pycyaneus*, applying it to deal with masses of wounded people. This extract, referred to as pyocyanase, was used until the 1910s. Pyocyanase nearly disappeared from the literature after 1914. Alsberg and Black (1913) isolated a phenolic compound from a *Penicillium* strain consistent with its color reaction. The compound has been given the name "mycophenolic acid." Oxford and Raistrick (1933) studied its chemical structure. Birkinshaw et al. (1948) reported the definitive structure of these compounds. Alsberg and Black (1913) isolated penicillic acid from moldy maize. This substance inhibited growth of *E. coli*. Birkinshaw et al. succeeded to determine its structure (Birkinshaw, Oxford, and Raistrick 1936; Birkinshaw, Raistrick, and Ross, 1952). Both of these antibiotics were clinically tested.

German bacteriologist Paul Ehrlich searched for dyes that stain and kill bacteria. Ehrlich advanced artificial arsenic compounds mainly based on pro-capsules salvarsan (salvation arsenic). They have been used to deal with *Treponema pallidum*, the causative agent of syphilis. Salvarsan has been superseded by the sulfonamide pro-drug Prontosil. Gerhard Domagk, a German chemist, discovered it. Domagk and colleagues effectively continued the work started by Ehrlich. Domagk was a bacteriologist at Bayer Division of IG Farben, which is a group of German dye manufacturers. He tested azo-related dyes for activity against bacteria in a program started in 1927. In 1932, KI-730 (subsequently named Prontosil) was recognized and has been submitted for testing, but it has been devoid of activity *in vitro*. However, it has been still under investigation for their *in vivo* activity using mice infection by *Streptococcus pyogenes*. It proved to work well *in vivo*, and its discovery was announced in 1935. Domagk's daughter contracted streptococcal blood poisoning by pricking herself with a needle. Domagk injected her with "Prontosil Rubrum" and saved her life. In 1935, sulphanilamide, marketed under the name Prontosil, was used in World War II (Lewis 2013). The sulfanilamide portion of the Prontosil molecule is responsible for its antibacterial effect. It inhibits essential enzyme activity in affected bacteria. Domagk won the Nobel Prize in Medicine in 1939 (Domagk 1935; Sneader 2005).

In 1922, Alexander Fleming isolated the substance lysozyme from tears. He observed that tears contain something that kills bacteria and causes no physical damage to the body. In 1928, Fleming isolated and identified the mold *Penicillium notatum*, which produces penicillin as an antibiotic, but it was not used therapeutically until 1940. In 1939, Ernst Chain and Howard Florey continued the research on it and later conducted a human trial. Penicillin was purified by Norman Heatley, Howard Florey, Ernst Chain, and co-workers at Oxford (Abraham et al. 1941). Dorothy Hodgkin solved the β-lactam shape of penicillin in 1945 (Hodgkin 1949). Fleming, Florey, and Chain shared in 1945 the Nobel Prize for their work leading the era of antibiotics.

In 1930, René Dubos found an oligopeptide from a soil-derived *Bacillus* that is able to decompose *Streptococcus pneumoniae* type III capsular polysaccharide. The oligopeptide is able to treat mice with pneumococcal peritonitis (Avery 1930). The importance of Dubos's work resulted from the first deliberate search for compounds produced by soil microbes that have been able to kill pathogenic bacteria. He made a unique experiment by feeding Gram-positive bacteria at intervals to a large sample of mixed soils. He had a hope to find microbes that were capable of destroying the bacteria. The product has been an alcohol-soluble compound able to inhibit the growth of Gram-positive bacteria. It has been given the name tyrothricin. The alcohol extract has been a mixture of two compounds: tyrocidin and gramicidin. It has been isolated from the *Bacillus brevis* oligopeptide gramicidin that widely inhibited Gram-positive bacterial species (Dubos 1941). Gramicidin exhibited potent toxicity for humans, except for local treatment (Herrell 1941).

Many scientists have described the antagonism between microbes. In the 1930s, Selman Waksman in the United States proved that soil microbes can produce antimicrobial compounds (Durand, Raoult, and Dubourg 2019). Waksman was the first to use the noun "antibiotic" as a term in 1942. Waksman described an antibiotic as a compound made by a microbe to kill other different microbes (Waksman, Schatz, and Reynolds 2010). The main distinguishing point between antibiotics and other antimicrobial agents is that antibiotics are produced by microorganisms. Antibiotics that can kill bacteria are called bactericidal, and antibiotics that inhibit the growth of bacteria are called bacteriostatic (Walsh 2003). Using his platform, in the 1940s Waksman discovered numerous major antibiotics and antifungals, such as actinomycin (from *Streptomyces* spp.) (Waksman, Geigerm, and Reynolds 1946), streptomycin (from *Streptomyces griseus*) (Jones 1944), neomycin (from *Streptomyces fradiae*) (Waksman and Lechevalier 1949), fumigacin (from *Aspergillus fumigatus*), and clavacin (from *Aspergillus clavatus*) (Waksman, Horning, and Spencer 1942). Actinomycin, neomycin, and streptomycin are still in use today.

Many of streptomycetes' natural products are active in antagonism to microbes such as bacteria, fungi, and viruses, as well as nematodes and insects. They have additionally been advanced as anticancer and immunosuppressant drugs (Newman and Cragg 2016). One story that can be addressed is the story of streptomycin; in 1943, Albert Schatz (a graduate in Waksman's lab) discovered streptomycin, which is active against Gram-negative bacteria, particularly against *Mycobacterium tuberculosis* (Waksman, Schatz, and Reynolds 2010). It has been active in animal models of TB and was used to cure the disease by 1946. The Merck company originally was the official partner in

Waksman's lab. The demand encouraged Merck to give licenses to other pharmaceutical companies to manufacture streptomycin. A thousand kilograms per month were produced.

Waksman developed numerous culture techniques and strategies, which were named as "Waksman platform," in order to find antagonism between bacterial species (Lewis 2012). In 1952, Waksman was awarded the Nobel Prize for his work with actinomycetes and the discovery of streptomycin. During this golden era, out of many bacterial and fungi species, more than twenty classes of antibiotic were discovered.

Later, the industry examined different *in vitro* synthetic molecules mostly based on the knowledge of known mechanisms of active antibiotics. Alternatively, a completely new structure proved its antibacterial activity. Few new classes of antibiotics have been discovered: nitrofuran (1953), quinolones (1960), sulphonamides (1961), and oxazolidinones (1987). Modifying and improving already known molecules have also been carried out. Daptomycin, known in 1986, was commercialized in 2001, and linezolid, known since 1955, was commercialized in 2003. A new cephalosporin, named cefiderocol, has been found to be active against carbapenem-resistant Gram-negative bacteria (Falagas, Skalidism, and Vardakasm 2017). Usually, the first step is promising when the synthetic compound shows antimicrobial activity. But in contrast to naturally produced antibiotics, the synthetic ones need more clinical investigation to prove their safety. Hemisynthetic compounds from natural products have been also developed, such as ketolides (derivatives from macrolides) and metronidazole (derivatives from a natural product of *Streptomyces* sp.). Of the 20 pharmaceutical companies that financed antibiotic discovery in the 1980s, only 5 were left by 2015 (Blaskovich, Butlerm, and Cooper 2017).

Antibiotic resistance is an increasing global threat to human health and to the clinical treatment of disease (Allen et al. 2010). Antibiotic resistance genes (ARGs) have been detected in the last decade in all environments, including natural, developed, and clinical habitats. Anthropogenic activities, including the clinical use of antibiotics, are widely regarded as the main drivers of the dissemination of ARGs (Allen et al. 2010; Zhu et al. 2017; Zhang et al. 2020). It is highly depreciable to conduct research, install industrial production lines, and do many marketing efforts to seal a new antibiotic, and after a short time, bacterial pathogens develop resistance to it and start to spread the resistance factor to other species. For that there is an urgent need to solve the problem of antibiotic resistance. The best way is by understanding its mode of action correctly. Resistance to the antibiotics itself was first claimed to the activity of the R factor only. Today our understanding has become clearer. The early non-proved suggestions have become facts. One should not neglect the opinions, remarks, and observations of any scientist. Prestigious publisher should accept publishing scientists' believes that not proved experimentally yet at least when evidences prove that there is something could build on it. Some microbes are awkward; *P. aeruginosa* is an example (Adekunle et al. 2020; Afshari et al. 2020; Algammal et al. 2020). They could carry many virulence factors beside their ability to produce alginate. Isolates from the same geographical area contain diverse strains. One nearly could not find two similar strains; while they are being either highly mutated or highly diverse. Nutrients are supposed to be contributed to an epigenetic-like-factor effect on the similarity and differences of the microbes (Amara and Shibl 2015b). The microbes are different genetically. For that, they are reacting differently if they existed in different environments. The different types of the microbial cells' macromolecules are sensitive to the different environmental factors, so the overall cell structure is changed by the change in the nutrients and the environment as proposed.

β-lactamase is a clear example. It includes the use of protons or electrons in the side chains residue of the catalytic amino acids (serine β-lactamase) or importing some help from the ions around, such as the z^{2+}metallo β-lactamase. The biological system needs a case-by-case study as proposed by Hussein and Amara (2006) where each microbial strain is different genetically and environmentally (react differently in different environments) (Amara and Hussain 2006; Hussein and Amara 2006). Using a simple inhibition zone experiment is not enough. *P. aeruginosa* firstly could show sensitivity in low antibiotic concentration that turns out to be resistance in higher dosage as described by Amara and Hussain (2006) and Hussein and Amara (2006). That could be explained by the role of the high concentrations in elevating new or non-working yet specific resistant mechanisms. Those different issues involved in drawing a complete image of antibiotic resistance include many physical, chemical, biological, economic, personal, environmental, and strain-based features and the like. This chapter is concerned with differentiation between genetically based and structure-based resistance, the different resistance mechanisms, the effect of various physical and chemical biological factors in the resistance, some weak points concerning our misuses of antibiotics, alternatives to antibiotics, and solutions to control antibiotic resistance. Lessons from antibiotic discovery, production, use in different applications, modification, and the like are included. The main aim is to enlarge the image of antibiotic resistance, spot the factors that cause it, and propose solutions.

2.2 MAJOR POINTS CONCERNING ANTIBIOTIC RESISTANCE

2.2.1 Why Antibiotic Resistance Is a Natural Phenomenon

Antibiotic resistance is usually defined as the ability of bacteria and other microorganisms to develop resistance to antibiotics (survive in the existence of one antibiotic or more the one) that they once were sensitive to. This definition is tricky because it does not mention that the antibiotic-producing microbes could be resistant naturally (to their own products). It gives an opinion that antibiotic resistance is something new that developed only because of (and after) treatment with antibiotics. For antibiotic-producing microbes, antibiotic resistance is a natural phenomenon that is essential for their survive and does not harm. Other points are seldom discussed, like the power of hydrocarbon-degrading microbes in degrading antibiotics (e.g., *Pseudomonas aeruginosa*). The resistance to different antibiotics protects many microbes from eradication.

There are identified mechanisms that support a certain bacterial strain to survive against an antibiotic. Such mechanisms can include one or more than one, such as gene transformation, transduction, conjugation, mutation, target change, efflux, and other types of gene transfer and the like. Additionally, many naturally existing microbial structures have diverse responses to the antibiotic and give to some full protection and for others enough time to develop resistance. *P. aeruginosa* is an example of a nearly full and diverse protective microbe. It is a unique bacterial strain with extra different properties such as being able to degrade hydrocarbons such as disinfectants (Russell 1998). Using disinfectants in hospitals is responsible for their dominance (Hussein and Amara 2006). Additionally, those microbes are responsible for gene transfer and protecting other microbes from antibiotics under their alginate biofilm (Amara and Hussain 2006; Amara 2011). Moreover, Amara (2011) recommended using simple practices to control *P. aeruginosa*, like keeping places dry. Dry environments could help control this pathogen in hospitals (or at least block their reproduction process). In fact, this opportunistic pathogen could be a safe platform (for the researchers) to understand many of the different antibiotic resistance mechanisms. It has different virulence factors to survive and is usually dominant in hospitals.

New infected patients, when hospitalized, bring more pathogenic strains that might harbor additional new antibiotic resistance genes. For that *P. aeruginosa*, *Streptococcus* sp., *Staphylococcus* sp., *Klebsiella* sp., *Shigella* sp., and the like are antibiotic resistant strains that existed with multiple antibiotic resistance factors. We are responsible for the dominance of those microbes in hospitals.

Acquiring new antibiotic resistance genes is an optional additional chance for the bacteria that are self-protective against antibiotics. One should not focus only on gene transfer but study the antibiotic resistance as a phenomenon. What happened is that we invited the microbes that harbor the resistance genes to supply the other microbes with additional resistance factors. The hospitals are the most expected places for exchanging resistance genes between pathogenic microbes and other ones. That made them dominant with additional antibiotic resistance genes. In other words, the bacterial strain which is naturally sensitive, upon their acquiring resistance factor, will turn to be resistant. And the one which is already resistant to one type of antibiotic upon their acquiring a different resistance factor will be resistant to more than one antibiotic. Moreover, the one which has the resistance gene in the chromosome upon their acquiring a mobile resistance factor will be able to resist higher dosage of the antibiotic(s). Outside the area of the transfer of antibiotic resistance genes, one could summarize some evidence, which proves that resistance to antibiotics is a natural phenomenon in the following points:

1 Antibiotic-producing microbial strains are antibiotic resistant microbes. They contain genes or have developed mechanisms for antibiotic resistance that could resist their own product (the antibiotic).
2 Bacterial strains that survived (in their native ecosystem) and co-existed with these antibiotic-producing strains probably have antibiotic resistant mechanisms.
3 Most eukaryotic cells and viruses are not affected by the antibiotics.
4 Spores are mostly resistant to antibiotics.
5 Many types of antibiotics are bacteriostatic. So, not all of the microbes that have been treated with a bacteriostatic agent are correctly killed.
6 Microbes that are sensitive to certain antibiotic could resist it by activating certain mechanisms of resistance (e.g., production of exopolysaccharide) which could give them the chance and time to develop the correct resistance mechanism or acquire a resistance factor.
7 Synthetic antibiotics show less resistance by microbes, which means that these compounds are new to the bacterial biological system (e.g., the bacterial macromolecules). If they are produced from a modified natural antibiotic, the resistance might be elevated after mutating the naturally existing resistance-related genes. So, the microbe will be resistant faster for both the natural antibiotic and the synthetic one which was built based on its structure than for that one which completely synthetic and not match existing biological structure.
8 Certain microbes are able to protect themselves against harsh environmental condition by producing unusual protecting agents such as alginates, capsules, spores, and the like. In such a case, microbes will have more chances and more time to develop or acquire resistance genes or resistance mechanisms. Additionally, their co-existence with other microbes will enable them to protect themselves under their biofilms.
9 For sensitive microbes, an antibiotic could do their appropriate activity only if the correct concentration is used.

2.2.2 Microbial Load, Number, Adaptation, and Other Issues

Did we become sick because pathogenic microbes entered our body? The answer is yes. Nevertheless, pathogenic microbes enter our body every day! What is the difference? What is the misunderstood point(s)? Opportunistic pathogens and immune-compromised patients are special issues. However, they also highlight that microbial pathogenicity as a phenomenon still needs more study. Why have microbes become suddenly so virulent and epidemic in certain areas or even against certain individual(s)? The microbial load, number, and host tolerance are important issues. Travelers, after eating local cooked foods, are getting diarrhea. In contrast, native individuals are not affected! Our bodies show different tolerance to the local microbes, which should attract our focus. In addition, the second or third exposure is usually less effective.

Microbes search for food and produce products that help them to degrade such food to simpler components that could inter to their own cells. If they find proteinaceous food, first they will start to produce proteases. In their first attack, they will be programmed for this type of food. They might use more virulence factors to adapt the process. Meanwhile, introducing a simpler carbon source (like sugar) will change their interest. That means they attack our tissue because they need to feed themselves. If they find instead an easier carbon source, they will stop attacking us and will utilize it. The Arab world considered honey as one of the three illness solutions where honey is the first solution, cupping therapy is the second, and ironing with fired iron (until it becomes red) is the last solution and not recommended (only as the last choice when there is no other solution). Nowadays, honey is proving to have potent antimicrobial activity (Mandal and Mandal 2011).

Other issues are included, such as our daily practices. For example, the way we chew food in the mouth is different. Mouth saliva (containing lysozyme) could degrade microbes. Correct lysis to the bacteria happens if correct chewing time is applied. Even the way of food chewing could make a health difference between individuals. The microbes (e.g., bacterial pathogen) that enter our gastrointestinal system (GIS) might harbor antibiotic resistance genes carried on mobile elements (plasmids). Our mouth lysozyme will probably lysis them. However, if we eat fast or eat food contaminated by a large number of such microbes, they might pass into our GIS. Based on the number of the harmful microbes passed, their side effect is different. In such cases, their plasmids that carry resistance genes will be exchanged with other bacterial strains. Lay people observed early the importance of chewing food. Other simple practices can save us, such as cleaning hands, using the right hand for eating foods and the left one for handling items that are expected to be contaminated with microbes, cutting the nails regularly, and the like.

2.2.3 Our Practices That Support the Elevation of Antibiotic Resistance

What has happened to lead to distributing antibiotic resistance? A very high percentage of antibiotic resistance could be existed, like in cases of MRSA (methicillin-resistant *Staphylococcus aureus*). Such accumulated resistance genes and resistance mechanisms are a result of a long history of treatment with different antibiotics, acquiring new resistance genes, and selection due to the survive against the antibiotic stresses. Many factors could adapt the conditions that favor the elevation of one or more than one of the antibiotic resistance mechanisms. However, there are general pathways that could lead to that. They range between acquiring new resistance genes, developing new mutated genes able to resist new antibiotics, and developing new mechanisms to get rid of the antibiotics or their potency. During our attempt to kill pathogens with different antimicrobial agents, most sensitive microbes are killed. Sensitive pathogenic or sensitive beneficial ones are included. Only resistant ones, or those that can acquire or develop resistance (in the correct time), can survive. After a certain time, they will dominate their environment. The same strategy is applied to short active peptides, disinfectants, and some other antimicrobial agents.

2.2.3.1 Excessive Antibiotic Production and Use in Non-Human Treatment

The first claimed cause for the existence of antibiotic resistance is the genes that existed naturally in the microbes that produce antibiotics. Meanwhile, microbes produce antibiotics to protect themselves or to antagonize other surrounding microbes. In other words, in nature, one could not found a water of lack left naturally contains abnormal level of antibiotics. In contrast, when the antibiotics are extensively produced and applied in different fields, the antibiotic resistance problem has been elevated. One could find these changes in modernized regions, area around hospitals, or industries that produce antibiotics.

In nature, the war between microbes never stops; but only starts if there is a real requirement for it. Antibiotics are one type of weapons that are used by microbes. Our use of antibiotics in various applications is the second major cause of the development of antibiotic resistance. Apparently, there are more claimed causes for antibiotic resistance development than the human use of antibiotics for different treatment purposes. Antibiotics are used in many other applications such as in food preservation, agriculture (e.g., growing poultry and livestock animals), and the like (Podlacha et al. 2021). The excessive use of antibiotics was started early after the discovery of antibiotics. For example, in the 20 years following the discovery of antibiotics, antibiotics have been used as food preservatives (Mahalmani et al. 2019; Kasimanickam, Kasimanickam, and Kasimanickam 2021).

Excessive use of antibiotics in various applications – linked by one or another link with the various food chains – accelerates the rate of developing antibiotic resistance (Devirgiliis, Barile, and Perozzi 2011; Founou, Founou, and Essack 2016; Kumar, Arnipalli, and Ziouzenkova 2020). Most of the used antibiotics will finally reach nature. That will put additional stress on the microbial ecosystem. Did we establish correct regulation for the use of antibiotics in our backyard? Even the environment is the final collecting pool, but we will consume them again by one or other means (e.g., as agriculture products). We are mostly the last consumer of many of the food chains contaminated by antibiotics.

Those who practice biology know that the existence of an antibiotic in the cultivation media is an essential step to keep the antibiotic genes carried by plasmid in a recombinant strain. In contrast, sub-cultivation of a recombinant strain that harbors the antibiotic resistance gene in media does not contain antibiotics that will lead to plasmid curing. For that, the wide existence of antibiotic resistance genes everywhere is a correct response to the wide existence of different types of antibiotics in our ecosystems. That is because most bacterial strains that acquire antibiotic resistance genes on plasmids will have limited chance to cure them naturally. In another word, we put the bacterial strains under continuous need to have the resistance genes.

2.2.4 The Spreading and Mobility of the Resistant Genes

The continuous emergence of bacteria resistant to nearly all known antibiotics is a cause for serious concern, and this phenomenon has increased the search for new and more effective antibiotics. Although about 2,000 antibiotics have been discovered, only a few dozen antibiotics survive (Lengeler, Drews, and Schlegel 1999). Additional concern is the side effects associated with most of the antibiotics discovered (particularly the synthetic ones). The correct characterization and full understanding of the mode of action of antibiotics are the key to safeguarding the human healthcare system.

The spreading and mobility of the antibiotic resistance genes and mutated gene transfer among microbes could happen mainly as follows:

2.2.4.1 Chromosome Replication and Recombination-Mediated Resistance

Chromosome-mediated resistance occurs most probably because of spontaneous mutation (no new genes are involved). This is caused by a mutation in the gene that codes for either the target of the drug or the transport system in the membrane of the cell wall, which controls the entry of drugs into cells. The frequency of chromosomal mutation is much less than the plasmid-mediated resistance. One should not neglect that the genes of the antibiotic resistance exist either in the main genomic DNA or as a mobile element (the plasmid). The chromosomal gene is less expressed than that carried on plasmids. Antibiotic resistance gene carried by the chromosome probably existed as a single copy. Mutation in the chromosomal antibiotic resistance gene could turn the produced protein to be able to resist more than one antibiotic. Additionally, as described prior, in relaxed growth, microbes could get rid of the plasmid when there is no need for their genes. In contrast, genes built in the chromosomes could not be cured or corrupted easily, but under very specific conditions, such as inserting of transposon elements inside the gene or due to the treatment of chemical compounds that have an effect on the DNA. Phages also could cut the antibiotic resistance genes from the chromosome, but this will not be recognized because phage infection will destroy the bacterial mother cell. Meanwhile, the main way to transfer the chromosomal resistance genes is by the bacterial replication process. Each single cell of an antibiotic resistance strain, if given a chance to replicate safely, will be a source for the antibiotic resistance gene in its new ecosystem.

The recombination in microbes is a combining of genes or characters different from what they were in the wild microbial cells. It has been claimed that direct recombination might play a bigger role than expected in many natural gene exchanges like sickle cell anemia and antibiotic resistance (Valen et al. 2012; Amara 2017).

2.2.4.2 Plasmid-Mediated Resistance (Transformation and Conjugation)

The plasmid is a piece of DNA, usually circular (also called a mobile chromosomal element), functioning as part of the genetic material of a cell, not integrated with the chromosome and replicating independently, but transferred to subsequent generations. In bacteria, plasmids often carry the genes for antibiotic resistance. Plasmid-mediated drug resistance in bacteria occurs by the transfer of plasmids that contain the resistance genes to other bacterial cells. It is mediated by plasmids (that carry antibiotic resistance factors or RFs). These circular, usually, double-stranded DNA molecules can carry the genes responsible for resistance against a variety of antibiotics. These RFs may carry one or even two or more of the resistant genes. Plasmid-mediated resistance plays a very important role in antibiotic usage in clinical practice. This is because a high rate of transfer of plasmids from one bacterium to another bacterium takes place by conjugation. Additionally, plasmids are multi-copy, which means that the gene will be expressed in high amount of active proteins.

2.2.4.3 Transposons-Mediated Drug Resistance

Drug resistance is also mediated by transposons that often carry the drug resistance genes. Transposons are a segment of DNA that can become integrated at many different sites along a chromosome (especially a segment of bacterial DNA that can be translocated as a whole). Transposons at the end of DNA are able to move from one site of the bacterial chromosome to another and from the bacterial chromosome to plasmid DNA. Many RFs are carried in one or more transposons. Their copies can be inserted at random sites within chromosomes. A transposon has almost identical sequences at each end and inverted repeat sequences (that run in the opposite direction). This end encodes for transposase, that catalyzes its insert in the chromosome. If transposons are inserted inside a gene, this gene will become unable to translate the correct active protein.

2.2.4.4 Bacteriophage-Mediated Resistance

Bacteriophage is a virus that is obligatory parasitic (reproduces itself in bacteria). Bacteriophage uses the bacterium's machinery and energy to produce more phage until the bacterium is destroyed and the phage is released to invade surrounding bacteria. When a phage infects a bacterium, it either can destroy its host or be incorporated in the host genome in a state of lysogenicity. Lysogenic phages are able to re-infect the bacterial cells and adapt a lytic cycle. In some cases, they take a part of the bacterial chromosome with it. The antibiotic resistance gene can be transferred due to the temperate phages (Pires et al. 2016). Phages are involved in transferring different antibiotic resistance genes (Deng, Li, and Zhang 2018; Mathieu et al. 2019; Zhang et al. 2019). The temperate phages in bacterial cells are problematic in both nature and in specific areas (e.g., the human body). They are responsible for encoding different virulence factors, including ARGs, bacterial toxins adhesion, serine protease, haemolysin, thermostable enterotoxin, and special catalase system (Alexa et al. 2001; Hayashi 2001; Kutter et al. 2010; Vondruskova et al. 2010; Balcazar 2014).

2.2.4.5 Fungal Networks and Their Role in Dissemination and Enrichment of ARGs

Fungi in soil had significant influences on the diversity and abundance of a broad spectrum of bacteria with ARGs through soil microcosms (Nazir et al. 2017). In 2017, Nazir et al. suggested that fungi can also participate in global ARGs. They identified unique ARGs specially found in the mycosphere of different fungi via qPCR (qualitative polymerase chain reaction) assay. Fungi in ecological and taxonomical features exerted different selection pressures on ARGs, which originate from the same source. Some soil fungi could support different antibiotic resistance bacteria enriched in the mycosphere and unexpectedly transported to far places. Different fungi exhibit different effects to assist horizontal gene transfer potential under fungal selection pressure (Nazir et al. 2017).

2.2.4.6 Gene Mutation Followed by One of the Aforementioned Mechanisms

Excessive exposure to compounds or nutrients could enhance mutations (Amara 2013a). A new mutation in the β-lactamase gene can lead to multiple resistance to antibiotics. For example, β-lactamase is the RF against the β-lactam antibiotics. The plasmid-encoded TEM lactamase is the most prevalent in the Gram-negative gut bacteria. Discovery of a new β-lactam antibiotic leads to inducing new resistance. Venkatachalam et al. (1994) have examined TEM lactamase variants.

They induce site-directed mutation in the active site pocket of the enzyme. The resistance has been introduced to natural isolates. The resulted modified enzymes show increased resistance. Extended spectrum cephalosporins such as cefotaxime and ceftazidime are reported. Mutants have been selected for 100-fold higher ceftazidime resistance than the wild type. All mutants had a serine substitution at position 238, lysine or arginine at position 240, and a small amino acid at position 241 (Venkatachalam et al. 1994). This is an example of what could happen in the lab. After the resistant gene is mutated and becomes resistant to additional antibiotics, one of the aforementioned transformation methods might transfer it to other bacterial species.

2.3 DIFFERENT WAYS TO CLASSIFY ANTIBIOTIC RESISTANCE

For understanding the different types of antibiotic resistance, it is wise to first understand the way antibiotics themselves have been classified. Antibiotics are subjected to different steps concerning their classifications. As in the introduction part, the first belief was that only fungi are able to produce antibiotics, and then antibiotics were identified as bacterial products as well. The source of the microbe is also another thing that distinguishes between different microbes able to produce antibiotics. However, from the first days, they have been recognized for their inability to attack eukaryotic cells such as human cells (no cell wall) or fungal cells. For example, the fungal cell wall contains a long polymer of D-glucose (not peptidoglycan as in bacteria); thus, fungi are insensitive to antibiotics, such as penicillin, that inhibit peptidoglycan synthesis (Amara 2021). After that they have been recognized for their ability to attack different kinds of macromolecules such as the DNA, RNA, protein, peptidoglycan layer, ribosomes, some metabolites, and the like. The type of substrate, or the macromolecules that the antibiotic attacks, is place-dependent. In other words, the antibiotics that attack certain macromolecules should find a way to be able to attack them in their specific places in the cell. For example, DNA- and RNA-attacking antibiotics should reach the cytoplasm; hence, they might be prevented from reaching their target by the cell wall or the outer layer of different kinds of macromolecules.

Antibiotic resistance classification is mostly based on the opposite mechanism of the antibiotic. The resistance to antibiotics in its simple mean is a mechanism that blocks, deactivates, destroys, or fraud (or the like) the antibiotic mechanism. They generally are based on a gene that produces a protein able to either destruct or modify the active site of the antibiotic or protect/change the substrate structure. For example, the antibiotic target site can be modified so the antibiotic will be useless. Alternatively, the target site could be protected. Moreover, the microbial cells have different types of natural structures and macromolecules that are helping in resisting antibiotics, including the cell wall, capsules, exopolysaccharide outer layer, etc. Additionally, some microbes are able to perform unusual mechanisms to mediate resistance to antibiotics, such as slowing their consumption to the antibiotics. Being naturally resistant or acquiring additional foreign resistance genes is another phenomenon.

The ARGs in their nature either naturally existed in the antibiotic-producing microbes or were transmitted by one or another mechanism of the gene transfer mechanisms (as mentioned previously). Additionally, certain new mechanisms could be developed.

So, bacterial resistance to antibiotics could be classified by different means. For example, one classification is based on their main activity against the antibiotics, such as (a) production of enzymes, (b) production of altered enzymes, (c) synthesis of modified targets, (d) alteration of permeability of the cell wall, (e) alteration of metabolic pathways, and (f) efflux pump and many others. Many factors could differentiate between the different mechanisms of resistance. The time needed for elevating resistance to the antibiotic could also be used as some microbes show faster resistance developing for the new antibiotics such as *Staphylococcus* spp. Additionally, it has been remarked that exposure to natural antibiotics is followed by resistance in a shorter time than compared with the synthetic ones. Synthetic antibiotics take more time to develop resistance. In addition, synthetic structures that have been modified from a naturally existing structure could show resistance faster than a synthetic one that has no reference/similarity in the nature. Additionally, it might be important to highlight that some microbes are potent hydrocarbon-degrading machines, where the different antimicrobial agents (including antibiotics) are food for them (with some differences).

Another classification could be based on the genetic basis. For example, is the resistance based on one gene, so one protein is able to accomplish the process, or is it based on more than one gene, so more than one protein collaborates to perform the resistance? Or they are based on an already-existing structure (e.g., the cell wall, capsules, etc.) or an existing mechanism to get rid of any toxic materials (e.g., efflux)? Additionally, there are extra-fine differences that could also be used to differentiate between some branches of different antibiotic resistance mechanisms such as the need/no need for the divalent cations to finalize the effect of the antibiotic-degrading enzymes, such as the need for zinc by ampicillinase.

In fact, one who is interested in studying antibiotic resistance should be aware of these differences. The genetic base of the resistance will be used in the following part. The classification will be either based on the activation of one gene/multiple genes or mediated through an already existing structure or macromolecules.

2.4 CLASSIFICATION BASED ON THE GENETIC BASIS

It might be interesting to classify antibiotic resistance mechanisms as either being based on the activation of one gene or multiple genes or based on already existing mechanisms or structures. That will put the major group, which is based on the direct enzymatic attack to the antibiotic together in one group. They are the most claimed agent in both antibiotic resistance and in transfer the resistant genes within microbes particularly those, which are carried, on extra chromosomal mobile elements (the plasmids).

The other group will include the resistance mechanisms that are controlled by many genes that give many proteins to perform antibiotic resistance. This does not mean that all the genes are involved at once, but the last effective reaction is mediated mostly by one protein produced by one gene (so they are still genetically based). Again, one should highlight that a protein might do not directly cause the effect of the antibiotic resistance alone but needs other factors, such as the divalent cations. Even so, in such cases, it will be considered in the frame of the direct gene to a direct protein that mediates antibiotic resistance. This classification might bring better understanding to the role of antibiotic resistance as well as to the effect of the power of microbes in developing resistance or the effect of an external factor like the geographic and epigenetic factors in the different resistance mechanisms (Amara and Shibl 2015b).

Some wise countries kept some selected antibiotics as a last option and did not allow their use in hospitals. This let these countries avoid the global antibiotic problem. The war against pathogens needs more understanding of the behavior of microbes and continuous evaluation and validation to our practices starting from daily activities, sanitization, the disinfection of hospitals, diagnosis, antibiotic tests, treatments, and the like. Any broken step could lead to the elevation of the resistance.

Based on the genetic basis, antibiotic resistance could be classified to:

1. One gene to one protein that could mediate antibiotic resistance.
2. More than one gene gives more than one protein each one mediates a step in antibiotic resistance mechanism.
3. Antibiotic resistance because of special microbial structure or macromolecules.

2.4.1 One Gene to One Protein That Could Mediate Antibiotic Resistance

The genetic basis of antibiotic resistance mediated by genetic change in bacteria that were antibiotic sensitive is most important in the development of antibiotic resistance in bacteria. The antibiotic resistant genes are the genes that encode for a variety of proteins (mostly enzymes) that destroy the antibiotics by degrading/modifying/blocking antibiotics or modifying the membrane transport system. For example, the genes that code for enzymes like β-lactamases that destroy the β-lactam ring are responsible for the anti-bactericidal action of β-lactam antibiotics (e.g., penicillins and cephalosporins).

2.4.1.1 Antibiotic Resistance Enzymes

Bacteria produce enzymes that inactivate antibiotics. For example, penicillin-resistant staphylococci produce β-lactamase that destroys the penicillins and cephalosporins by splitting the β-lactam ring of these antibiotics. Gram-negative bacteria resistant to aminoglycosides, mediated by a plasmid, produce adenylating, phosphorylating, or acetylating enzymes that destroy these antibiotics. Carbapenemases, from *Klebsiella pneumonia*, is able to break down carbapenems and most other β-lactams. Enzymatic inactivation of antibiotics occurs with numerous natural antibiotic classes. In contrast to the naturally produced antibiotics, it has not yet been observed as a major route of resistance development for the classes of synthetic antibacterial agents such as the sulfamethoxazole trimethoprim combination, the fluoroquinolones, or the oxazolidinones. Such lack of a direct gene that produces protein (enzyme) to degrade the synthetic antibiotic proves the natural origin of the ARGs for natural antibiotics. Either the antibiotic resistance microbes that mediate the resistance through an enzyme probably naturally harbored the related genes or one of the aforementioned transforming mechanisms has transferred these genes. Additionally, a synthetic antibiotic that was derived from natural structure (but subjected to a certain modification) might be able to resist if mutations that happened to the wild-type genes enable the enzyme to interact with the modified structure. In most cases, these modifications did not affect their resistance to the original natural antibiotic with the wild structure. The modified enzyme will be able to resist different antibiotics mostly from the same group.

2.4.1.2 Destroying/Modifying/Protecting Antibiotic Target Sites

Some antibiotics can single out and destroy specific parts (targets) of a bacterium. Bacteria change an antibiotic's target so the antibiotic can no longer fit and do its job. The *E. coli* mcr-1 gene adds a compound outside the cell wall so that the antibiotic colistin cannot latch onto it. An alternative to destroying or modifying the target site are proteins that are able to mediate target protection. The antibiotic resistance genes can be found in the bacterial DNA. But most of the clinically relevant resistance genes are carried by plasmids. Tetracycline resistance determinants include tetracycline resistance gene M (Tet(M)) and tetracycline resistance gene O (Tet(O)). Tet(M) existed in *Streptococcus* spp. and Tet(O) in *Campylobacter jejuni*. They are acqu ired by different bacterial species. They are found in numerous plasmids and in broad-range conjugated transposons (Connell et al. 2003). TetO and TetM act together with the ribosome and displace tetracycline from its binding site, a process which is Guanosine triphosphate (GTP) dependent. This interaction alters ribosomal conformation, preventing rebinding of antibiotics (Dönhöfer et al. 2012). TetO has also been shown to compete with tetracycline for the same ribosomal space. It alters the geometry of binding sites of antibiotics (Li et al. 2013). Certain bacteria produce modified targets against which the antibiotic has no effect. For example, a methylated 23S ribosomal RNA can result in resistance to erythromycin, and a mutant protein in the 50S ribosomal subunits can result in resistance to streptomycin. Penicillin resistance in *S. pneumoniae* and enterococci is caused by the loss or alteration of penicillin-binding proteins. These target changes may consist of:

1. Point mutations in genes that encode the target site.
2. Enzymatic alterations of binding sites (e.g., addition of methyl groups).
3. Replacement or bypass of the original target.

2.4.2 More Than One Gene Gives More Than One Protein Each One Mediates a Step in the Antibiotic Resistance Mechanism

2.4.2.1 Different Efflux Pumps

Bacteria have large numbers of efflux pumps, used physiologically to export specific metabolites and pump out foreign toxic substances. The *E. coli* efflux system that is able to pump tetracycline out of cytoplasm was described in the early 1980s and was the first to be described (McMurry, Petrucci, and Levy 1980). After that, many different classes of efflux pumps have been specified in Gram-negative and Gram-positive pathogens (Poole 2005). They can be categorized in five major families, including i) the major facilitator superfamily (MFS), ii) the small multi-antibiotic resistance family (SMR), iii) the resistant-nodulation-cell division family (RND), iv) the ATP-binding cassette (ABC), and v) the multi-antibiotic and toxic compound extrusion family (MATE). These families prove to be different in terms of structural conformation, energy sources, and range of substrates (Piddock 2006). The efflux pumps have been found to be responsible for conferring resistance to many groups of antibiotics including aminoglycosides, quinolones, etc. Pumps that evacuate antibiotics from cells are unregulated (efflux) (Morita 2012). Antibiotics are removed from bacteria via pumps embedded in their cell walls, which remove antibiotics that enter the cell. Pumps can be produced by certain *P. aeruginosa* bacteria to get rid of a variety of antibiotics, including fluoroquinolones, β-lactams, chloramphenicol, and trimethoprim. The up regulation of pumps (efflux) expels the antibiotic from the cell (Morita 2012). The active efflux is mediated by transmembrane proteins, both in the cytoplasmic membranes and in the outer membranes of Gram-negative bacteria. The transmembrane proteins are acting as pumps that export the antibiotics. Active efflux can be clinically relevant for β-lactam antibiotics, macrolides, the pristinamycin peptides, fluoroquinolones, and most classically the tetracyclines. Some efflux pumps have relatively narrow specificity (e.g., the tetracycline pumps). Others have broad tolerance and confer multiantibiotic resistance phenotypes.

2.4.2.2 Reducing/Preventing the Antimicrobial Molecule's Absorption and Cell Wall Re-Modeling

Vancomycin is an antibiotic that targets the bacterial cell wall, preventing its formation in the correct way. This antibiotic works by binding to a section of the building blocks of the cell wall named D-ala-D-ala (these are two unusual alanine amino acids). This D-ala-D-ala structure is normally essential to a bacterial cell wall. Cells contain D-ala-D-ala subunits in their cell wall. Upon being susceptible to vancomycin, the synthesis of cell wall containing D-ala-D-lactate by the enzyme VanH and VanA renders cells resistant to vancomycin. That kind of drastic change requires a major realignment in the cell. To pull it off, resistant cells must acquire five new genes (these genes come from other bacteria that are already resistant to the antibiotic). These genes produce proteins that sense the presence of vancomycin and then build and assemble the new cell wall structures. The structures or enzymes being mutated are typically essential for survival.

Many antibiotics used in medical treatment have intracellular bacterial targets or, in the case of Gram-negative bacteria, located in the cytoplasmic membrane (inner membrane). Hence, the compound must enter through the outer and/or cytoplasmic membrane in order to employ its antimicrobial effect. Bacteria have evolved ways to prevent antibiotics from reaching their intracellular or periplasmic targets by reducing the antimicrobial molecule's absorption. This process is especially evident in Gram-negative bacteria, since it limits the entry of chemicals from the surrounding environment. The outer membrane, in reality, serves as a first line of defense against the entry of a variety of hazardous chemicals, including antimicrobial agents. Hydrophilic compounds like β-lactams, tetracyclines, and certain fluoroquinolones are particularly impacted by changes in the outer membrane's permeability because they frequently employ water-filled diffusion channels known as porins to traverse them (Pagès, James, and Winterhalter 2008).

The prime case of the productivity of this normal boundary is the reality that vancomycin, a glycopeptide antimicrobial, is not dynamic against Gram-negative life forms because of the need of infiltration through the external layer. Moreover, there is the natural defenselessness of *Pseudomonas* and *Acinetobacter baumanii* to β-lactams (if compared to Enterobacteriaceae); to a diminished number and/or differential expression of porins (Hancock and Brinkman 2002). Beside natural ability to reduce the antibiotic penetration in some bacteria, some bacteria develop resistance to antibiotics by changing their permeability to the antibiotic in such a way that an effective intracellular concentration of the antibiotic is not achieved inside the bacterial cell. For example, *P. aeruginosa* develops resistance against tetracyclines by changing its porins that can reduce the amount of tetracycline entering the bacteria, thereby developing resistance to the antibiotics. Resistance to polymyxins is also associated with a change in permeability to the antibiotics. Streptococci have a natural permeability barrier to aminoglycosides. This can be partly overcome by the simultaneous presence of a cell wall–active antibiotic, e.g., penicillin.

2.4.2.3 Utilizing Pre-formed Metabolites to Avoid Blockage of Key Metabolic Pathways

Numerous antibiotics, such as sulfonamides and trimethoprim, have been shown to mimic the substrates required for bacterial cell metabolism. This deception causes bacterial enzymes to attach to antibiotics rather than the usual substrate. In particular, sulfonamides act like tetrahydro folic acid, which is required for folic acid synthesis in bacterial cells (Talaro and Chess 2008). Folic acid is essential for the metabolism of nucleic acids and amino acids. For this reason, sulfonamides mimic the substrates required for folic acid metabolism and ultimately interfere with the production of nucleic acids (DNA and RNA) and amino acids (Talaro and Chess 2008). Bacteria may develop resistance by altering metabolic pathways that bypass the reaction inhibited by the antibiotic. For example, certain sulfonamide-resistant bacteria do not require extracellular PABA (para aminobenzoic acid) but, like mammalian cells, can utilize preformed folic acid.

2.4.3 Antibiotic Resistance Because of Special Microbial Structure or Macromolecules

The different bacterial structure backbones (other than the antibiotic resistance genes) and different macromolecules monomeric/heteromeric compositions are playing different roles in antibiotic resistance. The antibiotics should interact with the existing macromolecules or could be transferred by them through some structures in the cell barrier to the bacterial preplasmic space or to the cytoplasm. The non-genetic basis of resistance plays a less important role in the development of antibiotic resistance as follows:

1 Certain bacterial species under ordinary circumstances are usually killed by penicillin. But these bacteria, if they lose their cell wall and become protoplast, become unsusceptible to the action of the cell wall–attacking antibiotics such as penicillin.
2 It is a common observation by molecular biologists to find sensitive colonies able to grow after a few days of using ampicillin. After transforming *E. coli* with ampicillin resistance genes carried on plasmids, only cells transformed by the plasmids could grow on ampicillin selective Petri dishes. Meanwhile, ampicillin is a bacteriostatic antibiotic. Not all the used recombinant cells (that did not acquire the plasmid) are killed – many are still live. They appear suddenly after days. They take more time to grow, probably after the antibiotic in the Petri dishes is degraded by the correct transformants that harbor the ampicillin resistance gene (degrade the antibiotic in the surrounding area). Their ability to survive in the presence of the ampicillin for days proves that the bacterial cells could rebuild their cell wall (or reconstruct it) if not severely damaged or use another mechanism to survive.

3 In certain conditions, such as in the abscess cavity, bacteria can be walled off, which prevents antibiotics from penetrating effectively into bacteria. Surgical drainage of pus, however, makes these bacteria again susceptible to the action of antibiotics.
4 Presence of foreign bodies (such as surgical implants and catheters) and penetration injury caused by splinters and sharpeners make successful antibiotic treatment more difficult, probably due to the activity of incorrect concentrations of antibiotics.
5 Non-replicating bacteria in their resting stage are less sensitive to the action of cell wall inhibitors such as penicillin and cephalosporins. This is particularly true for certain bacteria such as *Mycobacterium tuberculosis*, which remains in the resting stage in tissues for many years, during which it is insensitive to antibiotics. However, when these bacteria begin to multiply, they become susceptible to antibiotics. This phenomenon has been observed also in aged cultures.
6 Hidden under the live or dead cells (biomass) and biofilms as proposed by Amara (2011) is another possibility of acquiring antibiotic resistance gene(s) or developing resistance mechanism(s). Viable microbes are targets for free mobile elements or DNA free fragments in their surrounding area. Ruptured microbes (or partially raptured) are the source of such free elements (Amara 2011). The process may possibly be given the name "static natural transformation." It is expected to happen during the treatment with cell wall–targeting antibiotics.
7 Presence of factors that are able to induce mutation in certain microbes, such as H_2O_2, which induces alginate overproduction (e.g., *P. aeruginosa*) can protect microbes from being killed by the antibiotics. The alginate will form a layer/biofilm that protects the microbe as a biofilm. Alginate contributes to an effective role in antibiotic resistance (Amara, Salem-Bekhit, and Alanazi 2013a; El-Shanshoury et al. 2018a). Alginate biofilm allows mechanical protection of the viable microbes.
8 The microbes build a unique strategy upon being part of the biofilm while they reduce their biological activity as well as the amount of the consumed nutrient. Very slow growth in its simple explanation means that the cells eat less. So, the effective dosage of any antimicrobial agent (calculated based on active growth) will not reach the cells.
9 The geographic factors could affect the ARGs themselves or the way of their transformation. That has been proposed by Amara and Shibl (2015). It has been proposed that the food has an epigenetics-like mechanism (Amara and Shibl 2015b). Or, at least it enriches some microbes over others and spreads resistance genes in their ecosystem.
10 Other human activities could also accelerate antibiotic resistance. For example, the use of antimicrobial agents during plant growth could contaminate the tissue. This affects food quality and the microbial activity in our bodies as well as the animals that might consume such plants. Captive birds are often overtreated with antibiotics. This can lead to an increase in antibiotic resistance. Captivity as well is known to have a negative effect on immunological responses.

2.5 SUGGESTED TACTICS TO CONTROL ANTIBIOTIC RESISTANCE

Even late, some tactics could be introduced to control antibiotic resistance. They are mainly extracted from different experiences and ideas concerning the troubles faced by the use of antibiotics and the expected purposes that lead to the elevation of resistance. Some of them are as follows:

1 Reducing the dependence on antibiotics in different fields by using different alternatives to antibiotics (as summarized in the following section).
2 Using the physical conditions that control microbial growth in different fields, such as a dry environment (preventing water and moisture from reaching the microbes).

3 Use antibiotics different than those applied to human infections in different applications. That means keeping one or more classes to be used in non-human-treatment applications such as agriculture, poultry, and farm stock (Jeżak and Kozajda 2022).
4 Keep some antibiotics from different groups out of pharmacies as a last choice to be used in hospitals only (in emergence cases).
5 Classifying microbial infections based on their species and installing an update database for their expected properties and their ability to develop resistance.
6 For multi-antibiotic-resistance microbes (have different types of antibiotic resistance mechanisms), a mixture of antibiotics should be used.
7 Patients should be isolated in separate areas to prevent the distribution of resistance genes, particularly in hospitals.
8 Re-evaluate the different strategies for disinfecting and sanitizing important places such as hospitals. As reported in this review and others, some microbes are hydrocarbon degrading, so disinfectants are food for them. The relationship between the microbes that live in the community should be observed. If water is the last exposing material, the water will be evaporated. But if disinfectant is the last applied agent, they will be the subject of the growth of some microbes that are able to utilize them as nutrient (particularly fungi). Such microbes will have disinfectant-degrading genes as well as they might have antibiotic resistance genes.
9 Some tactics could suppress the antibiotic resistance enzymes and improve the use of antimicrobial agents (such as antibiotics and disinfectants). An example of that is chelating agents, which could be used in different superficial treatment with antibiotics or as disinfectants (Amara and Hussain 2006; Hussein and Amara 2006). Curing the divalent cations will help both in interrupting the antibiotic resistance enzymes that need them to finalize their activity and disturbing or preventing the biofilm from being formed in correct structure (less ridge).
10 Removing any source of antibiotics that contaminate the ecosystem to decrease the probability that the microbes use the plasmids that carry the resistance factors will keep them active. The absence of the antibiotic(s) will encourage the microbes to get rid of them (e.g., plasmid curing). Microbes, if they find no need to keep the resistance genes, they mostly will get rid of them.

2.6 ALTERNATIVES TO ANTIBIOTICS

The alternatives to antibiotics are so important in the war against microbes. They are huge, applicable, and some exist in our foods as nutraceutical compounds. Most of them are safe, such as acids in dairy products and vegetables. Some still need research to enable safe use, such as bacteriophages. Other are safe microbes in their nature, such as probiotics, which could control pathogens and produce different types of antimicrobial products and byproducts (Amara and Shibl 2015a). Some ancient folk remedies could be used particularly in emergency cases where the medicine might not be available, such as ash that contains carbonic acid and might contain some heavy metals in safe quantities, which could be used for superficial treatment. Volatile oil is a wonderful natural compound that if used probably will give many positive effects against microbes and will protect against them. Treatment with metals, particularly the noble ones, by using and applying them to our skin will improve the health generally and will protect against different infections. Lysozymes and short active peptides are alternatives to antibiotics and prove to be efficient. Production of vaccines for microbes that usually infect humans will be another good choice. One promising tactic is using microbial ghosts either for Gram-negative bacterial strains by applying the bacteriophage *E* lysis gene strategy or by applying the sponge-like protocol which uses the critical concentration of some chemical compounds to evacuate microbes such as viruses, fungi, yeast, bacteria, spores, and the like (Amara and El-Baky 2022).

In some cases, some nutraceutical foods are so efficient for particular cases. For example, lemon juice might be enough for throat infection! The lemon contains chemical ingredients that kill the *Streptococcus pyrogen* in an efficient way (Adil, Mh, and Hpv 2019). The lemon juice (also helps in absorbing vitamin K), garlic, onion, and the like in a salad dish protect from different kinds of pathogens. Chowing food correctly will apply the correct dosage of the saliva lysozyme that will lysis different types of microbes (Allgrove, Oliveira, and Gleeson 2014). Out of a huge number of antibiotic alternatives, some have been summarized in Table 2.1.

2.7 FUTURE PROSPECTIVE

From much of the literature, one could observe a pessimistic sense concerning antibiotics. In fact, that is not correct, because antibiotics are produced by wild microbes and wild microbes have survived over the ages. The chances that the original wild microbes (either those which are able to produce antibiotics or those which are sensitive) turn to be resistance microbes is great. Meanwhile, by using better manipulation and a controlled use of antibiotics, many of our recent problems with antibiotics could be solved. The problem is not only human-based but also environmentally based. Many antibiotic genes can be cured from nature spontaneously if the antibiotic did not exist for a long time in certain environments where certain microbes caring resistance genes on plasmids are existed. Our deep understanding to the antibiotics, their mode of actions, antibiotic resistance, and the different structures of the bacteria, as well as the different defending tactics that they developed/acquired or supported by the surrounding environment (e.g., biofilm), will enable better use of antibiotics, which finally will reduce the dilemma of antibiotic resistance. If we are starting to use antibiotics wisely and using their alternatives, global antibiotic resistance will be diluted over time. Humans made such problems, and they are the only ones able to solve it by stopping their unwise activity.

2.8 CONCLUSION

Antibiotics are wonderful drugs that are able to kill most bacteria. They have been applied for a long time safely and in many applications. But putting so much dependence on antibiotics leads to the elevation of the antibiotic resistance. Antibiotic resistance is a natural phenomenon as the antibiotic producer strains harbor genes and have different mechanisms that enable them to survive against their own products. Genes are transferred by different mechanisms starting from normal recombination to plasmid transformation as well as through phages and transposon elements and the like. Additionally, quite a number of bacteria are able to conjugate. Those mechanisms facilitate the spreading of ARGs. Additionally, some microbes have some extra genetic elements than the other one. For example, *Pseudomonas* spp., *Staphylococcus* spp., *Streptococcus* spp., and the like have more genes if compared with *E. coli*. These microbes have additional natural mechanisms to resist antibiotics and any other toxic compounds. Even some of them are hydrocarbon degrading. Hydrocarbon compounds are considered as food for them. For that, they dominate in hospital, waiting for new microbial visitors to gain new resistant genes! Each cycle of disinfectants will kill other microbes but keep the worst. Using additional strategies like drying the hospital environment is proposed to control them. Synthetic antibiotics take a long time to develop resistance. In contrast, synthetic antibiotics from a modified existed structure will need only an additional mutation in the related genes.

It is important to understand the different mechanisms involved in the resistance of antibiotics and which one is based on genes and which one is based on the basic bacterial structure. Meanwhile, sometimes the basic structure is supported by a gene reaction to apply for a specific type of resistance like modifying the cell wall structure or altering the substrate site.

There are quite enough alternatives to antibiotics, including microbial antagonism, antimicrobial natural products such as plant extracts or volatile oils, probiotics, bacteriophages, vaccines, antibodies, etc., as in Table 2.1.

TABLE 2.1
Some Alternatives to Antibiotics; Types and Mode of Action

Alternative way	Antimicrobial agent	Mode of action
Direct antagonism	**Microbes against microbes**	Antagonism is defined as the relation between opposing principles, forces, or factors in our context. It means using microbes against microbes. It was used blindly since ancient times. For example, in the Egyptians folk, if one takes a swap by their fingers from the area around the teeth bottom after being woke up and apply it to the infected skin (e.g., infection on the face), this infection will disappear after a few days. Using sand, mud, spoilage foods, and the like to antagonist microbes was reported by many authors before and during the Pasteurian area (as prior)
Noble metals (silver, gold, platinum, etc.)	**Silver, gold, platinum**	Silver, gold, platinum, and the other noble metals are metals that are resistant to corrosion or oxidation. Many have been using silver and/or gold instruments and other noble metals in surgeries (self-sterile). Europeans' nobles have been healthier than the lay peoples. Historically, the explanation has been given to their pure blood "blue blood." Today, the explanation is thought to be the silver jewels that were touching their skin for a long time.
Plants		The Egyptian folk and belief was that lemon juice is poisonous killer. Crud lemon juice is used traditionally against throat infections. It gives a fast-curing reaction against the throat infection with *Streptococcus pyrogenes*. The juice of one lemon applied to the mouth every two to three hours for three times is enough to get rid of such infection. Arabs use branches of *Salvadora presica* to clean their teeth. It is also used for wound treatment. Onion and garlic are used widely to protect/treat various kinds of microbial infections particularly against the common cold.
Ashes and wood perfumes	**Carbonic acids and heavy metals**	The ash from fired dry woods and plants are well known for their antimicrobial activity. Also, the perfumes come out during the fire process of some woods. They were used by early humans to disinfect injury, preserve animals' lather, and the like. In wild nature, wood, stones, and sand are widely existed. In such cases, ashes could be used to disinfect injury. Ashes consist of carbonic acids and concentrated heavy metals (if existing). This mixture will kill microbes. Humans discovered its sanitization properties and used it without any knowledge about the microbe itself. Some still practice the treatment with ashes nowadays while it is safe particularly if applied to superficial injuries.
Volatile oils	**Aromatic compounds**	Volatile oils from different plants are well known for treating microbial infections, in different sanitization applications, and as perfume since ancient times.
Probiotics and prebiotics		Prebiotics are foods and drinks that can favor the presence of beneficial microorganisms in the intestines (Amara and Shibl 2015a). Using a product that contains microbes to control other microbes goes back far in human history (used blindly). Meanwhile, the name "probiotic" was invented by modern civilization. The main source of most probiotics strains is traditionally old and present in various fermented food. Parker (1974) described probiotics as the other half of the antibiotic. Besides their beneficial effect on food, some species of probiotics are able to inhibit the growth of pathogens (Parker 1974).

(*Continued*)

TABLE 2.1 *(Continued)*
Some Alternatives to Antibiotics; Types and Mode of Action

Alternative way	Antimicrobial agent	Mode of action
		One critical part in our body that could lead to degrade our health is gastrointestinal microflora (GIM). If the pathogenic microbes in the GIM increase in number or become dominant, they become a source for toxins and undesirable micro- and macromolecules. Each meal will become a new load of such undesirable structures. In such cases, unexpected illness starts to appear, such as eczema.
		After medical treatment with antibiotics, or antibiotics that could affect the GIS microflora, it is advised to use therapeutic probiotics under the supervision of a physician (Trapecar et al. 2011; Dinleyici et al. 2012; Magnuson et al. 2013; Koutsoumanis, Lianou, and Sofos 2014; Jeong, Heo, and Lee 2017; Laulund et al. 2017; Murphy et al. 2017; Cebrián et al. 2019; Agriopoulou et al. 2020; Zugic Petrovic et al. 2020; Iseppi, Lauková, and Sabia 2021).
Probiotic byproducts	**H_2O_2, CO_2, ethanol, lactic acid** **Acetic acid, propionic acid, bacteriocin**	Probiotics produce different kinds of antimicrobial agents such as acids, H_2O_2, CO_2, ethanol, and the like. The most common acids produced as byproducts of the fermentation processes are lactic acid (Carr, Chill, and Maida 2002; Leroy 2004), acetic acid (Papalexandratou et al. 2011; Sengun 2011), and propionic acid (Thierry 2015). As an example about bacterial acid activity, lactic acid permeabilizes Gram-negative bacteria by disrupting the outer membrane (Alakomi et al. 2000; Suskovich et al. 2010; Papalexandratou et al. 2011; Sengun 2011). Acetic acid producers such as *Acetobacter aceti* and propionic acid producers such as *Propionibacterium* spp. include *Propionibacterium freudenreichii*, *P. jensenii*, and *P. acidipropionici* (Woskow and Glatz 1991; Brede et al. 2007; Liu et al. 2015, 2016; Piwowarek et al. 2018).
		Bacteriocin are ribosomal synthesized peptides originally defined as proteinaceous compound affecting the growth or viability of closely related organisms. Bacteriocin is similar to antibiotics but contains amino acids and/or modified amino acids. They are variants in their types (O'Shea et al. 2013; Furtado et al. 2015; Shyntum et al. 2019). Bacteriocin is applied in the control of foodborne pathogenic and spoilage bacteria (Mokoena 2017; Dicks et al. 2018; Nazari and Smith 2020; Pogány Simonová, Chrastinová, and Lauková 2020). They are largely used in food preservation and have been reported as safe products to preserve food (Gautam and Sharma 2009).
Bacteriophages	**Lytic bacteriophages**	Another natural, safe bacterial attacker is the lytic bacteriophage (wild and genetically engineered) (Pires et al. 2016). Phages are able to infect and kill many of the prokaryotic microbes. Some bacteriophages are strain specific and can be used to replace some antibiotics. The capacity to engineer phages genetically with new properties for therapeutic use is potentially advantageous. Direct treatment with a cocktail of bacteriophages is applied to control different bacterial species (Loc Carrillo et al. 2005; Gutiérrez et al. 2016; Mancuso, Shi, and Malik 2018; D'Accolti et al. 2021). Bacteriophages are effective if they directly interact and kill disease-causing bacteria. They should be administered around the time of infection. They will work only when bacteria are actually present in abundance and causing infections. In the absence of bacteria, phages mostly become inactive in the animal (Podlacha et al. 2021).
Vaccines	**Dead, attenuated, and degraded microbes; similar safe microbes**	Vaccines consist of a suspension of weakened or dead pathogenic cells injected in order to stimulate the production of antibodies.

		Examples of some pathogens that have been produced as vaccines are Newcastle virus (Tatár-Kis et al. 2020), genotype VII Newcastle disease virus (Wang et al. 2021), a *Salmonella* vector vaccine delivering three *Clostridium perfringens* antigens protecting poultry from necrotic enteritis (Wilde et al. 2019), *Campylobacter jejuni* vaccine (Zeng, Xu, and Lin 2010), airborne gumboro vaccine virus (Zhao et al. 2012), attenuated *Salmonella Typhimurium* live vaccine, and many others. Some factors can influence the choice of vaccination method such as the user's health status, innate immunity, and vaccination costs (Rauw et al. 2009). In some countries (Belgium, the Netherlands, and Germany), vaccination of farm animals against some diseases is mandatory. In France, only long-lived poultry (laying and breeding poultry) are vaccinated (Rauw et al. 2009).
Antimicrobial peptides	**More than 1,200 antimicrobial peptides**	Antimicrobial peptides are active peptides with antimicrobial activity. The advantages of active peptides are their activity on the general spectrum that includes most of the Gram-positive and Gram-negative bacteria, as well as their bactericidal and fast action, target-based resistance, and low immunogenicity. More than 1,200 antimicrobial peptides have been discovered from various origins, from plants to invertebrates and animals (Boman 2003; El-Baky, Elkhawaga, et al. 2021; El-Baky and Amara 2023; El-Baky, Amara, and Redwan 2023).
Microbial ghosts	**All kinds of microbes**	Bacterial ghosts have been prepared traditionally using the bacteriophage cloned *E* lysis gene under a control of the activity of a heat-sensitive promoter carried on suitable plasmids to induce pores in the cloned microbes. A new method called the Sponge-Like Protocol was invented in 2013 (Amara, Salem-Bekhit, and Alanazi 2013a) and shows acceptance as a universal protocol for microbial evacuation as well as in filed controlling pathogens (El-Baky, Rahman, et al. 2021). The described protocol is simple, inexpensive, and forward. However, certain experiments should be conducted first to determine the minimum inhibition concentration and minimum growth concentration of the used chemical compounds for each microbe. The new protocol succeeded to evacuate each of the prokaryotic and the eukaryotic microbes. H_2O_2 has been recommended by El-Baky and Amara (2014) to control Newcastle virus and other viruses, particularly in water (El-Baky and Amara 2014). The Sponge-Like Protocol might represent perfectly the ability of both chemical compounds and enzymes to do similar targeted processes. The used chemical compound (NaCl, NaOH, SDS, $CaCO_3$, $NaHCO_3$, and H_2O_2) might also collaborate with some of the microbe macromolecules to cause self-degradation to its cell wall (Amara, Salem-Bekhit, and Alanazi 2013a, 2013b; Amrara et al. 2014; Amara 2015b, 2016; Menisy et al. 2017; El-Baky et al. 2018; Sheweita et al. 2019). Or they function by activating some other ions (Amara 2011). The Sponge-Like Protocol is designed for evacuating the microbial cells to mimic what happened naturally (Amara, Salem-Bekhit, and Alanazi 2013a, 2013b; Amara 2015b; Menisy et al. 2017). Quite enough microbes have been evacuated either by the bacteriophage *E* lysis gene-based protocol or by the Sponge-Like Protocol. The most critical point in the Sponge-Like Protocol is to determine the MIC (minimum inhibitory concentration) and MGC (minimum growth concentration) of the selected chemical compounds (e.g., SDS, NaOH, $NHCO_3$ and the like) or enzymes (e.g., lysozyme).

(*Continued*)

TABLE 2.1 *(Continued)*
Some Alternatives to Antibiotics; Types and Mode of Action

Alternative way	Antimicrobial agent	Mode of action
Control the divalent cations	**EDTA and the like**	There are many ways to deactivate enzymes. Enzymes mostly responsible for inducing resistance and degrading, modifying, and blocking the antibiotic (Amara 2015a). Nevertheless, the charges and ions are localized in a highly specific orientation around their specific target. Organization, coordination, step-fitting, and function/structure/specificity are critical issues in the interaction of the charged macromolecules (e.g., active amino acids residue that interacts with divalent cations) and their substrates (Amara 2013b). Eliminating single fine steps from the process (e.g., the cations) will corrupt all the processes (Amara 2005a, 2005b, 2011, 2015b; Hussein and Amara 2006). It might also be interesting to report that controlling some types of divalent cations will interfere with important structures and macromolecules such as exopolysaccharide (Amara 2022). That interferes with a protective structure like biofilm made from alginate (Amara 2011; Amara et al. 2013; El-Shanshoury et al. 2018b; El-Shanshoury et al. 2018a).
Drug delivery	**Liposomes, microspheres, polymeric carriers, nanoplexes, polymeric carriers, bacterial ghosts, and the like**	The treatment of intracellular bacterial pathogens (as well as some other microbes) (Alomrani et al. 2014) is still limited because of the spreading of antibiotic resistance (Czyż et al. 2014). Most antibiotics are hydrophilic and cannot invade the plasma membrane of infected cells spontaneously. Therefore, intracellular infections remain ineffective despite standard treatment. Increasing the therapeutic dose and duration of treatment frequently results in severe side effects. Conventional antimicrobial compounds have difficulty penetrating, accumulating, or remaining in mammalian cells in the presence of intracellular bacterial pathogens (Gomarasca et al. 2017). The drug delivery for antimicrobials reached an important advancement in recent years, particularly the nanoparticle-based strategies (Zhang et al. 2022). Various developed drug carriers have been improved, including liposomes, microspheres, polymeric carriers, and nanoplexes. Despite these new approaches, the delivery of antibiotics still faces numerous obstacles (Chifiriuc et al. 2016). Different organs, tissues, and subcellular sections have different transport barriers for antibiotics, as well as different pathophysiological conditions, which interfere with the delivery of antibiotics (Jager et al. 2019). Numerous bacterial strains normally multiply within the cells, causing a diverse chronic infection. At the intracellular level, several pathogens can invade and multiply within host cells as macrophage and epithelial cells – for example, *Shigella*, *Legionella pneumophila*, *Listeria monocytogenes*, and methicillin-resistant *Staphylococcus aureus*, among others (Johnston et al. 2017). Intracellular infection is frequently connected with a sum of recurrent or chronic infection as pulmonary infections, rhinosinusitis, endocarditis, and osteomyelitis (Hodgson et al. 2015).

Geographic area, personality, microbial strains, food, and the like all are proving to have an effect on antibiotics, microbes, and our responses. Those factors and others should not be neglected. A catastrophic side effect comes from treating all microbes as one unit such as treating a patient with a broad-spectrum antibiotic or while disinfecting hospitals with a group of disinfectants daily and every time. Specialists should play more roles in optimizing the beneficial effects and reducing the side effect of the antimicrobial agents used here and there. Out of the different kinds of antimicrobial drugs, the most important one is the antibiotic.

REFERENCES

Abraham, E. P., E. Chain, C. M. Fletcher, A. D. Gardner, N. G. Heatley, M. A. Jennings, and H. W. Florey. 1941. Further observations on penicillin. *Lancet* 238 (6155):177–189.

Adekunle, C., A. Mustapha, G. Odewale, and O. Richard. 2020. Detection of antibiotic resistance genes among multiple drug resistant *Pseudomonas aeruginosa* isolated from clinical sources in selected health institutions in kwara state. *Infect Disord Drug Targets* 21 (6):e170721187999. doi: IDDT-EPUB-111517 [pii] 10.2174/1871526520666201116103625.

Adil, M., B. Mh, and R. Hpv. 2019. [Impact of citral and phloretin, alone and in combination, on major virulence traits of streptococcus pyogenes. *Molecules* 24 (23):4237.

Afshari, H., M. Maleki, M. Hakimian, R. A. Tanha, and M. Salouti. 2020. Immunogenicity evaluating of the SLNs-alginate conjugate against *Pseudomonas aeruginosa*. *J Immunol Methods* 488:112938. doi: S0022-1759(20)30232-5[pii]10.1016/j.jim.2020.112938.

Agriopoulou, S., E. Stamatelopoulou, M. Sachadyn-Król, and T. Varzakas. 2020. Lactic acid bacteria as antibacterial agents to extend the shelf life of fresh and minimally processed fruits and vegetables: Quality and safety aspects. *Microorganisms* 8 (6):952. doi: 10.3390/microorganisms8060952.

Alakomi, H. L., E. Skyttä, M. Saarela, T. Mattila-Sandholm, K. Latva-Kala, and I. M. Helander. 2000. Lactic acid permeabilizes gram-negative bacteria by disrupting the outer membrane. *Appl Environ Microbiol* 66 (5):2001–2005.

Alexa, P., K. Štouračová, J. Hamřík, and I. Rychlík. 2001. Gene typing of the colonisation factor K88 (F4) in enterotoxigenic *Escherichia coli* strains isolated from diarrhoeic piglets. *Veterinární Medicína* 46 (2):46–49. doi: 10.17221/7851-vetmed.

Algammal, A. M., M. Mabrok, E. Sivaramasamy, F. M. Youssef, M. H. Atwa, A. W. El-Kholy, H. F. Hetta, and W. N. Hozzein. 2020. Emerging MDR-*Pseudomonas aeruginosa* in fish commonly harbor oprL and toxA virulence genes and blaTEM, blaCTX-M, and tetA antibiotic-resistance genes. *Sci Rep* 10 (1):15961. doi: 10.1038/s41598-020-72264-410.1038/s41598-020-72264-4[pii].

Allen, Heather K., Justin Donato, Helena Huimi Wang, Karen A. Cloud-Hansen, Julian Davies, and Jo Handelsman. 2010. Call of the wild: Antibiotic resistance genes in natural environments. *Nature Reviews Microbiology* 8 (4):251–259. doi: 10.1038/nrmicro2312.

Allgrove, J. E., M. Oliveira, and M. Gleeson. 2014. Stimulating whole saliva affects the response of antimicrobial proteins to exercise. *Scand J Med Sci Sports* 24 (4):649–655.

Alnaimat, S., N. S. Alharbi, S. A. Alharbi, S. H. Salmen, A. Chinnathambi, B. O. Al-Johny, and M. Wainwright. 2015. Mycelium of fungi isolated from mouldy foods inhibits staphylococcus aureus including MRSA – A rationale for the re-introduction of mycotherapy? *Saudi J Biol Sci* 22 (5):600–603.

Alomrani, A. H., G. A. Shazly, A. A. A. F. Amara, and M. M. Badran. 2014. Itraconazole-hydroxypropyl-$$-cyclodextrin loaded deformable liposomes: In vitro skin penetration studies and antifungal efficacy using candida albicans as model. *Colloids Surf B Biointerfaces* 121:74–81.

Alsberg, G. L., and O. F. Black. 1913. *Contributions to the Study of Maize Deterioration; Biochemical and Toxicological Investigations of Penicillium puberulum and P. stoloniferum*. Vol. 270 (pp. 1–47). Nabu Press, Washington.

Amara, A. A. 2005a. Simple slot lysis and plasmid curing for determination the transfer of resistant factors within the pathogenic microbial strains. In proceeding of tagungsband zum 2. Gemeinsamen kongress der DGHM und VAAM Göttingen, 25–28.9.2005. in Göttingen, Germany. *BioSpectrum* 74.

Amara, A. A. 2005b. New strategy for controlling resistant strain of *Pseudomonas aeruginosa*. In proceeding of tagungsband zum 2. Gemeinsamen kongress der DGHM und VAAM Göttingen, 25–28.9.2005. in Göttingen, Germany. *BioSpectrum* 74.

Amara, A. A. 2011. Opportunistic pathogens and their biofilm "food for thought." In science against microbial pathogens: Communicating current research and technology advances: Edit by Mendez-Vilas. *Formatex Microbiology Series No 3* 2:813–824.

Amara, A. A. 2013a. The inevitability of balanced lives: Genes and, foods in action and, interactions. *IIOAB J* 4 (2):1–27.

Amara, A. A. 2013b. Pharmaceutical and industrial protein engineering: Where we are? *Pak J Pharm Sci* 26 (1):217–232.

Amara, A. A. 2015a. *Kostenlose Viral Ghosts, Bacterial Ghosts, Microbial Ghosts and More*. Schüling Verlage (Schüling Press), Münster Germany.

Amara, A. A. 2015b. *Saccharomyces cerevisiae* ghosts using the sponge-like re-reduced protocol. *SOJ Biochemistry* 2 (1):1–4. doi: 10.15226/2376–4589/2/1/00107.

Amara, A. A. 2016. Lysozymes, proteinase K, bacteriophage E lysis proteins, and some chemical compounds for microbial ghosts preparation: A review and food for thought. *SOJ Biochemistry* 2 (1):1–16. doi: 10.15226/2376–4589/2/1/00110.

Amara, A. A. 2017. α globin hemoglobinopathy as a case study: Mutants and prototype mutants: Are they absent α0, reduced α+ or recombinant? *Cellular and Molecular Medicine* 3 (2):10.

Amara, A. A. 2021. Chapter 9: Improving Animal Immunity to Prevent Fungal Infections with Folk Remedies and Advanced Medicine. In *Fungal Diseases in Animals from Infections to Prevention*. Eds: A. Gupta and N. Pratap Singh (pp. 127–167). Springer Publishing, Cham.

Amara, A. A. 2022. The Role of Divalent Cations in Antibiotic Sensitivity: A Molecular Aspect. In *Biomolecules from Natural Sources: Advances and Applications* (pp. 252–277) Eds, Gupta, V. K., Sarker, S. D., Sharma, M., Pirovani, M. E., Usmani, Z., Jayabaskaran, C. (eds.), Wiley and Sons Ltd. USA/UK.

Amara, A. A., and Z. M. Hussain. 2006. Offline controlling of *Pseudomonas aeruginosa* resistant to protein inhibitor antibiotics using combination of EDTA and Na-citrate or disinfectant(s). *Microbiology (China)* 33:95–100.

Amrara, A. A., A. J. Neama, A. Hussein, E. A. Hashish, and S. A. Sheweita. 2014. Evaluation the surface antigen of the *Salmonella typhimurium* ATCC 14028 ghosts prepared by SLRP. *Sci World J* 2014:840863. doi: 10.1155/2014/840863.

Amara, A. A., M. M. Salem-Bekhit, and F. K. Alanazi. 2013a. Plackett Burman randomization method for Bacterial ghosts preparation form *E. coli* JM109. *Saudi Pharmaceutical Journal* 22 (3):273–279. doi: 10.1016/j.jsps.2013.06.002.

Amara, A. A., M. M. Salem-Bekhit, and F. K. Alanazi. 2013a. Sponge-like: A new protocol for preparing bacterial ghosts. *Sci World J* 2013:1–7. doi: 10.1155/2013/545741.

Amara, A. A., M. M. Salem-Bekhit, and F. K. Alanazi. 2013b. Preparation of bacterial ghosts for *E. coli* JM109 using sponge-like reduced protocol. *Asian J Biol Sci* 6 (8):363–369. doi: 10.3923/ajbs.2013.363.369.

Amara, A. A., and A. M. Shibl. 2015a. Role of probiotics in health improvement, infection control and disease treatment and management. *Saudi Pharmaceutical Journal* 23 (2):107–114.

Amara, A. A., and A. M. Shibl. 2015b. A hypothesis about the variation within *β*-lactamases: An epigenetic-like mechanism. *IIOAB J* 6 (1):1–11.

Amara, A. A., A. M. Shibl, H. Radwan, A. Al-Omrani, and Al-K. F. Tawfik. 2013. Towards safer alginate production from a strain of pseudomonades. *Asian Journal of Biological Sciences* 6 (8):376–381. doi: 10.3923/ajbs.2013.376.381.

Amara, A. A. A. F., and N. A. El-Baky. 2022. Evacuation of yeast and fungal cells using the sponge-like protocol: A new approach for fungal control. *J Appl Pharm Sci* 12 (4):185–186.

Avery, O. T., and R. Dubos. 1930. The specific action of a bacterial enzyme on pneumococci of type III. *Science* 72 (1858):151–152. doi: 10.1126/science.72.1858.151.

Balcazar, J. L. 2014. Bacteriophages as vehicles for antibiotic resistance genes in the environment. *PLoS Pathogens* 10 (7):e1004219.

Birkinshaw, J. H., A. Bracken, E. N. Morgan, and H. Raistrick. 1948. Studies in the biochemistry of micro-organisms. The molecular constitution of mycophenolic acid, a metabolic product of *Penicillium brevi-*compactum Dierckx. Part 2. Possible structural formulae for mycophenolic acid. *Biochem J* 43 (2):216.

Birkinshaw, J. H., A. E. Oxford, and H. Raistrick. 1936. Studies in the biochemistry of micro-organisms: Penicillic acid, a metabolic product of *Penicillium puberulum* Bainier and *P. cylopium* Westling. *Biochem J* 30 (3):394–411. doi: 10.1042/bj0300394. PMID: 16746037; PMCID: PMC1263039.

Birkinshaw, J. H., H. Raistrick, and D. J. Ross. 1952. Studies in the biochemistry of micro-organisms. The molecular constitution of mycophenolic acid, a metabolic product of *Penicillium brevicompactum* Dierckx. Part 3. Further observations on the structural formula for mycophenolic acid. *Biochem J* 50 (5):630.

Blaskovich, M. A. T., M. S. Butlerm, and M. A. Cooper. 2017. Polishing the tarnished silver bullet: The quest for new antibiotics. *Essays Biochem* 61 (1):103–114. doi: 10.1042/EBC20160077.

Boman, H. G. 2003. Antibacterial peptides: Basic facts and emerging concepts. *J Intern Med* 254 (3):197–215.

Brede, D. A., S. Lothe, Z. Salehian, T. Faye, and I. F. Nes. 2007. Identification of the Propionicin F Bacteriocin Immunity Gene (pcfI) and development of a food-grade cloning system for *Propionibacterium freudenreichii*. *Appl Environ Microbiol* 73 (23):7542–7547. doi: 10.1128/aem.01023–07.

Carr, F. J., D. Chill, and N. Maida. 2002. The lactic acid bacteria: A literature survey. *Critical Reviews in Microbiology* 28 (4):281–370.

Cebrián, R., M. E. Rodríguez-Cabezas, R. Martín-Escolano, S. Rubiño, M. Garrido-Barros, M. Montalbán-López, M. J. Rosales, M. Sánchez-Moreno, E. Valdivia, M. Martínez-Bueno, C. Marín, J. Gálvez, and M. Maqueda. 2019. Preclinical studies of toxicity and safety of the AS-48 bacteriocin. *J Adv Res* 20:129–139. doi: 10.1016/j.jare.2019.06.003.

Chifiriuc, Mariana Carmen, Alina Maria Holban, Carmen Curutiu, Lia-Mara Ditu, Grigore Mihaescu, Alexandra Elena Oprea, Alexandru Mihai Grumezescu, and Veronica Lazar. 2016. Antibiotic Drug Delivery Systems for the Intracellular Targeting of Bacterial Pathogens. In *Smart Drug Delivery System*. InTechOpen, Rijeka.

Connell, S. R., D. M. Tracz, K. H. Nierhaus, and D. E. Taylor. 2003. Ribosomal protection proteins and their mechanism of tetracycline resistance. *Antimicrob Agents Chemother* 47 (12):3675–3681.

Czyż, Daniel M., Lakshmi-Prasad Potluri, Neeta Jain-Gupta, Sean P. Riley, Juan J. Martinez, Theodore L. Steck, Sean Crosson, Howard A. Shuman, and Joëlle E. Gabay. 2014. Host-directed antimicrobial drugs with broad-spectrum efficacy against intracellular bacterial pathogens. *mBio* 5 (4):e01534–14. doi: 10.1128/mbio.01534–14.

D'Accolti, M., I. Soffritti, S. Mazzacane, and E. Caselli. 2021. Bacteriophages as a potential 360-degree pathogen control strategy. *Microorganisms* 9 (2):261. doi: 10.3390/microorganisms9020261.

Deng, Yu, Bing Li, and Tong Zhang. 2018. Bacteria that make a meal of sulfonamide antibiotics: Blind spots and emerging opportunities. *Environmental Science & Technology* 52 (7):3854–3868. doi: 10.1021/acs.est.7b06026.

Devirgiliis, C., S. Barile, and G. Perozzi. 2011. Antibiotic resistance determinants in the interplay between food and gut microbiota. *Genes Nutr* 6 (3):275–84. doi: 10.1007/s12263–011–0226-x.

Dicks, L. M. T., L. Dreyer, C. Smith, and A. D. van Staden. 2018. A review: The fate of bacteriocins in the human gastro-intestinal tract: Do they cross the gut – blood barrier? *Front Microbiol* 9:2277. doi: 10.3389/fmicb.2018.02297.

Dinleyici, Ener Cagri, Makbule Eren, Metehan Ozen, Zeynel Abidin Yargic, and Yvan Vandenplas. 2012. Effectiveness and safety of *saccharomyces boulardiifor* acute infectious diarrhea. *Expert Opinion on Biological Therapy* 12 (4):395–410. doi: 10.1517/14712598.2012.664129.

Domagk, G. 1935. Ein beitrag zur chemotherapieder bakteriellen infektionem. *Deut Med Wochenschr* 61:250–258.

Dönhöfer, A., S. Franckenberg, S. Wickles, O. Berninghausen, R. Beckmann, and D. N. Wilson. 2012. Structural basis for TetM-mediated tetracycline resistance. *Proc Natl Acad Sci USA* 109 (42):16900–16905.

Dubos, R. J., and Hotchkiss, R. D. 1941. The production of bactericidal substances by aerobic sporulating bacilli. *J Exp Med* 73 (3):629–640.

Durand, G. A., D. Raoult, and G. Dubourg. 2019. Antibiotic discovery: History, methods and perspectives. *International Journal of Antimicrobial Agents* 53:371–382.

El-Baky, A. N., and A. A. Amara. 2014. Newcastle disease virus (LaSota strain) as a model for virus ghosts preparation using H2O2 bio-critical concentration. *International Science and Investigation Journal* 3 (5):38–50.

El-Baky, N. A., and A. A. Amara. 2023. Natural Antimicrobial Therapeutic Peptides: Milk Lactoferricin and *Spirulina platensis* Peptides. In *Antimicrobials in Pharmaceutical and Medicinal Research* (pp. 19–48). CRC Press.

El-Baky, N. A., A. A. Amara, and E. M. Redwan. 2023. Nutraceutical and Therapeutic Importance of Clots and Their Metabolites. In *Nutraceuticals: Sources, Processing Methods, Properties, and Applications* (pp. 241–268). Elsevier.

El-Baky, N. A., M. A. Elkhawaga, E. S. Abdelkhalek, M. M. Sharaf, E. M. Redwan, and H. R. Kholef. 2021. De novo expression and antibacterial potential of four lactoferricin peptides in cell-free protein synthesis system. *Biotechnology Reports* 29:e00583.

El-Baky, N. A., R. A. A. Rahman, M. M. Sharaf, and A. A. Amara. 2021. The development of a phytopathogenic fungi control trial: Aspergillus flavus and aspergillus niger infection in jojoba tissue culture as a model. *Scientific World Journal* 2021:2021.

El-Baky, N. A., M. M. Sharaf, E. Amer, H. R. Kholef, M. Z. Hussain, and A. A. Amara. 2018. Protein and DNA isolation from *Aspergillus niger* as well as ghost cells ormation. *SOJ Biochemistry* 4 (1):1–7. doi: 10.15226/2376–4589/4/1/00129.

El-Shanshoury, A. E., E. R. Kenawy, A. A. Amara, S. F. Mohamed, A. F. Salama, and S. Kishk. 2018a. Optimization of the culture medium using the burman and box-behnken experimental designs for an enhanced production of alginate by *azotobacter vinelandii*. *Biotechnologia* 99 (3):185–196. doi: 10.5114/bta.2018.77479.

El-Shanshoury, Abd El-Raheem R., El-Refaie Kenawy, Amro A. Amara, Samia F. Mohamed, Afrah F. Salama, and Shimaa Kishk. 2018b. Optimization of the culture medium using the burman and box-behnken experimental designs for an enhanced production of alginate by azotobacter vinelandii. *BioTechnologia* 99 (3):185–196. doi: 10.5114/bta.2018.77479.

Falagas, M. E., T. Skalidism, and K. Z. Vardakasm. 2017. Legakis NJHellenic Cefiderocol Study Group. Activity of cefiderocol (S-649266) against carbapenem-resistant Gram negative bacteria collected from inpatients in Greek hospitals. *J Antimicrob Chemother* 72:1704–1708. doi: 10.1093/jac/dkx049.

Founou, L. L., R. C. Founou, and S. Y. Essack. 2016. Antibiotic resistance in the food chain: A developing country-perspective. *Front Microbiol* 7:1881. doi: 10.3389/fmicb.2016.01881.

Furtado, D. N., S. D. Todorov, M. Landgraf, M. T. Destro, and B. D. Franco. 2015. Bacteriocinogenic *Lactococcus lactis* subsp. lactis DF04Mi isolated from goat milk: Application in the control of listeria monocytogenes in fresh minas-type goat cheese. *Braz J Microbiol* 46 (1):201–206. doi: 10.1590/s1517–838246120130761.

Gautam, N., and N. Sharma. 2009. Bacteriocin: Safest approach to preserve food products. *Indian J Microbiol* 49 (3):204–211. doi: 10.1007/s12088-009-0048-3.

Gomarasca, Marta, Thaynan F. C. Martins, Lilo Greune, Philip R. Hardwidge, M. Alexander Schmidt, and Christian Rüter. 2017. Bacterium-derived cell-penetrating peptides deliver gentamicin to kill intracellular pathogens. *Antimicrobial Agents and Chemotherapy* 61 (4):e02545–16. doi: 10.1128/aac.02545-16.

Gutiérrez, D., L. Rodríguez-Rubio, B. Martínez, A. Rodríguez, and P. García. 2016. Bacteriophages as weapons against bacterial biofilms in the food industry. *Front Microbiol* 7:825. doi: 10.3389/fmicb.2016.00825.

Hancock, R. E., and F. S. Brinkman. 2002. Function of *Pseudomonas* porins in uptake and efflux. *Annu Rev Microbiol* 56 (1):17–38.

Hayashi, T. 2001. Complete genome sequence of enterohemorrhagic *Eschelichia coli* O157:H7 and genomic comparison with a laboratory strain K-12. *DNA Research* 8 (1):11–22. doi: 10.1093/dnares/8.1.11.

Herrell, W. E., and D. Heilman. 1941. Experimental and clinical studies on gramicidin. *J Clin Invest* 20 (5):583–591. doi: doi:10.1172/JCI101251.

Hodgkin, D. C. 1949. X-ray crystallographic investigation of the structure of penicillin. *Adv Sci* 6:85–89.

Hodgson, K., J. Morris, T. Bridson, B. Govan, C. Rush, and N. Ketheesan. 2015. Immunological mechanisms contributing to the double burden of diabetes and intracellular bacterial infections. *Immunology* 144 (2):171–185.

Hussein, M. Z., and A. A. Amara 2006. Case-by-case study using antibiotic-EDTA combination to control *Pseudomonas aeruginosa*. *Pak J Pharm Sci* 19 (3):236–243.

Hutchings, M. I., A. W Truman, and B. Wilkinson. 2019. Antibiotics: Past, present and future. *Current Opinion in Microbiology* 51:72–80.

Iseppi, R., A. Lauková, and C. Sabia. 2021. Editorial: Bacteriocin-producing probiotic bacteria: A natural solution for increasing efficiency and safety of livestock food production. *Front Microbiol* 12:675483. doi: 10.3389/fmicb.2021.675483.

Jager, Nynke G. L., Reinier M. van Hest, Jeffrey Lipman, Jason A. Roberts, and Menino O. Cotta. 2019. Antibiotic exposure at the site of infection: Principles and assessment of tissue penetration. *Expert Review of Clinical Pharmacology* 12 (7):623–634. doi: 10.1080/17512433.2019.1621161.

Jeong, Do-Won, Sojeong Heo, and Jong-Hoon Lee. 2017. Safety assessment of *Tetragenococcus halophilus* isolates from doenjang, a Korean high-salt-fermented soybean paste. *Food Microbiology* 62:92–98. doi: 10.1016/j.fm.2016.10.012.

Jeżak, K., and A. Kozajda. 2022. Occurrence and spread of antibiotic-resistant bacteria on animal farms and in their vicinity in Poland and Ukraine-review. *Environ Sci Pollut Res Int* 29 (7):9533–9559.

Johnston, Daniel G. W., Jay Kearney, Zbigniew Zasłona, Michelle A. Williams, Luke A. J. O'Neill, and Sinéad C. Corr. 2017. MicroRNA-21 limits uptake of listeria monocytogenes by macrophages to reduce the intracellular niche and control infection. *Frontiers in Cellular and Infection Microbiology* 7:201. doi: 10.3389/fcimb.2017.00201.

Jones, D., H. J. Metzger, A. Schatz, and S. A. Waksman. 1944. Control of gram-negative bacteria in experimental animals by streptomycin. *Science* 100 (2588):103–105. doi: 10. 1126/science.100.2588.103.

Kasimanickam, V., M. Kasimanickam, and R. Kasimanickam. 2021. Antibiotics use in food animal production: Escalation of antimicrobial resistance: Where are we now in combating AMR? *Med Sci (Basel)* 9 (1):14. doi: 10.3390/medsci9010014.

Koutsoumanis, K., A. Lianou, and J. Sofos. 2014. Food safety: Emerging pathogens. *Encyclopedia of Agriculture and Food Systems*:250–272. doi: 10.1016/b978-0-444-52512-3.00049–8.

Kumar, S. B., S. R. Arnipalli, and O. Ziouzenkova. 2020. Antibiotics in food chain: The consequences for antibiotic resistance. *Antibiotics (Basel)* 9 (10):688. doi: 10.3390/antibiotics9100688.

Kutter, Elizabeth, Daniel De Vos, Guram Gvasalia, Zemphira Alavidze, Lasha Gogokhia, Sarah Kuhl, and Stephen Abedon. 2010. Phage therapy in clinical practice: Treatment of human infections. *Current Pharmaceutical Biotechnology* 11 (1):69–86. doi: 10.2174/138920110790725401.

Laulund, S., A. Wind, P. M. F. Derkx, and V. Zuliani. 2017. Regulatory and safety requirements for food cultures. *Microorganisms* 5 (2):28. doi: 10.3390/microorganisms5020028.

Lengeler, J. W., G. Drews, and H. G. Schlegel. 1999. *Biology of the Prokaryotes*. 7th Ed. Cambridge University Press, Cambridge. Thieme (Stuttgart and New York 1999) and Blackwell Science.

Leroy, F., and L. De Vuyst. 2004. Lactic acid bacteria as functional starter cultures for the food fermentation industry. *Trends in Food Science & Technology* 15:67–78.

Lewis, K. 2012. Antibiotics: Recover the lost art of drug discovery. *Nature* 40 (485):439. doi: 10.1038/485439a.

Lewis, K. 2013. Platforms for antibiotic discovery. *Nat Rev Drug Discovery* 12 (5):371–387. doi: 10.1038/nrd3975.

Li, W., G. C. Atkinson, N. S. Thakor, U. Allas, C. C. Lu, K. Y. Chan, T. Tenson, K. Schulten, K. S. Wilson, V. Hauryliuk, and J. Frank. 2013. Mechanism of tetracycline resistance by ribosomal protection protein Tet(O). *Nat Commun* 4:1477.

Liu, L., N. Guan, G. Zhu, J. Li, Hd Shin, G. Du, and J. Chen. 2016. Pathway engineering of *Propionibacterium jensenii* for improved production of propionic acid. *Sci Rep* 6:19963. doi: 10.1038/srep19963.

Liu, L., X. Zhuge, Hd Shin, R. R. Chen, J. Li, G. Du, and J. Chen. 2015. Improved production of propionic acid in *Propionibacterium jensenii* via combinational overexpression of glycerol dehydrogenase and malate dehydrogenase from klebsiella pneumoniae. *Appl Environ Microbiol* 81 (7):2256–2264. doi: 10.1128/aem.03572-14.

Loc Carrillo, C., R. J. Atterbury, A. El-Shibiny, P. L. Connerton, E. Dillon, A. Scott, and I. F. Connerton. 2005. Bacteriophage therapy to reduce campylobacter jejuni colonization of broiler chickens. *Appl Environ Microbiol* 71 (11):6554–6563. doi: 10.1128/aem.71.11.6554-6563.2005.

Magnuson, B., I. Munro, P. Abbot, N. Baldwin, R. Lopez-Garcia, K. Ly, L. McGirr, A. Roberts, and S. Socolovsky. 2013. Review of the regulation and safety assessment of food substances in various countries and jurisdictions. *Food Addit Contam Part A Chem Anal Control Expo Risk Assess* 30 (7):1147–1220. doi: 10.1080/19440049.2013.795293.

Mahalmani, V. M., P. Sarma, A. Prakash, and B. Medhi. 2019. Positive list of antibiotics and food products: Current perspective in India and across the globe. *Indian J Pharmacol* 51 (4):231–235. doi: 10.4103/ijp.IJP_548_19.

Mancuso, F., J. Shi, and D. J. Malik. 2018. High throughput manufacturing of bacteriophages using continuous stirred tank bioreactors connected in series to ensure optimum host bacteria physiology for phage production. *Viruses* 10 (10):537. doi: 10.3390/v10100537.

Mandal, M. D., and S. Mandal. 2011. Honey: Its medicinal property and antibacterial activity. *Asian Pac J Trop Biomed* 1 (2):154–160.

Mathieu, Jacques, Pingfeng Yu, Pengxiao Zuo, Marcio L. B. Da Silva, and Pedro J. J. Alvarez. 2019. Going viral: Emerging opportunities for phage-based bacterial control in water treatment and reuse. *Accounts of Chemical Research* 52 (4):849–857. doi: 10.1021/acs.accounts.8b00576.

McMurry, L. M., R. E. J. Petrucci, and S. B. Levy. 1980. Active efflux of tetracycline encoded by four genetically different tetracycline resistance determinants in *Escherichia coli Proc Natl Acad Sci USA* 77 (7):3974–3977.

Menisy, M., A. Hussein, G. A. Abd El Kader, S. Sheweita, and A. A. Amara. 2017. *Klebsiella pneumoniae* ghosts as vaccine using sponge like reduced protocol. *Cellular & Molecular Medicine: Open Access* 03 (02):11. doi: 10.21767/2573-5365.100034.

Mokoena, M. P. 2017. Lactic acid bacteria and their bacteriocins: Classification, biosynthesis and applications against uropathogens: A mini-review. *Molecules* 22 (8):1255. doi: 10.3390/molecules22081255.

Morita, Y., J. Tomida, and Y. Kawamura. 2012. MexXY multidrug efflux system of *Pseudomonas aeruginosa*. *Fronteirs in Microbiology* 3:1–13.

Murphy, D., A. Ricci, Z. Auce, J. G. Beechinor, H. Bergendahl, R. Breathnach, J. Bureš, J. P. Duarte Da Silva, J. Hederová, P. Hekman, C. Ibrahim, E. Kozhuharov, G. Kulcsár, E. Lander Persson, J. M. Lenhardsson, P. Mačiulskis, I. Malemis, L. Markus-Cizelj, A. Michaelidou-Patsia, M. Nevalainen, P. Pasquali, J. Rouby, J. Schefferlie, W. Schlumbohm, M. Schmit, S. Spiteri, S. Srčič, L. Taban, T. Tiirats, B. Urbain, E. Vestergaard, A. Wachnik-Święcicka, J. Weeks, B. Zemann, A. Allende, D. Bolton, M. Chemaly, P. S. Fernandez Escamez, R. Girones, L. Herman, K. Koutsoumanis, R. Lindqvist, B. Nørrung, L. Robertson, G. Ru, M. Sanaa, M. Simmons, P. Skandamis, E. Snary, N. Speybroeck, B. Ter Kuile, H. Wahlström, K. Baptiste, B. Catry, P. S. Cocconcelli, R. Davies, C. Ducrot, C. Friis, G. Jungersen, S. More, C. Muñoz Madero, P. Sanders, M. Bos, Z. Kunsagi, J. Torren Edo, R. Brozzi, D. Candiani, B. Guerra, E. Liebana, P.

Stella, J. Threlfall, and H. Jukes. 2017. EMA and EFSA joint scientific opinion on measures to reduce the need to use antimicrobial agents in animal husbandry in the European Union, and the resulting impacts on food safety (RONAFA). *Efsa j* 15 (1):4666. doi: 10.2903/j.efsa.2017.4666.

Nazari, M., and D. L. Smith. 2020. A PGPR-produced bacteriocin for sustainable agriculture: A review of thuricin 17 characteristics and applications. *Front Plant Sci* 11. doi: 10.3389/fpls.2020.00916.

Nazir, R., J. P. Shen, J Wang, T., H. W. Hu, and J. Z. He. 2017. Fungal networks serve as novel ecological routes for enrichment and dissemination of antibiotic resistance genes as exhibited by microcosm experiments. *Sci Rep*. 7 (1):15457.

Newman, D. J., and G. M. Cragg. 2016. Natural products as sources of new drugs from 1981 to 2014. *J. Nat Prod* 79 (3):629–661.

O'Shea, E. F., P. M. O'Connor, O. O'Sullivan, P. D. Cotter, R. P. Ross, and C. Hill. 2013. Bactofencin A, a new type of cationic bacteriocin with unusual immunity. *mBio* 4 (6):e00498–13. doi: 10.1128/mBio.00498–13.

Oxford, A. E., and H. Raistrick. 1933. Studies in the biochemistry of micro-organisms: A note on the mechanism of the production of phenolic acids from glucose by *Penicillium brevi*-compactum Dierckx. *Biochem J* 27 (5):1473.

Pagès, J. M., C. E. James, and M. Winterhalter. 2008. The porin and the permeating antibiotic: A selective diffusion barrier in gram-negative bacteria. *Nat Rev Microbiol* 6:893–903.

Papalexandratou, Z., G. Vrancken, K. De Bruyne, P. Vandamme, and L. De Vuyst. 2011. Spontaneous organic cocoa bean box fermentations in Brazil are characterized by a restricted species diversity of lactic acid bacteria and acetic acid bacteria. *Food Microbiol* 28 (7):1326–1338.

Parker, R. B. 1974. Probiotics, the other half of the antibiotic story. *Anim. Nutr. Health* 29:4–8.

Piddock, L. J. 2006. Clinically relevant chromosomally encoded multidrug resistance efflux pumps in bacteria. *Clin Microbiol Rev* 19 (2):382–402.

Pires, D. P., S. Cleto, S. Sillankorva, J. Azeredo, and T. K. Lu. 2016. Genetically engineered phages: A review of advances over the last decade. *Microbiol Mol Biol Rev* 80 (3):523–543.

Piwowarek, K., E. Lipińska, E. Hać-Szymańczuk, M. Kieliszek, and I. Ścibisz. 2018. *Propionibacterium spp.* – source of propionic acid, vitamin B12, and other metabolites important for the industry. *Appl Microbiol Biotechnol* 102 (2):515–538. doi: 10.1007/s00253-017-8616-7.

Podlacha, M., L. Grabowski, K. Kosznik-Kawśnicka, K. Zdrojewska, M. Stasiłojć, G. Węgrzyn, and A. Węgrzyn. 2021. Interactions of bacteriophages with animal and human organisms-safety issues in the light of phage therapy. *Int J Mol Sci* 22 (16):8937.

Pogány Simonová, M., Ľ. Chrastinová, and A. Lauková. 2020. Autochtonous strain *Enterococcus faecium* EF2019 (CCM7420), its bacteriocin and their beneficial effects in broiler rabbits – A review. *Animals (Basel)* 10 (7):1188. doi: 10.3390/ani10071188.

Poole, K. 2005. Efflux-mediated antimicrobial resistance. *J. Antimicrob. Chemother*. 56 (1):20–51.

Rauw, F., Y. Gardin, B. T. van den, and B. Lambrecht. 2009. La vaccination contre la maladie de newcastle chez le poulet (Gallus gallus). *Societe et Environnement* 13:587.

Russell, A. D. 1998. Bacterial resistance to disinfectants: Present knowledge and future problems. *Journal of Hospital Infection* 43:S57–S68.

Sengun, Y. I., and S. Karabiyikli. 2011. Importance of acetic acid bacteria in food industry. *Food Control* 22 (5):647–656.

Sheweita, S. A., A. M. Batah, A. A. Ghazy, A. Hussein, and A. A. Amara. 2019. A new strain of *Acinetobacter baumannii* and characterization of its ghost as a candidate vaccine. *Journal of Infection and Public Health* 12 (6):831–842. doi: 10.1016/j.jiph.2019.05.009.

Shyntum, D. Y., N. P. Nkomo, N. L. Shingange, A. R. Gricia, D. Bellieny-Rabelo, and L. N. Moleleki. 2019. The impact of type VI secretion system, bacteriocins and antibiotics on bacterial competition of *Pectobacterium carotovorum* subsp. brasiliense and the regulation of carbapenem biosynthesis by iron and the ferric-uptake regulator. *Front Microbiol* 10. doi: 10.3389/fmicb.2019.02379.

Sneader, W. 2005. *Drug Discovery: A History*: Wiley, Hoboken.

Suskovich, J., B. Kos, J. Beganovich, A. Pavunc, K. Habjanic, and S. Matosich. 2010. Antimicrobial activity – the most important property of probiotic and starter lactic acid bacteria. *Food Technol Biotechnol* 48 (3):296–307.

Talaro, K. P., and B. Chess. 2008. *Foundations in Microbiology*. 8th Ed. McGraw Hill, New York.

Tatár-Kis, T., E. A. Fischer, C. Cazaban, E. Walkó-Kovács, Z. G. Homonnay, F. C. Velkers, V. Palya, and J. A. Stegeman. 2020. A herpesvirus of turkey-based vector vaccine reduces transmission of newcastle disease virus in commercial broiler chickens with maternally derived antibodies. *Vaccines (Basel)* 8 (4):614. doi: 10.3390/vaccines8040614.

Thierry, A., F. Valence, S. M. Deutsch, S. Even, H. Falentin, Y. Le Loir, G. Jan, and Valerie Gagnaire. 2015. Strain-to-strain differences within lactic and propionic acid bacteria species strongly impact the properties of cheese – A review. *Dairy Science & Technology* 95 (6):895–918. doi: doi: 10.1007/s13594-015-0267-9.

Trapecar, Martin, Thomas Leouffre, Morgane Faure, Henrik E. Jensen, Per E. Granum, Avrelija Cencic, and Simon P. Hardy. 2011. The use of a porcine intestinal cell model system for evaluating the food safety risk of *Bacillus cereus* probiotics and the implications for assessing enterotoxigenicity. *APMIS* 119 (12):877–884. doi: 10.1111/j.1600–0463.2011.02797.x.

Valen, D. Van, D. Wu, Y. J. Chen, H. Tuson, P. Wiggins, and R. Phillips. 2012. A single-molecule Hershey-Chase experiment. *Curr Biol* 22 (14):1339–1343.

Venkatachalam, K. V., W. Huang, M. LaRocco, and T. Palzkil. 1994. Characterization of TEM-1 β-Lactamase mutants from positions 238 to 241 with increased catalytic efficiency for ceftazidim. *J. Boil Chem* 269:23444–43450.

Vondruskova, H., R. Slamova, M. Trckova, Z. Zraly, and I. Pavlik. 2010. Alternatives to antibiotic growth promoters in prevention of diarrhoea in weaned piglets: A review. *Veterinární Medicína* 55 (5):199–224. doi: 10.17221/2998-vetmed.

Waksman, S. A., W. B. Geigerm, and D. M. Reynolds. 1946. Strain specificity and production of antibiotic substances: VII. Production of actinomycin by different actinomycetes. *Proc Natl Acad Sci U S A* 32:117–120.

Waksman, S. A., E. S. Horning, and E. L. Spencer. 1942. The production of two antibacterial substances, fumigacin and clavacin. *Science* 96 (2487):202–203. doi: 10.1126/science.

Waksman. S. A., and H. A. Lechevalier. 1949. Neomycin, a new antibiotic active against streptomycin-resistant bacteria, including tuberculosis organisms. *Science* 109:305–307. doi: 10.1126/science.109.2830.305.

Waksman, S. A., A. Schatz, and D. M. Reynolds. 2010. Production of antibiotic substances by actinomycetes. *Ann. NY Acad. Sci.* 1213 (1):112–124.

Walsh, C. 2003. *Antibiotics: Actions, Origins, Resistance*. 1st Ed. ASM Press, Washington, DC.

Wang, Z., X. Zhao, Y. Wang, C. Sun, M. Sun, X. Gao, F. Jia, C. Shan, G. Yang, J. Wang, H. Huang, C. Shi, W. Yang, A. Qian, C. Wang, and Y. Jiang. 2021. *In vivo* production of HN protein increases the protection rates of a minicircle DNA vaccine against genotype VII newcastle disease virus. *Vaccines (Basel)* 9 (7):723. doi: 10.3390/vaccines9070723.

Wilde, S., Y. Jiang, A. M. Tafoya, J. Horsman, M. Yousif, L. A. Vazquez, and K. L. Roland. 2019. Salmonella-vectored vaccine delivering three *Clostridium perfringens* antigens protects poultry against necrotic enteritis. *PLoS One* 14 (2):e0197721. doi: 10.1371/journal.pone.0197721.

Woskow, S. A., and B. A. Glatz. 1991. Propionic acid production by a propionic acid-tolerant strain of Propionibacterium acidipropionici in batch and semicontinuous fermentation †. *Appl Environ Microbiol* 57 (10):2821–2828.

Zeng, X., F. Xu, and J. Lin. 2010. Development and evaluation of CmeC subunit vaccine against *Campylobacter jejuni*. *J Vaccines Vaccin* 1 (3):112. doi: 10.4172/2157–7560.1000112.

Zhang, Anyun, Douglas R. Call, Thomas E. Besser, Jinxin Liu, Lisa Jones, Hongning Wang, and Margaret A. Davis. 2019. β-lactam resistance genes in bacteriophage and bacterial DNA from wastewater, river water, and irrigation water in Washington State. *Water Research* 161:335–340. doi: 10.1016/j.watres.2019.06.026.

Zhang, Qi, Zhenyan Zhang, Tao Lu, W. J. G. M. Peijnenburg, Michael Gillings, Xiaoru Yang, Jianmeng Chen, Josep Penuelas, Yong-Guan Zhu, Ning-Yi Zhou, Jianqiang Su, and Haifeng Qian. 2020. Cyanobacterial blooms contribute to the diversity of antibiotic-resistance genes in aquatic ecosystems. *Communications Biology* 3 (1). doi: 10.1038/s42003-020-01468-1.

Zhang, Yuning, Shiyu Lin, Jingyuan Fu, Wei Zhang, Gang Shu, Juchun Lin, Haohuan Li, Funeng Xu, Huaqiao Tang, Guangneng Peng, Ling Zhao, Shiqi Chen, and Hualin Fu. 2022. Nanocarriers for combating biofilms: Advantages and challenges. *Journal of Applied Microbiology* 133 (3):1273–1287. doi: 10.1111/jam.15640.

Zhao, Y., A. J. A. Aarnink, R. Dijkman, T. Fabri, M. C. M. de Jong, and P. W. G. Groot Koerkamp. 2012. Effects of temperature, relative humidity, absolute humidity, and evaporation potential on survival of airborne gumboro vaccine virus. *Appl Environ Microbiol* 78 (4):1048–1054. doi: 10.1128/aem.06477–11.

Zhu, Yong-Guan, Yi Zhao, Bing Li, Chu-Long Huang, Si-Yu Zhang, Shen Yu, Yong-Shan Chen, Tong Zhang, Michael R. Gillings, and Jian-Qiang Su. 2017. Continental-scale pollution of estuaries with antibiotic resistance genes. *Nature Microbiology* 2 (4). doi: 10.1038/nmicrobiol.2016.270.

Zugic Petrovic, T. D., P. D. Ilic, M. Z. Grujovic, K. G. Mladenovic, S. D. Kocic-Tanackov, and L. R. Comic. 2020. Assessment of safety aspect and probiotic potential of autochthonous *Enterococcus faecium* strains isolated from spontaneous fermented sausage. *Biotechnol Lett* 42 (8):1513–1525. doi: 10.1007/s10529-020-02874-5 10.1007/s10529-020-02874-5 [pii].

3 Aspects of Antimicrobial Resistance in Agriculture

Mechanism, Types, and Prospects

Sakshi Patel, Ameeta Sharma, Neha Batra, and P.C. Trivedi

3.1 INTRODUCTION

Antimicrobials are synthetic, natural (obtained from bacteria or fungi), or chemical in origin and prevent microbial development with little to no harm to the host (Linares-Otoya *et al.*, 2017). First antibiotic, penicillin was discovered accidentally by Alexander Fleming after he observed that *Penicillium notatum* secreted a chemical that prevented *Staphylococcus* from growing. The era of antibiotics began with this discovery. By broadening the range of antibiotic activity at low dosages, decreasing the development of resistance, and enhancing the general use of antibiotics, these medicines exhibited extraordinary efficacy. Over time, these substances started to have complicated and pleiotropic downstream effects on bacterial cells, driving microorganisms to develop coping mechanisms. As a result, the novel process of creating synthetic antibiotics did not succeed in creating agents that are longer lasting and more effective than antibiotics derived from natural sources. Soon, novel approaches to producing additional narrow-spectrum antibiotics and related diagnostics using natural antibiotics, including their semi-synthetic forms, may be required (Brown and Wright, 2016).

Antimicrobial resistance (AMR) develops when bacteria, viruses, fungi, and parasites rapidly adapt and multiply in the presence of drugs that once negatively affected them. AMR is seen as a danger to public health systems not just in underdeveloped nations but throughout the world. The inability of antibiotics to treat infectious infections indicates an uncertain future for healthcare (Chokshi *et al.*, 2019). AMR infection causes significant infections and protracted hospital stays, rising healthcare expenses, increased prices for second-line medications, and unsuccessful treatment outcomes (Shrestha *et al.*, 2018). The scenario in the agricultural ecosystem seems to be more challenging since different farming practices and niches may individually have a different impact on the persistence and proliferation of antibiotic resistance (Mafiz *et al.*, 2018).

Antibiotics, antibiotic resistance bacteria (ARB), and antibiotic resistance genes (ARGs) are distributed throughout the intricate structure of the agricultural ecology, finally reaching the food chain. The discharge of wastewater effluents to receptive watersheds, the application of biosolids and manure to the soil, and the irrigation of agricultural products with effluents have all been encouraged as a result of the use of antibiotics in healthcare facilities, farming, and aquaculture, and eventually back to humans. This leads to unpredictable health consequences in humans. In an account of DNA fingerprinting's evidence of a connection between ARB isolated from an agricultural source and a sick person, the World Health Organization (WHO) stated that ARB on farms can infect fruits and vegetables.

3.2 THE DYNAMICS OF ANTIBIOTIC RESISTANCE

3.2.1 Overuse of Antibiotics

In the past, antibiotics have been used to cure diseases caused by bacteria in major fauna, flora, and ornamental plants. Oxytetracycline and streptomycin are still widely employed for this reason

DOI: 10.1201/9781003269380-3

today. To combat the fire blight brought on by the bacterium *Erwinia amylovora*, streptomycin is sprayed on the surfaces of plants like apples, pears, and ornamental plants in the US. Global antibiotic usage by livestock alone was estimated to be 63,151 tons in 2010, and if this problem is not resolved, it is expected to increase by 67% by 2030 (Van Boeckel *et al.*, 2015). The majority of antibiotics consumed by humans (between 50% and 70%) are beta-lactam antibiotics like penicillin, cephalosporins, and carbapenems. Antibiotics are widely used in medical, veterinary, and agricultural platforms for therapeutic, prophylactic, metaphylactic, and growth-promoting purposes.

Contrary to popular belief, the hospital is not the only place where antimicrobials can be found in the environment through community sewage. Furthermore, the use of antibiotics during breeding results in meat with higher protein content and lower fat content. Additionally, fruit production and beekeeping both require antibiotics. Antibiotic use and the development of antibiotic resistance might well be related, as per epidemiological studies. Antibiotic resistance, therefore, begins to develop as a result of excessive antibiotic use. Typically, when antibiotics are overused, they kill drug-sensitive bacteria, leaving resistant bacteria to spread by natural selection. Despite the concerns regarding their abuse, antibiotics are overprescribed globally, and a lack of control encourages this overuse by making antibiotics more widely available, more easily accessible, and more reasonably priced (Michael *et al.*, 2014).

3.2.2 Incorrect Antibiotic Prescriptions

Antibiotic resistance in bacteria can also be caused by incorrect prescription of antibiotics. Studies have shown that the selection of the antibiotic and the length of the antibiotic therapy are inappropriate in roughly 30% to 50% of clinical cases and treatment regimens (Luyt *et al.*, 2014). Inadequately given antibiotics frequently have negative therapeutic effects and could cause patients to experience clinical problems (Lushniak, 2014). By encouraging gene alterations like irregular gene expression, mutagenesis, and horizontal gene transfer (HGT), the concentration of subtherapeutic antibiotics may cause antibiotic resistance. Antibiotic-induced alterations in gene expression may result in virulence, but greater mutagenesis and HGT may promote the emergence of antibiotic resistance. Studies have demonstrated that tazobactam and piperacillin cause extensive proteome mutations in *Bacteroides fragilis* at sublethal concentrations (Viswanathan, 2014).

3.2.3 Extensive Use of Antibiotics in the Agriculture Sector

Research has shown that the environmental microbiome is impacted by the use of antibiotics in agriculture. Tetracyclines and streptomycin are applied to fruit trees as insecticides in the western and southern parts of the US. Antibiotic resistance exposes environmental microorganisms to inhibitory substances and changes the ecology of the environment. The total amount of antibiotics used in agriculture worldwide is thought to be between 63,000 and 240,000 tons, although this estimate isn't exact due to insufficient data collecting and surveillance of antibiotic resistance in some nations (O'Neill, 2015). The consumption of agricultural antibiotics is also predicted to climb globally by 67% by 2030 as a result of the anticipated increase in the number of animals required for the production of food. Additionally, it is predicted that by 2030, the consumption of antibiotics in China, South Africa, Brazil, and Russia will also have doubled due to the expected 13% population growth of these countries (Kuppusamy *et al.*, 2018).

3.3 ANTIBIOTIC RESISTANCE AND ITS MOLECULAR MECHANISM

The development of antibiotic resistance results from stable genetic alterations such as mutations, transformation, conjugation, and transduction (Holmes *et al.*, 2015). There are two ways that antibiotic resistance might develop: intrinsic and acquired. As a result of bacteria's evolutionary adaptability to their environment, altered preexisting genes found in their chromosomes result in reduced antibiotic affinity to their targets (Ventola, 2015). During this evolutionary process, bacterial cells develop genetic flaws in their genes (plasmids or chromosomes) and pass these genes to daughter cells via vertical gene transfer, leading to intrinsic or innate resistance (Founou *et al.*, 2016). Through HGT between the same or different bacterial species, acquired resistance happens through conjugation, transformation, or transduction. Instead of mutation, it might happen due to the acquisition of movable extrachromosomal DNA elements such as transposons, integrons, and plasmids. An example of acquired resistance is the development of the genes that produce beta-lactamase, an enzyme that breaks down beta-lactam medicines like cephalosporins and penicillin. Bacteria transform when they take up free DNA from the environment containing the resistance genes; transduction occurs when the DNA is transferred via bacteriophages, which are viruses that affect bacteria; and conjugation occurs when the transfer of genes takes place via conjugative pili.

Antibiotic resistance mechanisms such as changing cellular targets for antibiotics, preventing antibiotic entry into cells, removing antibiotics from cells, degrading antibiotics, and overexpressing membrane transporters are also being outlined with the routine discovery of vectors and genes that encode antibiotic resistance. Targets of antibiotics include functional proteins like ribosomal and enzyme proteins. A specific interaction occurs between an antibiotic and its target site. Antibiotics may lose their effectiveness if certain target areas are mutated. Antibiotics are also rendered ineffective because of the lower influx of medications due to variations in the porins penetration in the outer cell membrane. Additionally, some bacteria contain specific proteins called efflux pumps or plasma translocases which they use to remove significant amounts of medicines from their cells, thereby preventing their buildup (Kumarasamy *et al.*, 2010). Inactivation of antibiotics can also occur by chemical modification or enzymatic cleavage. Another way of developing drug resistance by bacteria is through use of an alternative metabolic process. Due to this, the utilization of substrates occurs from the surrounding environment which in turn reduces the antibiotic effect.

3.4 ANTIBIOTIC RESISTANCE IN AGRICULTURE

Agriculture soil is a rich source of antibiotic resistance, which has significant effects on public health and food safety. The soil is thought of as a reservoir of ARB and ARGs because of the diverse microbial community, which includes bacteria that create antibiotics in the soil and is a result of natural or anthropocentric behaviors like applying manure. Globally, manure and manure-based soil have yielded several antibiotics in high quantities of about 1,000 g/kg. It is hypothesized in this context that the soil may serve as an "under-recognized reservoir" for antibiotic resistance that has been previously developed or is still establishing in disease-causing bacteria and that this reservoir must be taken into account while evaluating the risk factors which contribute to the global spread of antibiotic resistance (Gatica and Cytryn, 2013). The feces and urine of grazing animals; the use of sewage sludge, biosolids, and dung during soil amendment; and the use of recycled water for irrigation are additional significant pathways of ARB, ARG, and antibiotic transmission to the soil environment. Recent research has revealed that it is challenging to entirely remove all antibiotics present inside manure, although composting can lower antibiotic levels in dung before it is applied to the soil (Ray *et al.*, 2017). This increases the number of ARB and ARGs in the soil and applies deliberate pressure to the native community of microbes (Chen *et al.*, 2017)

Once antimicrobials are released into the environment, their behavior and fate are typically affected by a variety of factors, including soil characteristics and climate conditions, as well as their physicochemical properties like water solubility, sorption, volatility, lipophilicity, chemical properties, and sequestration capacities. Since antibiotics must be in an aqueous state for quick uptake, sorption is the primary process that influences the movement and dissemination of antimicrobials in the environment. This process also controls the availability of antibiotics for plant uptake (Albero *et al.*, 2018). The sorption of antibiotics to soil regulates both the absorption of antibiotics by plants and their transport within the soil environment. This process is influenced by several factors, such as soil characteristics, steric configuration, and amphoteric and amphiphilic properties of the antibiotic. For instance, tetracyclines' capacity to bind to doubly charged cations enhances their capacity to form robust interactions with soil particles. It was found that samples of topsoil obtained from farms after the second day of applying animal slurry contained 9.5 g/kg of chlortetracycline while samples taken from the soil below 80 cm contained a lower concentration of about 0.7 g/kg (Iwu *et al.*, 2020). The amount of antibiotics in the soil that are transferred to the surrounding environment is further increased by surface runoff and transmission made possible by particles. Sulfonamides experienced a 0.5% runoff loss (Stoob *et al.*, 2007).

Small levels of antibiotics present in the soil can be absorbed by plants grown there, but these amounts are sufficient to have phytotoxic effects on plant growth and development (Bártíková *et al.*, 2016). Antibiotic absorption depends on the characteristics of the plant as well as the antibiotic. ARGs from the soil resistome may enter the food chain via contaminated food crops or groundwater and in certain situations make their way to human pathogens, having an impact on human health. Researchers have demonstrated that several agricultural practices, such as spreading animal manure on farms, might increase the soil's intrinsic resistance (Popowska *et al.*, 2012). ARGs have also been found on the surface of fruits and vegetables, which is relevant because the bacteria group found on plant surfaces can be altered through a variety of antibiotic-resistant strains detected in the soil (Durso *et al.*, 2012).

The choice and the consumption pattern of antibiotics show geographic variation across the continents and are influenced by food and animal species, regional production patterns, types of production systems, intensive or extensive farming, the purpose of farming (commercial, industrial, or domestic), lack of clear legislative framework or policies on the use of antibiotics, growth, and the socioeconomic status of the population (Van *et al.*, 2015). In general, in developing countries, the way farmers obtain (over the counter) and utilize (multidrug practices) these antibiotics, as well as the existence of existing determinants, may have an impact on the amount and rate of antibiotic consumption in the farming sector. The factors currently in place include a high incidence or level of infections, a severe lack of state growth and management plans, a lack of planning for the husbandry zones, and minimal hygienic practices in livestock husbandry combined with the existence of an integrated agricultural system (Bashahun and Odoch, 2015). Numerous authors have shown that farmers use antibiotics carelessly and blamed this behavior on ignorance of how to use them wisely, the potential side effects of abusing them, neglect of manufacturer's guidelines and antibiotic withdrawal times, and the lack of veterinary care.

Furthermore, inexperienced farmers depend on the guidance and counsel of seasoned farmers and local medicine vendors for medication administration. Rich farmers, on the other hand, tend to use several antibiotics since they have the financial means to do so. Additionally, only some animals are sent to the laboratory for diagnosis to determine the causative agent and evaluate the antibiotic susceptibility tests before applying antibiotics (Kim *et al.*, 2014). As a result, some of these developing nations continue to utilize antibiotics that have been outlawed in industrialized nations, such as chloramphenicol, tylosin, and TCN (a powdered combination of oxytetracycline, chloramphenicol, and neomycin). As a result, these medications have been linked to human aplastic anemia development, carcinogenicity, mutagenicity (chloramphenicol), and the worsening of kidney illness (neomycin) (Berendsen *et al.*, 2010).

3.5 ANTIBIOTIC RESISTANCE IN MANURE

Due to the overuse of antibiotics in the animal environment, animal manure acts as an important reservoir of ARGs, which raises the level of antibiotic resistance in agricultural soil, especially after the application of antibiotic- and ARG-rich animal manure, posing issues for the environment and public health (Zhang *et al.*, 2019). When agricultural soil is amended, manure that has been dumped on it contains a sizeable amount of ARGs. Manure is a well-known "hot zone" for bacteria, and these bacteria contain ARGs in their mobile genetic elements (MGEs). Manure has been found to contain high amounts of antibiotics ranging from 91 to 136 mg/kg dry matter of tetracyclines and sulfonamides, which are then released into other environmental niches such as soil sediment, surface water, and groundwater. Additionally, manure has been found to contain several heavy metals, including Cu and Zn, which are thought to be the main causes of antibiotic resistance. The reason for this is that they are employed as food supplements in animal husbandry.

A significant portion of these substances are disposed into the aquatic environment via wastewater treatment plants (WWTPs), and effluents are released into agricultural soil via sludge due to the frequent use of antibiotics, which has led to their constant release into WWTPs, which were not initially designed to treat substances of this nature (Ezzariai *et al.*, 2018). Because it is inexpensive and safe to dispose of, animal dung is favored over chemical fertilizers to increase soil fertility and nutrient cycling during the production of organic crops. Animal dung is typically obtained from "restricted animal feeding facilities," where it can be obtained in big enough quantities to cover a large area of farmland. Antibiotics are utilized in this type of environment to both treat illnesses and help animals flourish. Only a small portion of these substances are absorbed by the animals; the remainder (30%–90%) is expelled in the manure (Negreanu *et al.*, 2012). Long-term use of animal manures on fields and the direct release of animal waste through drainage systems may result in the transfer of residues of antibiotics, and ARGs released from livestock into neighboring bodies of water, such as rivers, streams, lakes, and agricultural soil habitats, pose a threat to the environment and public health (Fang *et al.*, 2018). In one case, it was discovered that fertilizer of organic origin increased the amount of tetracycline in the soil by 20 mg/kg. Furthermore, the research by (Peng *et al.*, 2015) revealed that prolonged application of compost and fresh manure to arable soil enhanced the number of genes for tetracycline resistance with the dominant tet G genotype, strikingly similar to those seen in pathogenic bacteria (Peng *et al.*, 2015). Unbroken antibiotics may favor resistant bacterial strains in the soil when manure is applied to the soil (Wang *et al.*, 2015).

ARB and ARG distribution onto food crops designed for human or animal use can be accelerated by applying manure to agricultural soil used for food crop cultivation. Fortunately, modern technological advancements like aerobic composting and anaerobic digestion significantly minimize antibiotic and ARG residues in manure before they are implemented into the soil (Sun *et al.*, 2016). Because of its ease of use, effectiveness, and economic viability, aerobic composting is one of the commonly accepted techniques utilized during manure production. However, in a few situations, changing treatment methods, manure kinds, or places affects how well antibiotics and ARGs are eliminated from the manure. Composting is undoubtedly a cost-effective and environmentally friendly way to treat manure. A variety of thermophilic, mesophilic, and thermo-tolerant bacteria are present in the process, which is characterized by a high concentration of microbial flora. As a result, if the procedure is not well handled, it could result in the transfer of ARGs and ARB to the soil (Zhang *et al.*, 2018).

Similar to the effect of physicochemical condition of organic materials on the degradation process during composting, antibiotic persistence and movement is also influenced by their physicochemical characteristics. It was suggested that humic compounds aid in the elimination of antibiotics from manure by sequestering themselves inside the organic and inorganic matrix, which makes them harder for different antibiotics to extract throughout the 40 days of broiler dung composting (Ho *et al.*, 2013). Similarly to this, Wang *et al.* (2016) reported that at 2.4 log copies/g dry matter, tetracyclines ARGs were largely eliminated. After 42 days of manure composting, Selvam *et al.* (2012) demonstrated a 100% elimination of tetracyclines, fluoroquinolones, and sulfonamides from ARGs (Selvam *et al.*, 2012).

3.6 ANTIBIOTIC RESISTANCE IN CROP PRODUCTION

Antibiotics, ARBs, and ARGs are constantly being accumulated on agricultural fields, where they are then exposed to plant crops due to the compounds' physicochemical qualities, ambient conditions, and absorption capacities (Zhang *et al.*, 2017). Manure is widely regarded as a rich source of antibiotics and ARGs for soil, and its application encourages the evolution of ARB within the soil microbiome, which in turn affects the population of bacteria in plants growing above and below the soil. Therefore, changes in the microbial population of the soil can affect the microbiome of plants grown in manure-enriched soil. ARGs from the upper surfaces of many fruits and vegetables have been found in several investigations. Wang *et al.* (2015) retrieved ARGs from the phyllosphere, leaf endophytes, and roots of lettuce cultivated on manure-amended soil. In comparison to their utilization in medical contexts, antimicrobials are often used directly on plants in very limited quantities.

Antimicrobials applied to plants, however, are released into soil and water systems, impacting the resistance of humans, animals, and the environment (Thanner *et al.*, 2016). Antimicrobials from the soil can occasionally be passively absorbed by plants through their roots or by the movement of water. It has been demonstrated that crops like rice, carrot, cabbage, corn, spinach, cucumber, lettuce, tomato, potato, and wheat can absorb and accumulate medications like oxytetracycline, trimethoprim, amoxicillin, sulfamethoxazole, ofloxacin, ciprofloxacin, and tylosin from the growth medium via their roots. These plant crops are consumed by humans and, in certain circumstances, animals, which leads to the emergence of antibiotic resistance. The complicated relationship between the agroecosystem and the food chain poses an increasing concern about antibiotic resistance contaminating food crops.

3.7 TECHNIQUES FOR DETECTING ANTIMICROBIAL RESISTANCE

3.7.1 Antibiotic Susceptibility Testing

Common methods for identifying antibiotic resistance focus on the phenotypic analysis of bacterial growth in the presence of several antibiotics. A crucial component of medical microbiology has always been the isolation of pure cultures of pathogens. To identify drug-resistant microorganisms and their sensitivity pattern to frequently used antibiotics during therapy, clinical and research microbiology laboratories conduct antibiotic susceptibility testing (AST) (Karkman *et al.*, 2018). In this method of detection, the pattern of growth of a particular quantity of microorganisms implanted in a growth medium containing a serial dilution of test antibiotics is observed. The most widely used AST techniques include broth dilution, automated systems, disc susceptibility testing, and E-test strips. The approach of phenotypic susceptibility testing has the drawbacks of being inefficient in terms of time and being unable to determine the precise mechanism of antibiotic resistance. Inherent culture bias results from the inability of this technology to be applied to environmental bacteria that are viable but uncultivable (Franklin *et al.*, 2016). Whole-genome sequencing is currently advised and used for the identification of antibiotic resistance due to the increasing accessibility and affordability of next-generation sequencing, and the outcomes have been outstanding, especially in particular organisms like *Staphylococcus aureus* (Holden *et al.*, 2013).

3.7.2 PCR

The most widely used molecular diagnostic technique is the polymerase chain reaction (PCR). This is to enable it to identify not only pathogens but also the genes that determine their virulence and resistance. Real-time PCR methods that recognize pathogens and screen for resistance determinants include the Verigene system (Nanosphere), which confirms up to five species and four genera of Gram-negative bacteria as well as the GeneXpert system, which conducts real-time PCR in single-use disposable cartridges; and the Film Array Blood Culture Identification Panel system (Biofiber Diagnostics), which can identify up to 11 species and 15 genera of Gram-positive bacteria as well as ARGs like van A/B, KPC, and mec A with up to 99% specificity. Without culturing, ARGs in environmental DNA are analyzed using quantitative PCR (qPCR). The drawback of this procedure is that to compare results, you must extract the DNA of numerous distinct bacteria from various environmental sample matrices (Franklin *et al.*, 2016).

Utilizing droplet digital PCR (ddPCR) is another method of overcoming such constraints. The ddPCR is a recent technology that accurately and sensitively measures the number of nucleic acids in samples. Before amplification, the ddPCR divides the PCR reaction into thousands of separate droplets, unlike the qPCR, so that each droplet's quantification of the nucleic acid happens independently (Martinez-Hernandez *et al.*, 2019). The ability to precisely quantify nucleic acids in samples devoid of standard curves results from this. It generates more dependable data even in the presence of impurities that can prevent the Taq polymerase and/or primer annealing from working (Taylor *et al.*, 2017).

3.7.3 Microarrays

This technique makes use of image analysis to identify target molecules that have hybridized to a specific probe mounted on a solid basis. As long as the probes – in this case, oligonucleotides – are attached sufficiently close to one another, microarrays can be used to detect a sufficient number of resistance genes in a single assay. Several microarrays in the market can identify various resistance genes, such as Extended Spectrum Beta-lactamases, Amp C (Ampicillin Class C beta-lactamases) and carbapenems These include Check-MDR CT102, Check-MD CT103, and Check-MDR CT103 XL (Checkpoints Health BV) (March-rosselló, 2017)

3.7.4 Immuno-Chromatographic Methods

Antibiotic-hydrolyzing bacterial enzymes are found using this method. This procedure applies a few drops of the diluent containing the test microorganisms to one end of a nitrocellulose strip. By using capillarity, the bacteria move to the other end of the strip, and if the test is successful, a clear band is seen on the test area of the strip where an antigen-specific bacterial antibody is attached. The immune-chromatography technology used to detect OXA-48 and KPC carbapenems with over 100% specificity and sensitivity is called the Coris Bio Concept. This technology is inexpensive, does not require technical expertise, yields result in two minutes, and is easily available in the marketplace (March-rosselló, 2017).

3.7.5 Metagenomics

Without any previous information on the target resistance genes, metagenomics can be utilized to capture the DNA sequences of an entire community. Since it is not limited to only a few genes of interest, it has been used in a variety of contexts to identify ARGs and capture the whole resistome of the environment (Karkman *et al.*, 2018). Delineation of discovered ARGs in metagenomics is based on confirmed genes that are listed in open ARG databases. The most trustworthy ones are those that have been experimentally verified, such as Resfams, a current version of CARD, and hidden Markov model–based databases. The majority of platforms used for metagenomics sequencing typically produce short reads that offer little information about the sequenced genes. However, the combination of short readings with longer, overlapping DNA segments, known as contigs, may be able to reveal information on the phylogeny and genetic location of the genes. Metagenome data can be used to reconstruct entire or partial genomes. This information is crucial for determining the degree of hazards connected with ARGs in the environment (Hultman *et al.*, 2015).

3.7.6 Functional Metagenomics

Functional metagenomics deals with the laboratory cloning and expression of DNA from the environment in a host. The advantage of this method over PCR and metagenomics is that it doesn't call on any prior knowledge of ARGs. In this method, DNA from the environment is typically cloned in large pieces (10–200 kb) in a lab host, such as *E. coli*, and then the host's resistance to various antibiotics is evaluated. Using silicon analysis, mutagenesis, or subcloning, resistant clones are further tested for resistance genes. This procedure's drawback is that it requires a lot of effort and time because it is difficult to clone and express genes in a particular host. By identifying the expressed proteins in a high-throughput manner and comparing them to a confirmed strain without having to identify the cloned DNA and alleged new resistance determinants, a combination of proteomics and functional metagenomics will help reduce the difficulties of screening potential clones that contain each DNA segment (Fouhy *et al.*, 2015)

3.7.7 MALDI-TOF Mass Spectrometry

By analyzing proteins, MALDI-TOF mass spectrometry (MS) can quickly identify bacteria, yeast, and filamentous molds. Antibiogram by MALDI-TOF MS determines in about three hours whether a bacterium is harboring an enzyme that breaks down antibiotics, like ESBLs and carbapenems. The test bacterium is incubated with the test antibiotic for a while in this experiment, following which the medium is centrifuged and the supernatant is examined using the MALDI-TOF MS. If the test is successful, the antibiotic-conforming peak will vanish and be replaced by new peaks that correspond to the antibiotic's separated metabolites. Only the peak consistent with the antibiotic will emerge if the test is negative. Using this method, it is possible to predict resistance to clindamycin and chloramphenicol by identifying the methylation of 16S ribosomal RNA that methyltransferases carry out (Savic *et al.*, 2009)

In general, molecular methods are preferred over culture-based methods for monitoring antibiotic resistance in the environment because they avoid cultural bias, provide a better method for tracking ARB/ARG sources within a system of agriculture, and provide additional information from the extracted DNA, RNA, or proteins present in samples collected from the environment (Franklin *et al.*, 2016).

3.8 EFFECTS OF ANTIMICROBIAL RESISTANCE ON HUMAN HEALTH AND ECONOMY

The emergence and distribution of resistance to an antibiotic in the agricultural environment pose a worldwide health emergency. Antibiotic resistance is predicted to result in an increase in mortality rate from 700,000 to 10 million per year and an increase of US$100 trillion by 2050 (Jasovský *et al.*, 2016). Therefore, it is crucial to comprehend how antibiotic resistance develops in the agricultural ecosystem and transfers to humans. Through direct contact with plant materials or livestock which are usually considered ARB reservoirs, humans become infected with bacteria that are resistant to antibiotics that originate from the agricultural environment. This could happen if you consume meat, food, or water that has been contaminated with microorganisms that are resistant to antibiotics. In a similar vein, HGT via the food chain can also spread genes that code for antibiotic resistance to human infections. Once these infections have colonized the human body and antibiotics have been given, antibiotic-resistant lineages, which these medications have naturally selected for, will start to arise and spread (Chang *et al.*, 2015). Agriculture increases AMR primarily through three mechanisms: (i) infecting people through livestock products, (ii) spreading resistant strains through the food chain, and (iii) transferring ARGs from environmental matrices to the human body. A study by Hassan *et al.* found that antibiotic residues in the liver and kidneys of fish, poultry, and dairy products such as meat, milk, and eggs can exacerbate damage to bone marrow and reproductive organs in humans and drive AMR through transmission down the food chain (Hassan *et al.*, 2021).

Infections with antimicrobial-resistant pathogens leave healthcare institutions, the government, patients, and their families at a financial and clinical disadvantage. Antibacterial-resistant microorganisms cause around two million healthcare-associated infections in Americans every year, which ultimately cause 99,000 mortalities. Caretakers are often forced to distribute antibiotics that are extremely harmful and costly to their patients when first-line and second-line antibiotics are in short supply due to the significant financial burden that antibiotic-resistant infections typically place on a country's healthcare system (Lushniak, 2014). Additionally, people who contract illnesses brought on by microorganisms resistant to antibiotics must stay in the hospital longer, visit the doctor more frequently, go through prolonged recovery periods, and may even be more likely to develop long-term disabilities (CDC, 2013).

3.9 STRATEGIES TO COMBAT ANTIMICROBIAL RESISTANCE

Most researchers are leaning toward the isolation of antibiotic-degrading microorganisms, process improvement, and other cost-effective substances in their effort to increase the elimination efficacy of antimicrobials and ARGs from manure. One of the methods involved adding a bulking agent to animal dung before composting it. This improved the manure's humidity and carbon-to-nitrogen ratio, which in turn raised the compost's required temperature. It is believed that adding a bulking agent to manure during composting will accelerate the breakdown of antibiotics and limit ARGs from spreading throughout the manure (Zhang *et al.*, 2019). As a result, a bulking agent is used in current research on the fate of antibiotics and ARGs in manure during composting (Chai *et al.*, 2016).

Another strategy is the proper handling and disposal of animal waste. The treatment of manure and WWTP effluents needs to be upgraded with more modern methods. On the majority of farms and municipal WWTPs, better waste management should be implemented along with government restrictions. Another practical way to lessen the negative effects of antibiotics on agricultural ecosystems is to control the sources of antibiotic residues by cleaning organic manure and urban wastewater (Rad *et al.*, 2022). It is widely known that techniques like anaerobic fermentation and composting can significantly lower the level of antibiotics present in manure.

Zhang *et al.* reported that erythromycin persulfate residues activated by heat may limit the spread of MGEs; as a result, heating is a potentially effective way to reduce ARGs in sewage sludge (Zhang *et al.*, 2020). In an analysis by Xu *et al.*, thermophilic anaerobic digestion of urban sewage sludge resulted in a 29.59% decrease in *Proteobacteria* and a 17.65% decrease in *Chloroflexi*, as well as more successfully eliminated tetracycline-, macrolide-, and fluoroquinolone-resistant genes (Xu *et al.*, 2020). Tetracyclines, macrolides, lincosamides, ARGs, and MGEs were all reduced by 94% using thermal hydrolysis treatment of wastewater sludge (Sun *et al.*, 2016). According to research by Liao *et al.*, hyperthermophilic composting eliminated ARGs and MGEs by 89% and 49% more than conventional composting techniques (Liao *et al.*, 2018). By reducing the activity of pathogenic bacteria and ARGs, composting improves the soil's nutrients and beneficial microbiota. During composting, probiotic species increased from 5.77% to 7.12% whereas the diversity of potentially harmful bacteria decreased from 37.18% to 3.43% (Deng *et al.*, 2020). ARG production can be reduced by the microbial electrolysis cell system, an innovative contamination-control method. Hua *et al.* showed that electrolysis is a useful method to increase the efficiency of antibiotic breakdown by examining the effects of microbial electrolysis cells on the decomposition of erythromycin in wastewater (Hua *et al.*, 2019).

The launch of public media–based awareness programs will encourage the education of people, including children, farmers, doctors, veterinarians, and researchers, on antibiotic resistance stewardship approaches. These programs could include documentaries, lectures, campaigns, online discussion forums, conferences, advertisements, short films, poster presentations, animations, and lectures (Tasho and Cho, 2016). To educate the public about antibiotics and spur governments to take action, WHO established the first World Antibiotic Awareness Week in November 2015. Such activities will lessen the abuse of antibiotics generally, which will lessen their environmental distribution.

3.10 CONCLUSION AND FUTURE PROJECTIONS

In the sphere of agriculture, antibiotics – whether they are produced naturally, semi-synthetically, or synthetically – are extremely important. In the lives of humans and animals, these compounds are equipped with bacteriostatic (inhibitory) or bactericidal activity required for prophylactic and therapeutic purposes to prevent and treat diseases. Despite these positive results, using antibiotics has negative side effects in addition to their subtherapeutic use as growth promoters for the crop over a lengthy period. These could result in antibiotic pollution, leaving antibiotic residues in products derived from plants that humans consume. It can cause direct toxicity, promote growth of microorganisms exhibiting antibiotic resistance, and therapeutic failure in clinical circumstances. However, the rate of antibiotic use and antibiotic resistance is alarming; considering the actual data. The information represented here is derived from the results obtained by several researchers who have previously researched antibiotic use and antibiotic resistance.

It is essential to switch policies for regulating antibiotic consumption towards precision agriculture strategies that prevent the overuse of agricultural inputs like fertilizers and pesticides, as well as technologies like bioremediation and microbial electrolysis that remove antibiotic residues before joining the food supply, given the growing global population and, consequently, the increasing demand for livestock production, antibiotics, and countering the spread of AMR.

The precautionary use of antibiotics in clinical and agricultural settings across all nations or countries appears to require calling for cooperation or teamwork from people, farmers, veterinarians, consumers, and local vendors of pharmaceutical products to combat the rising antibiotic resistance of bacteria. As a result, a campaign of awareness is necessary to inform both farmers and the general public about the risks associated with the careless use, poor handling, and storage of antibiotics as well as the consequent antibiotic resistance and its effects on public health. Pharmacists and veterinary officers should strictly abide by the rules governing the prescription of antibiotics used in animal production. They should also make sure the farmers follow the dosage amount, duration, and administration of the treatment and withdrawal periods of the antibiotics prescribed for a disease, as well as the diagnosis of the disease's causative agent and antibiotic susceptibility testing against it, before the prescription and use of the antibiotics. As a result, the government should maintain a constant and thorough monitoring program in coordination with the public authorities from the concerned health department to track both the use of antibiotics in the agriculture sector and the antibiotic resistance patterns of well-known bacteria that cause foodborne illnesses. To keep a record of the drug and its residual level in the majority of environmental samples, the government and public health sector should also make services for research and analysis of data gathered via surveillance studies available. For environmental toxins to be tracked from their source to their destination, which is typically the food chain, it is important to have a better understanding of transmission routes and hotspots. More research is nevertheless required to quantify exposure to antibiotics in humans, ARBs, and ARGs as well as the number of antibiotics released into the agricultural ecosystem. Further research is needed to fill in information gaps and uncover possible threats that environmental antibiotics, ARBs, and ARGs pose to people's health and the health of the larger ecosystem. Furthermore, researchers, farmers, veterinarians, medical professionals, and nongovernmental organizations must work together to stop the spread of antimicrobial resistance.

REFERENCES

Albero, B, Tadeo JL, Escario M, Miguel E, & Pérez RA. (2018). Persistence and availability of veterinary antibiotics in soil and soil-manure systems. *Sci Total Environ*. 1(643):1562–1570.

Bártíková, H, Podlipná R, & Skálová L. (2016). Veterinary drugs in the environment and their toxicity to plants. *Chemosphere*. 144:2290–22301.

Bashahun, DGM, & Odoch TA. (2015). Assessment of antibiotic usage in intensive poultry farms in Wakiso District, Uganda. *Livest Res Rural Dev*. 27:247.

Berendsen, B, Stolker L, de Jong J, Nielen M, Tserendorj E, Sodnomdarjaa R, Cannavan A, & Elliott C. (2010). Evidence of natural occurrence of the banned antibiotic chloramphenicol in herbs and grass. *Anal Bioanal Chem*. 397(5):1955–1963.

Brown, ED, & Wright GD. (2016). Antibacterial drug discovery in the resistance era. *Nature*. 529(7586):336–343. doi: 10.1038/nature17042. PMID: 26791724.

CDC (2013). Centers for disease control and prevention, office of infectious disease. *Antibiotic Resistance Threats in the United States, 2013 | Antibiotic/Antimicrobial Resistance | CDC*. [WWW Document]. Centre for Disease Control. Retrieved from www.cdc.gov/drugresistance/threat-report-2013/index.html.

Chai, R, Huang L, Li L, Gielen G, Wang H, & Zhang Y. (2016). Degradation of tetracyclines in pig manure by composting with rice straw. *Int J Environ Res Public Health*. 13(3):254.

Chang, Q, Wang W, Regev-Yochay G, Lipsitch M, & Hanage WP. (2015). Antibiotics in agriculture and the risk to human health: How worried should we be? *Evol Appl*. 8(3):240–247.

Chen, L, Lang H, Liu F, Jin S, & Yan T. (2017). Presence of antibiotics in shallow groundwater in the northern and southwestern regions of China. *Groundwater*. 56(3):451–457.

Chokshi, A, Sifri Z, Cennimo D, & Horng H. (2019). Global contributors to antibiotic resistance. *J Glob Infect Dis*. 11(1):36–42.

Deng, W, Zhang A, Chen S, He X, Jin L, Yu X, Yang S, Li B, Fan L, Ji L, Pan X, & Zou L. (2020). Heavy metals, antibiotics and nutrients affect the bacterial community and resistance genes in chicken manure composting and fertilized soil. *J Environ Manag*. 257:109980.

Durso, LM, Miller DN, & Wienhold BJ. (2012). Distribution and quantification of antibiotic resistant genes and bacteria across agricultural and non-agricultural metagenomes. *PLoS One*. 7(11):e48325.

Ezzariai, A, Hafidi M, Khadra A, Aemig Q, El Fels L, Barret M, Merlina G, Patureau D, & Pinelli E. (2018). Human and veterinary antibiotics during composting of sludge or manure: Global perspectives on persistence, degradation, and resistance genes. *J Hazard Mater*. 359:465–481.

Fang, H, Han L, Zhang H, Long Z, Cai L, & Yu Y. (2018). Dissemination of antibiotic resistance genes and human pathogenic bacteria from a pig feedlot to the surrounding stream and agricultural soils. *J Hazard Mater*. 357:53–62. doi: 10.1016/j.jhazmat.2018.05.066.

Fouhy, F, Stanton C, Cotter PD, Hill C, & Walsh F. (2015). Proteomics as the final step in the functional metagenomics study of antimicrobial resistance. *Front Microbiol*. 6:172. doi: 10.3389/fmicb.2015.00172.

Founou, LL, Founou RC, & Essack SY. (2016). Antibiotic resistance in the food chain: A developing country-perspective. *Front Microbiol*. 7:1881. doi: 10.3389/fmicb.2016.01881.

Franklin, AM, Aga DS, Cytryn E, Durso LM, McLain JE, Pruden A, Roberts MC, Rothrock MJ, Snow DD, Watson JE, & Dungan RS. (2016). Antibiotics in agroecosystems: Introduction to the special section. *J Environ Qual*. 45(2):377–393. doi: 10.2134/jeq2016.01.0023.

Gatica, J, & Cytryn E. (2013). Impact of treated wastewater irrigation on antibiotic resistance in the soil microbiome. *Environ Sci Pollut Res Int*. 20(6):3529–3538. doi: 10.1007/s11356-013-1505-4.

Hassan, MM, El Zowalaty ME, Lundkvist A, Jarhult JD, Khan Nayem MR, Tanzin AZ, Badsha MR, Khan SA, & Ashour HM. (2021). Residual antimicrobial agents in food originating from animals. *Trends Food Sci Technol*. 111:141–150.

Ho, YB, Zakaria MP, Latif PA, & Saari N. (2013). Degradation of veterinary antibiotics and hormone during broiler manure composting. *Bioresource Technology*. 131:476–484. https://doi.org/10.1016/j.biortech.2012.12.194.

Holden, MT, Hsu LY, Kurt K, Weinert LA, Mather AE, Harris SR, Strommenger B, Layer F, Witte W, de Lencastre H, Skov R, Westh H, Zemlicková H, Coombs G, Kearns AM, Hill RL, Edgeworth J, Gould I, Gant V, Cooke J, Edwards GF, McAdam PR, Templeton KE, McCann A, Zhou Z, Castillo-Ramírez S, Feil EJ, Hudson LO, Enright MC, Balloux F, Aanensen DM, Spratt BG, Fitzgerald JR, Parkhill J, Achtman M, Bentley SD, & Nübel U. (2013). A genomic portrait of the emergence, evolution, and global spread of a methicillin-resistant *Staphylococcus aureus* pandemic. *Genome Res*. 23(4):653–664. doi: 10.1101/gr.147710.112.

Holmes, AH, Moore LS, Sundsfjord A, Steinbakk M, Regmi S, Karkey A, Guerin PJ, & Piddock LJ. (2015). Understanding the mechanisms and drivers of antimicrobial resistance. *Lancet*. 387(10014):176–187. doi: 10.1016/S0140-6736(15)00473-0.

Hua, T, Li S, Li F, Ondon BS, Liu Y, & Wang H. (2019). Degradation performance and microbial community analysis of microbial electrolysis cells for erythromycin wastewater treatment. *Biochemical Engineering Journal*. 146:1-19.

Hultman, J, Waldrop MP, Mackelprang R, David MM, McFarland J, Blazewicz SJ, Harden J, Turetsky MR, McGuire AD, Shah MB, VerBerkmoes NC, Lee LH, Mavrommatis K, & Jansson JK. (2015). Multi-omics of permafrost, active layer and thermokarst bog soil microbiomes. *Nature*. 521(7551):208–212. doi: 10.1038/nature14238.

Iwu, CD, Korsten L, & Okoh AI. (2020). The incidence of antibiotic resistance within and beyond the agricultural ecosystem: A concern for public health. *Microbiology Open*. 9(9). https://doi.org/10.1002/mbo3.1035.

Jasovský, D, Littmann J, Zorzet A, & Cars O. (2016). Antimicrobial resistance-a threat to the world's sustainable development. *Ups J Med Sci.* 121(3):159–164. doi: 10.1080/03009734.2016.1195900.

Karkman, A, Do TT, Walsh F, & Virta MPJ. (2018). Antibiotic-resistance genes in waste water. *Trends Microbiol.* 26(3):220–228. doi: 10.1016/j.tim.2017.09.005.

Kim, J, Kang CI, Joo EJ, Ha YE, Cho SY, Gwak GY, Chung DR, Peck KR, & Song JH. (2014). Risk factor of community-onset spontaneous bacterial peritonitis caused by fluoroquinolone-resistant Escherichia coli in patients with cirrhosis. *Liver Int.* 34(5):695–699. doi: 10.1111/liv.12374.

Kumarasamy, KK, Toleman MA, Walsh TR, Bagaria J, Butt F, Balakrishnan R, Chaudhary U, Doumith M, Giske CG, Irfan S, Krishnan P, KumarAV, Maharjan S, Mushtaq S, Noorie T, Paterson DL, Pearson A, Perry C, Pike R, Rao B, Ray U, Sarma JB, Sharma M, Sheridan E, Thirunarayan MA, & Turton J. (2010). Emergence of a new antibiotic resistance mechanism in India, Pakistan, and the UK: A molecular, biological, and epidemiological study. *Lancet Infectious Diseases.* 10:597–602. https://doi.org/10.1016/S1473-3099(10)70143-2.

Kuppusamy, S, Kakarla D, Venkateswarlu K, Megharaj M, Yoon YE, & Lee YB. (2018). Veterinary Antibiotics (VAs) contamination as a global agro-ecological issue: A critical view. *Agricu, Ecosyst and Environ.* 257:47–59. https://doi.org/10.1016/j.agee.2018.01.026.

Liao, H, Lu X, Rensing C, Friman VP, Geisen S, Chen Z, Yu Z, Wei Z, Zhou S, & Zhu Y. (2018). Hyperthermophilic composting accelerates the removal of antibiotic resistance genes and mobile genetic elements in sewage sludge. *Environ Sci Technol.* 52(1):266–276.

Linares-Otoya, L, Linares-Otoya V, Armas-Mantilla L, Blanco-Olano C, Crüsemann M, Ganoza-Yupanqui ML, Campos-Florian J, König GM, & Schäberle TF. (2017). Diversity and antimicrobial potential of predatory bacteria from the peruvian coastline. *Mar Drugs.* 15(10):308. doi: 10.3390/md15100308.

Lushniak, BD. (2014). Surgeon general's perspectives. *Public Health Rep.* 129(3):220–221. doi: 10.1177/003335491412900302.

Luyt, CE, Bréchot N, Trouillet JL, & Chastre J. (2014). Antibiotic stewardship in the intensive care unit. *Crit Care.* 18(5):480. doi: 10.1186/s13054-014-0480-6.

Mafiz, AI, Perera LN, He Y, Zhang W, Xiao S, Hao W, Sun S, Zhou K, & Zhang Y. (2018). Case study on the soil antibiotic resistome in an urban community garden. *Int J Antimicrob Agents.* 52(2):241–250. https://doi.org/10.1016/j.ijantimicag.2018.05.016.

March-rosselló, GA. (2017). Rapid methods for detection of bacterial resistance to antibiotics. *Enfermedades Infecciosas Y Microbiologia Clinica.* 35:182–188. https://doi.org/10.1016/j.eimce.2017.02.007.

Martinez-Hernandez, F, Garcia-Heredia I, Lluesma Gomez M, Maestre-Carballa L, Martínez Martínez J, & Martinez-Garcia M. (2019). Droplet digital PCR for estimating absolute abundances of widespread pelagibacter viruses. *Front Microbiol.* 10:1226. doi: 10.3389/fmicb.2019.01226.

Michael, CA, Dominey-Howes D, & Labbate M. (2014). The antimicrobial resistance crisis: Causes, consequences, and management. *Front Public Health.* 2:145. doi: 10.3389/fpubh.2014.00145.

Negreanu, Y, Pasternak Z, Jurkevitch E, & Cytryn E. (2012). Impact of treated wastewater irrigation on antibiotic resistance in agricultural soils. *Environ Sci Technol.* 46(9):4800–4808. doi: 10.1021/es204665b.

O'Neill, J. (2015). Antimicrobials in agriculture and the environment: Reducing unnecessary and waste. *London, UK: The Review on Antimicrobial Resistance.* pp. 1–40.

Peng, S, Wang Y, Zhou B, & Lin X. (2015). Long-term application of fresh and composted manure increase tetracycline resistance in the arable soil of eastern China. *Sci Total Environ.* 506–507:279–286. doi: 10.1016/j.scitotenv.2014.11.010.

Popowska, M, Rzeczycka M, Miernik A, Krawczyk-Balska A, Walsh F, & Duffy B. (2012). Influence of soil use on prevalence of tetracycline, streptomycin, and erythromycin resistance and associated resistance genes. *Antimicrob Agents Chemother.* 56(3):1434–1443. doi: 10.1128/AAC.05766-11.

Rad, AK, Astaykina A, Streletskii R, Afsharyzad Y, Etesami H, Zarei M, & Balasundram SK. (2022). An Overview of antibiotic resistance and abiotic stresses affecting antimicrobial resistance in agricultural soils. *Int J Environ Res Public Health.* 19(8):4666. https://doi.org/10.3390/ijerph19084666.

Ray, P, Chen C, Knowlton KF, Pruden A, & Xia K. (2017). Fate and effect of antibiotics in beef and dairy manure during static and turned composting. *J Environ Qual.* 46(1):45–54. doi: 10.2134/jeq2016.07.0269.

Savic, M, Lovric J, Tomic TI, Vasiljevic B, & Conn GL. (2009). Determination of the target nucleosides for members of two families of 16S rRNA methyltransferases that confer resistance to partially overlapping groups of aminoglycoside antibiotics. *Nucleic Acids Res.* 37(16):5420–5431. doi: 10.1093/nar/gkp575.

Selvam, A, Xu D, Zhao Z, & Wong JW. (2012). Fate of tetracycline, sulfonamide and fluoroquinolone resistance genes and the changes in bacterial diversity during composting of swine manure. *Bioresour Technol.* 126:383–390. doi: 10.1016/j.biortech.2012.03.045.

Shrestha, P, Cooper BS, Coast J, Oppong R, Do Thi Thuy N, Phodha T, Celhay O, Guerin PJ, Wertheim H, & Lubell Y. (2018). Enumerating the economic cost of antimicrobial resistance per antibiotic consumed to inform the evaluation of interventions affecting their use. *Antimicrob Resist Infect Control*. 7:98. doi: 10.1186/s13756-018-0384-3.

Stoob, K, Singer HP, Mueller SR, Schwarzenbach RP, & Stamm CH. (2007). Dissipation and transport of veterinary sulfonamide antibiotics after manure application to grassland in a small catchment. *Environ Sci Technol*. 41(21):7349–7355. doi: 10.1021/es070840e.

Sun, W, Qian X, Gu J, Wang XJ, & Duan ML. (2016). Mechanism and effect of temperature on variations in antibiotic resistance genes during anaerobic digestion of dairy manure. *Sci Rep*. 6:30237. doi: 10.1038/srep30237.

Tasho, RP, & Cho JY. (2016). Veterinary antibiotics in animal waste, its distribution in soil and uptake by plants: A review. *Sci Total Environ*. 563–564:366–376. doi: 10.1016/j.scitotenv.2016.04.140.

Taylor, SC, Laperriere G, & Germain H. (2017). Droplet digital PCR versus qPCR for gene expression analysis with low abundant targets: From variable nonsense to publication quality data. *Sci Rep*. 7:2409. https://doi.org/10.1038/s41598-017-02217-x.

Thanner, S, Drissner D, & Walsh F. (2016). Antimicrobial resistance in agriculture. *mBio*. 7(2): e02227–15. doi: 10.1128/mBio.02227-15.

Van Boeckel, TP, Brower C, Gilbert M, Grenfell BT, Levin SA, Robinson TP, Teillant A, & Laxminarayan R. (2015). Global trends in antimicrobial use in food animals. *Proc Natl Acad Sci U S A*. 112(18):5649–5654. doi: 10.1073/pnas.1503141112.

Ventola, CL. (2015). The antibiotic resistance crisis: Part 1: Causes and threats. *P T*. 40(4):277–283. PMID: 25859123. PMCID: PMC4378521.

Viswanathan, VK. (2014). Off-label abuse of antibiotics by bacteria. *Gut Microbes*. 5(1):3–4. doi: 10.4161/gmic.28027. Epub 2014 Jan 29. PMID: 24637595. PMCID: PMC4049933.

Wang, FH, Qiao M, Chen Z, Su JQ, & Zhu YG. (2015). Antibiotic resistance genes in manure-amended soil and vegetables at harvest. *J Hazard Mater*. 299:215–221. doi: 10.1016/j.jhazmat.2015.05.028.

Wang, R, Zhang J, Sui Q, Wan H, Tong J, Chen M, Wei Y, & Wei D. (2016). Effect of red mud addition on tetracycline and copper resistance genes and microbial community during the full scale swine manure composting. *Bioresour Technol*. 216:1049–1057. doi: 10.1016/j.biortech.2016.06.012.

Xu, R, Zhang Y, Xiong W, Sun W, Fan Q, & Zhaohui Y. (2020). Metagenomic approach reveals the fate of antibiotic resistance genes in a temperature-raising anaerobic digester treating municipal sewage sludge. *J Clean Prod*. 277:123504.

Zhang, H, Li X, Yang Q, Sun L, Yang X, Zhou M, Deng R, & Bi L. (2017). Plant growth, antibiotic uptake, and prevalence of antibiotic resistance in an endophytic system of pakchoi under antibiotic exposure. *Int J Environ Res Public Health*. 14(11):1336. doi: 10.3390/ijerph14111336.

Zhang, J, Lin H, Ma J, Sun W, Yang Y, & Zhang X. (2019). Compost-bulking agents reduce the reservoir of antibiotics and antibiotic resistance genes in manures by modifying bacterial microbiota. *Sci Total Environ*. 649:396–404. doi: 10.1016/j.scitotenv.2018.08.212.

Zhang, R, Gu J, Wang X, Li Y, Zhang K, Yin Y, & Zhang X. (2018). Contributions of the microbial community and environmental variables to antibiotic resistance genes during co-composting with swine manure and cotton stalks. *J Hazard Mater*. 358:82–91. doi: 10.1016/j.jhazmat.2018.06.052.

Zhang, Y, Liu H, Dai X, Cai C, Wang J, Wang M, Shen Y, & Wang P. (2020). Impact of application of heat-activated persulfate oxidation treated erythromycin fermentation residue as a soil amendment: Soil chemical properties and antibiotic resistance. *Sci Total Environ*. 736:139668.

4 Application of Nanobiotechnology in Detection and Diagnosis of Phytopathogens

Riya Rachel Joji, Malavika Ram AK, Ajay Tomar and Meenakshi Rana

4.1 INTRODUCTION

Taniguchi coined the term “technology” for the branch of science that to a good extent manages the mix and utilization of nano-sized particles (1–100nm) of any material. Nanoparticles (NPs) were transformed into smaller called the nano-size which has a contrasting and distinguishable property. Nanoparticles have a higher surface area to volumetric ratio and have an expansion of their reacting nature stated by Dubchak et al., in 2010.

The vicinity of the nanoparticles is of 100 km when one gram gets transformed into the nano-form. As stated by Buffat and Borel, in 1906, gold NPs of size 2.5nm get dissolved at a much lower temperature of 300°C than a pure gold element which needs a very high temperature of 1,064°C. The slender film has lower retention capacity or the ability which is found higher in the materials moulded with the help of the nanoparticles. Wang et al., in 2011 stated that there is toxicity on the microbes like *Salmonella* where the original gold does not show the toxicity. In addition, the Ag NPs have hostile to microorganism and against parasitism whereas Ag in massive scale structure do not conduct so as stated by Sofi et al., in 2012. NPs are of assorted shapes and usually littler compared to microorganism cells. The NPs cannot be zoomed in by the optical magnifying instruments. The molecule size is significantly littler than associate degree infection molecule like respiratory disease infection ranging within 80–120nm distance across and ToMv infection with 300nm length and 10–18nm breadth. The nanoparticles have a purpose of dissipating lightweight beneath higher goals. The NPs move beyond locomotion and the speed shifts firmly within the molecule size. Nanoparticles can be of shapes of spheres, solid pole wrought.

Nanobiotechnology is that field within which nanotechnologies are utilized in completely different field areas. Final outcome is that the crossover field of nanobiotechnology that utilizes natural starting materials, organic arrange standards or has natural or clinical applications.

While biotechnology manages metabolic and different physiological procedures of natural subjects as well as the microorganisms, in mix with nanotechnology, nanobiotechnology will assume a necessary job in making and death penalty various valuable instruments within the investigation of life. In spite of the very fact that the coordination of nanomaterials with science has prompted the development of symptomatic gadgets, differentiate operators, logical apparatuses, treatment and medicine conveyance, biotechnology explore remains in its early stages.

Conventional sub-atomic indicative ways are loosely used in analysis facilities every place throughout the globe to acknowledge phytopathogenic with high level of affectability and pellucidity. In any case, the overwhelming majority of those systems cannot be applied within the field wherever salaries are very little. Moreover, the numerous expense and short rack half existence of

DOI: 10.1201/9781003269380-4

some atomic science reagents that limit the employment of customary sub-atomic ways in making nations. Nanotechnologies could have real arrangements against various commercial enterprise problems like disease management. Nano-based materials are going to be introduced, which is able to expand the viability of fungicides and chemicals, allowing simply minor parts to be used. In addition, nanodiagnostics and microfluidics provide novel devices to enhance the instance designing step that remaining elements laborious to include a scaled down stage. The sign improvement approaches might challenge those of target intensification. Location recognition of phytopathogens utilizing nano-based packs, nano-sensors, nano-biosensors, nanobarcodes and different versatile demonstrative frameworks can likewise facilitate the agricultural and nourishment trade to superintend numerous plant diseases. During this audit, we tend to portray the ideas and current condition of the technology application in plant pathology as well as nanodiagnosis by compact polymerase chain reaction (PCR), nanopore sequencing apparatuses, nanodiagnostic pack, gold nanoparticles, quantum dots (QDs), nanobarcodes and nano-sensors.

Nanotechnology contains a wide extent of utilization in medication, trade and commercial enterprise and might alter the total society, whenever victimized befittingly. In commercial enterprise, technology has potential extension to be used within the common plus abuse and preservation, as well as creation and assurance of yields and domesticated animals. Nanomaterials could likewise offer answers for mechanical and ecological difficulties within the regions of sun-based mostly vitality amendment and water treatment. Importance of technology has been acknowledged globally, and a number of nations have established specific councils/gatherings to assist and screen nano-mechanical headways and to saddle its interests for human beings. The Government of India has arranged nano-research throughout the Eleventh 5th Year Plan.

The targets chosen and the usage of the nano-sized particles are the two main vital purposes of the nanotechnique. The nano-sized particles in the range of 1–100nm with the blending of the matter with the nanoparticles.

The combination of NPs is however an easy procedure and needs specific experience and offices relying upon the strategy for integration. When the integration, characterization of nanoparticles is another vital advance that discovers fulfilment of the required molecule size of the fabric and their relative consistency.

Plant ailments area unit among the most issues limiting harvest profit. Associate degree initial move towards handling a plant below nursery and field conditions is to effectively distinguish the pathogens. Current advancements, as an example, quantitative polymerase chain reaction (Q-PCR), need a moderately large live of target tissue and depend upon various tests to exactly distinguish explicit phytopathogens. The fundamental harm of the customary symptomatic techniques is that they're tedious and want high affectability. Thus, straightforward techniques to boost the exactitude and rate of phytopathogens finding are needed. Engineering science, nanoparticles and QDs have developed as basic instruments for fast recognition of a particular natural marker with outrageous exactitude.

Biosensors, QDs, nanostructured stages, nano-imaging and nanopore DNA sequencing instruments will probably raise affectability, quality and speed of the infectious agent recognition, encourage high-through examination, and be utilized for wonderful perceptive and harvest assurance. Besides, nanodiagnostic unit hardware will while not a lot of a stretch and speedily acknowledge potential real plant pathogens, allowing specialists to assist within the turning away of pestilence ailments. The momentum survey manages the utilization of engineering science for quicker, additional sensible and precise demonstrative systems of plant diseases. Such a particular innovation might assist with coming up with associate degree acceptable coordinated the board framework which can alter crop things to unfavourably influence crop pathogens.

4.2 DETECTION OF PHYTOPATHOGENS

Bergeson, in 2010a stated that to identify the pathogens is a way to choose the specific chemicals or manures which indeed help in the eradication of the microbes from the field. Boonham et al., in 2008 stated that for the recognition of the plant parasitic bacteria, biomarkers are used as the tool, followed by Yao et al., in 2009 for the infections and Chartuprayoon et al., in 2010 for the growth indication. Baac et al., in 2006 stated that there is a restriction in the identification of the phytopathogens by the use of the sensors based on the nanoparticles. NPs will either be licitly adjusted to be used in infectious agent discovery or for associate degree associate degreealytic instrument to acknowledge mixes demonstrative to an unhealthy situation. Lopez et al., in 2009 stated that microarrays emitted by the nano-chips contains the fluorescent oligo rays that help in the distinguishing of the crossbred. Lopez et al., in 2009 stated that to recognize the single ester charge where its affectability and quality are detected by the nano-chips on the infection caused by the phytopathogens. Yao et al., in 2009 conducted an experiment in which he identified the pathogen named *Xanthomonas axonopodis pv. vesicatoria* causing spot of tomato using a light SiO NPs bound with the specific antibodies. Utilization of the nano-gold particles in the immunosensors to identify the Karnal bunt caused in wheat crop and the causal organism is *Tilletia indica* by using the SPR known as surface plasmon resonance stated by Singh et al., in 2010. The pathogens or any damage could easily be detected by the SPR technique as the viability is greater. The quick conclusion and the execution of the plant malady can be achieved by the use of the nanosenors in the analysis. Khan and Khan, in 2011b stated that variation in the pressure causes the changes in the physiology of the plants. Khan and Haque, in 2013 stated that jasmonates and SA helps in the guarding of the cells. The improvement was done by making up the fragile chemistry, adjusting the Ag and Cu NPs in screening of the SA and detecting the fungal growth example: *Sclerotium* concluded by Wang et al., in 2010. They are effective in the usage of the SA and comparing biosensors and detection techniques for the recognition of the phytopathogens, their screening and the changes in physiology of the plants.

4.3 MICROBIAL SYNTHESIS OF NANOPARTICLES

Khan, in 2000 used the biotechnology branch of science with the help of soil microbes and Gurunathan et al., in 2009 used the engineering branch. Khan and Anwer, in 2011a and El-Rfie et al., in 2012 confronted that there are microbes that forms nanoparticles within a substrate. Using microbes for the production of the nanoparticles with small and desired modifications. Gurunathan et al., in 2009 declared a union of Ag NPs through decrease of silver nanoparticles by the method of life afloat of *E. coli.*

Combination of Ag NPs were decontaminated by utilizing plant product thickness inclination natural process. Transmission electron microscopy (TEM), ultraviolet (UV) ray spectra and the light spectroscopic analysis are used in picturization of the pathogens. At 50 nm as a standard size for the identical distribution is attained by the use of TEM in the range of 420nm with highest absorbing capacity was observed. In 2011 Juliet and Manonmani used UV spectra analysis, FTIR and SEM techniques and synthesized Ag NPs using the *Escherichia coli* to be against the harmful microbial nutrients to fix the ability of SERRS and Ag nanoparticles which helps in delivering Ag-counter stain agent associated on the natural entities. *Fusarium solani* is used in the synthesis of Ag NPs stated by EL-Rafie et al., in 2012. The Ag nanoparticles in the compound AgNO focus on the ratio, the fixing of the biomass and also the concentration of H^+ ions. Using the pathogen *Fusarium solani* in the volume of 10g/100ml, 0.08g/100ml of AgNO with pH in the range of 10–12 and at 25°C which resulted in the production of 2,000ppm silver nanoparticles within the range of 9–15nm, which forms an ideal circumstance for the production of silver nanoparticles. In Portugal, Mourato et al., in 2011 studied about the production of silver and gold nanoparticles using the strains of the yeast that were extremophilic which were the by-product of the mine waste which was corrosive in nature. The metal blended with the nanoparticles from the yeast biomass resulted in the production of 20nm of silver nanoparticles and 40–100nm of gold particles.

Kalimuthu et al., in 2008 synthesized the production of the Ag NPs of size 50nm by using *Licheniformis* bacteria known as true bacteria. In 2008, Gade et al. successfully synthesized Ag NPs by using *Aspergillus niger.* Law et al., in 2008 used C-type Cytochromatic for the extracellular Ag which gives out silver iodide which is insoluble silver chloride or any other nanoparticles using *Geobacter sulfurreducens*. There was also a study in which the Ag NPs were synthesized by using a gram-positive bacteria named *Morganella morganii* stated by Abd et al., in 2013. Utilizing an extremophilic bacteria known as *Thermomonos pora* sp. has also been found useful in the production of Au NPs said by Ahmad et al., in 2003a and 2003b. Utilization of hydrogen in the binding with the Au NPs using a mesophilic microbe *Shewanella* by Konishi et al., in 2004. He et al., in 2000, created Au NPs of size in the range of 10–20nm using a microbe called *Rhodopseudomonas*. In 2000, Husse et al. declared that strain of a microbe called *Pseudomonas aeruginosa* afloat has a lesser effect on the Au nanoparticles. Woolfolk and Whiteley, in 1965 declared that strain of *Micrococcus lactilyticus* does the shrinking of the atomic number 92 (uranium). Lovely et al., in 1989 said that negatron uranium got reduced by the negatron element of donar which is developed by *Alteromonas putrefaciens*.

Lovley et al., in 1991 exhibited that *G. metallireducens* GS-15, developed anaerobically inside the sight of ethanoic acid derivation and U as negatron giver and negatron acceptor, severally, shrunk solvent U to insoluble U, oxidizing ethanoic acid derivation to CO. Di Gregorio et al., in 2005 stated that a vegetable named *Astragalus bisulcatus* has a sort of chemical element released by a strain of *Stenotrophomonas maltophilia*. In 1990, Losi and Frankenberger and later, in 1995, Tomei et al. studied non-obligatory non-oxygen required microbes, for example, *E. cloacea*; *Rhodospirullum rubrum*, which is known as a non-sulphur microbe; and *Desulfovibrio desulfuricans*, a microbe which in the absence of enough oxygen will reduce the element selenite to the chemical element and the storing within the living cell is obtained by the *E.coli*, discussed by Silverberg et al., in 1906 discussed the property of *Pseudomonas stutzeri* of reducing the selenite to the essential element and also was supported by Lortie et al., in 1992. *P. aeruginosa* has a property of biosynthesis or biodegrading of the selenite-rich soil from the rhizosphere area through biotransformation in the shape of the elemental oxy-anions to the shapeless allo-tropical element. Prasad et al., in 2000 said that there is production of the metallic elemental nanomaterials by a bacterial strain and at an optimum temperature the nanoparticles are produced of the size range of 40–60nm. Konishi et al., in 2000 said that a gram-negative bacterium named *Cynobacterium boryanum* is successful in deliverance of nanoparticles of the size range of 30–300nm. Yong et al., in 2002 discussed the bioreduction and biocrystallization of the metals to the metal NPs at nonpartial concentrated hydrogen ion which was achieved by an aerobic bacteria named *Desulfovibrio desulfuricans* which is also called microbe that feeds on salt. Windt et al., in 2005 from USA associated the reducing microbe named *S. oneidensis*, reduces the atomic number 46 palladium in to the nanoparticles palladium by the action of an actinomycetous *Thermomonos pora* which converts into an Au NPs of a sphere-like shape in the range of 8nm as discussed by Ahmed et al., in 2003a and 2003b. An alkali tolerant *Rhodococcus* sp. accumulation within the cells of the Au particles and converting them to the NPs with size of 5–15nm. *Anabaena*, *Calothrix* and *Leptolyngbya* are some of the eubacterial strains which are capable of the transformation of gold, silver and palladium and more of the NPs do the transformation within the elemental cells. In 2000, Brayner et al. discovered about the discharge of the nanoparticles and are sedimented by the organisms which are not considered as animals, plants, bacteria, fungi or viruses. Dameron et al., in 1989 used the fungus which is a yeast microbe known as *Glubrata* is used in the production of a monodispersed and peptidal bonding with the cadmium sulphide framed around a thiolate compound with a spot of size 2nm utilizing the plant-chelates. In 2002, Kowshik et al. used a microbe *Schizosaccharomyces pornbe* which got a shape of a polygon criss-crossed with the cadmium suphide blended NPs with the size of 1.5nm. Creation of varied polygonal shapes blended with the Au NPs in the size ranging from 1–100nm as described by Gericke and Pinches, in 2006. In 2009, Jha et al. used *S. cerevisiae* to produce the Au NPs in the size of 2–10nm. Bansal et al., in 2004 used *F. oxysporum*, Shankar et al., in 2003 worked on *Colletotrichum* sp., Fayaz et al., in 2010 used *Trichothecium* sp., *Trichoderma asperellum*, *T. viride*, Vigneshwaran et al., in 2006 used *Phaenerochaete chryosporium*, *F. solani* was used by Ingle et al. in the year 2009, *F. semitectum* was utilized by Basavaraja et al. in the year 2008, *Coriolus versicolor* was used and studied by Sanghi and Verma, in 2009, *Aspergillus niger* was used by Gade et al., in 2008, *Phoma glomerata* was studied by Birla et al., in 2009, genus *Penicillium brevicompactum* was used by Shaligram et al., in 2009, *Cladosporium cladosporioidis* by Balaji et al., in 2009, genus *Penicillium fellutanum* used in the research by Kathiresan et al. in the year 2009 and *Volvariella volvacea* was used by Philip, in 2009. The aforementioned examples of fungal and bacterial cultures and these scientists utilized these cultures for the production of the elemental NPs.

4.4 NANODIAGNOSIS AND NANO-PHYTOPATHOLOGY

The augmentation of sub-atomic medicine on a scale with nano-measurements could be an everlasting innovation for recognizing phytomicrobes. Nanomolecular indicative is that the utilization of nanobiotechnology technique to analyze the plant microbes and therefore referred to as nanodiagnostics. Specifically, many nano-devices and nano-systems used in the sequencing processes in the single particles of polymer. Examines with the employment of nano-size gadgets to explore polymer arrangements and analyze sickness have gotten more and more flexible and progressive.

It is necessary that new location procedures together with nano-biosensors for phytopathogens characteristic proof can in all probability be a foundation to the trend. Recognizing phytopathogens by means that of standard ways could take many days and during this manner analysts would like quick discovery devices that may offer leads in a few hours. To grow such discovery apparatuses, phytopathologists are regularly operating with nanotechnologists. Recently, created nanoparticles with non-common nano-scale qualities may introduce a big action in pathogens and contamination recognition. Engineering is to boot driving the development of lab-on-a-chip frameworks for recognizing harmfulness in water, looking at supplements in irrigation channels and its quality.

Nano-plant pathology could be a frontline science that utilizes nanotechnology for recognizing and identifying dominant plant diseases and also the plant microbes at a starting time, as a result of trim assurance from pandemic ailments. Gift day nanomolecular procedures are used for observance or comprehension of plant pathogens hereditary qualities, plant microorganism collaborations and quality exchange among plant microbes and additionally the host. Moreover, nanoparticles like nanosized silica-silver applied as antimicrobial and antifungal specialists. For the detection of any plant malady, we can use and observe the phytotoxin and identify it for early detection and diagnosis using the nanoparticles and also find the remedy for managing the maladies.

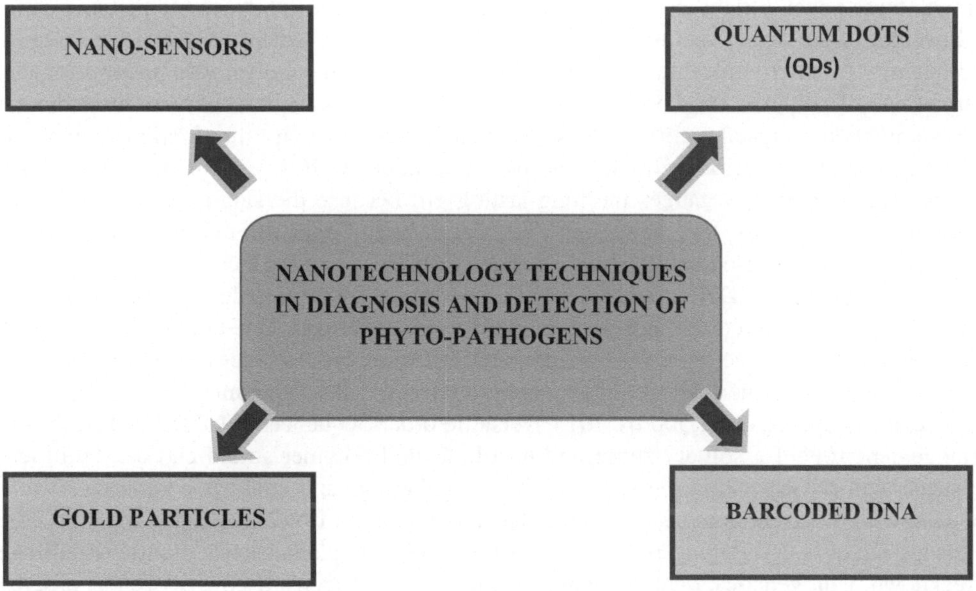

FIGURE 4.1 Techniques in nanotechnology.

4.5 TECHNOLOGIES IN THE DETECTION AND DIAGNOSIS OF PLANT PATHOGENS

4.5.1 Portable Diagnostic Equipment (Portable PCR Systems)

The ability in making a quick PCR amplicon in an exceedingly entirely mobile set-up won't simply allow phytopathologists to accomplish more and more fortunate and expected analysis centre, nonetheless additionally extensively improve alternative downstream applications, as an example, those requiring unmoved genomic instruments for identification. Palm PCR, created by a Korean company, could be a smart thermo-cycler meant for easy use, exactness and high productivity. Despite its very little size, this dominant contraption conveys astonishingly economical and brisk intensification for numerous varieties of polymer tests together with the plant order. The polymers are often increased inside 25 minutes to enough amounts for good agarose-gel identification. The versatile framework presents an exceptionally sensible and simple-to-know approach to perform numerous forms of PCR tests for each initiate and knowledgeable about scientists. Contrasted with alternative analytic testing frameworks, Twista quantitative and convenient in progress fluorometer could be a re-done contraption created for analyzing recombinase enzyme intensification (RPA) responses, and another insurgence in polymer medicine, just as consolidating advantage in speed, quality and utility with nice affectability and clearness. Twista RPA fluorometer can flexibly prompt fast analysis, allowing AN on-time acceptable treatment contrasted and customary biology measures (which is required) in any event hours and atomic examines that normally need incorporated gear.

4.5.2 Portable Genome Sequencer (Nanopore Sequencing System)

Numerous organizations are investigation utilizing nanopore technology. These are two principal challenges that have to be attended in any nanopore sequencing:

How to acknowledge the nucleotides because the strand goes through the nanopore.
How to management the speed of the polymer strand because it goes through the nanopore.

Even with these novel difficulties, the nanopore sequencing stages is believed to be basic and clear in lightweight of the actual fact that hypothetically long peruses are often created from an occasional quantity of macromolecule. The super-molecule nanopore catalyst was meant to manage a solitary strand of polymer, and because the polymer experiences the nanopore an immediate electronic examination is crystal rectifier. The super-molecule nanopore is embedded in an exceedingly chemical compound bilayer over the best purpose of a micro-well. Every micro-well incorporates a sensing element chip that gauges the long-lasting gift because the single particle goes through the nanopore. Nonetheless, the speed that the polymer strand goes through the nanopore is non-reasonably quick for precise identification. IBM and ROCHE along are increase another sequencing innovation delineated as "DNA transistors," which is used in the recording of the ester grouping because the format is gotten through the nanopore sensing element. One among the organizations that's spearheading sequencing advancements, Oxford Nanopore Advances, recently declared that they expect nanopore strand sequencing to own the choice to make a genomic map in an exceedingly short time at as expense of $1,500 by 2014. Versatile order Sequencer (FLUNKY) had the choice to arrangement 10kbof a solitary sense and hostile to find polymer strand and can build leading edge succession (NGS) within the span of diverse gatherings and analysis conditions. Nanopore stage dead within current demonstrative gear has the potential of breaking down the total order in minutes instead of hours. Engineered materials often applied to plant science inquire regarding thus on breakdown plant genetics, quality capability and phytopathogens discovery even as progress of yield species.

4.5.3 Nanodiagnostic Kit

Phytopathologists intend to contemplate the lab achievement in the field very soon that alludes to pressing modern estimating gadgets, reagents, power flexibly and different highlights that currently occupy the lab room into a package no bigger or heavier than a suitcase. An estimated unit is moved to a field where harvests are developing to look for pathogens that could taint and lessen the yield. This is a brisk and exact system. Nanodiagnostic pack gear can without much of a stretch and rapidly recognize potential genuine phytopathogens, permitting specialists to help ranchers in anticipation of disease pestilences from breaking out. For instance, a myco-sensor is a tetraplex serious immune response based examine in a dipstick design for the on-going recognition of ZEA, T-2/HT-2, Wear and FB1/FB2 phytotoxins on a similar single strip in corn, wheat, oat and grain tests at or underneath their individual European most extreme build-up limits (MRLs). A plan of multiplex stripe utilized all through the examination (myco-sensor) is depicted as the proposed immunoassay convention is quick, modest, simple to utilize and reasonable with the end goal of brisk screening of phytotoxins in seeds.

4.5.4 Loop-Mediated Isothermal Amplification (LAMP-PCR)

Circle intervened isotheral enhancement (Light) of deoxyribonucleic acid is a basic, savvy and fast strategy for the particular on location identification technique for evoluted DNA utilizing a lot of six oligonucleotide preliminaries with eight restricting locales hybridizing explicitly to various districts of an objective quality and a thermophilic DNA polymerase from *Geobacillus stearothermophilus* for DNA intensification. When the QDs are embedded onto the beads consisting of the small fragments of DNA blended with the Au NPs are also one of the few discoveries. The thiol-altered oligonucleotides (Au-nanoprobes) procedure disposes of the utilization of electrophoretic examination of Light PCR items, consequently speeding up, particularly the results and diminishing costs. Light can possibly execute early location of phytopathogens on a ranch rather than in a research centre or less well-resourced settings by sub-atomic strategies for observing inoculum levels noticeable all around and is creating models to foresee when does and from where a pathogen will initiate its infection in the farming fields.

Here an example to describe more about the use of PCR is as follows: A DNA is isolated from a microbe called *G. stearothermophilus* which thrives at a very high temperature and uses six oligonucleotides with eight restricted hybridizers and basically helps in intensifying a soil-borne pathogen named *Fusarium grraminearum* using Light PCR. The advantage of using this technique is that there is no need of assessing the object, which makes the diagnosis easy and simpler. If the result is favouring, then it would obtain us with a green fluorescent colour, whereas if the result turns out to be against, then the colour shown will be orange. If there is an outcome with the green I colour, then it would be said that the object that is the amplicons could be visible with the naked eye under an intensified light.

The technique was successful in obtaining a positive outcome on genus *Fusarium graminearum* and in some other species resulted negative. This technique is proved to be utilized to identify phytopathogenic microorganisms to rhizopherics. For example, *Phytophthora sojae*, *Phytophthora ramorum* and *Phytophthora kernoviae*, *Aspergillus flavus* and *Aspergillus parasiticus* and toxigenic *Fusarium*.

4.5.5 Gold Nanoparticles as Biosensors

Usage of nanoparticles in biosensors permits the flip of events and utilization of some novel sign-recognition strategies and equipment. Varied systems, as an example, counter acting agent matter, bond receptor, antimicrobial and integral deoxyribonucleic acid arrangement acknowledgements are found for a selected discovery between target phytopathogenic cells and bio-functionalized

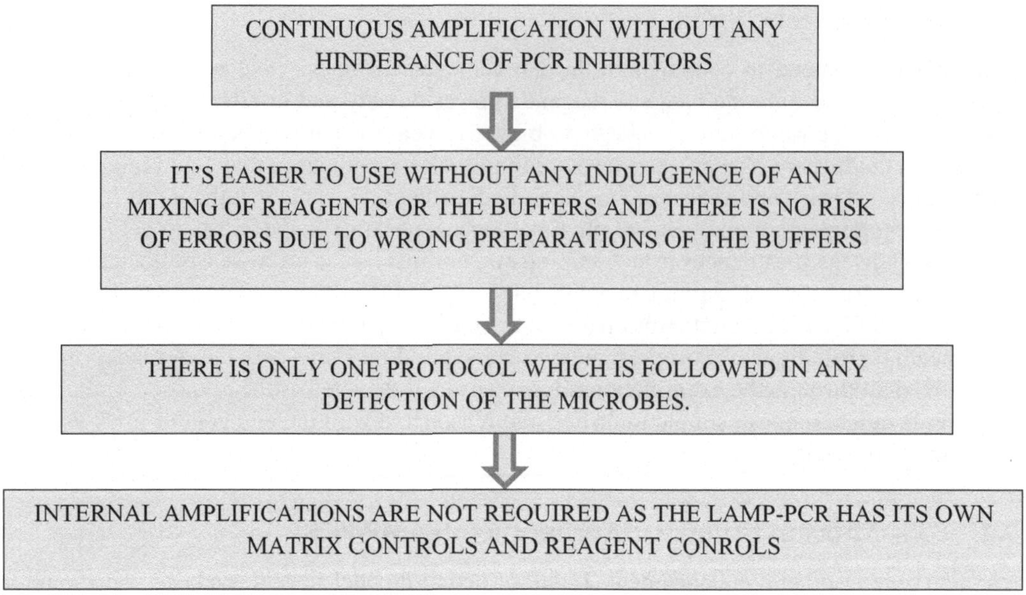

FIGURE 4.2 Advantages of LAMP-PCR.

nanoparticles. Gold nanoparticles area unit markers to be used in biosensors as some optical or electro-chemical methodology adjusted to acknowledge the microorganisms. Varied nanoparticles based mostly trials are performed to make biomolecular identification with deoxyribonucleic acid or super-molecule functionalized gold nanoparticles, used as associate in nursing objective express probes. Some nano-biosensors for the sub-atomic conclusions of borne pathogens and agro-psychological warfare specialists are found in later studies. These discovery techniques incorporate semiconducting chemical compound nano-wires, carbon nanotubes, nanoporous semiconductors and gold nanoparticles. Here, Dubertret et al. featured the capability of gold nanoparticles to travel concerning as visible light quenchers, and during this means, it can be used to settle vital drawbacks in these scientific atomic tests. For example, a deoxyribonucleic acid oligonucleotide can be intermingled, fluorescently named at its 5' finish and conjugated at the 3' finish with gold nanoparticles. These oligonucleotides are used in analytic methodology, particularly wherever deoxyribonucleic acid analysis and within the determination of the phytoplasmas connected with the flavescence dorée (FD) of grapevine. Fan et al., careful that the gold nanoparticles proficiently extinguish the visible light of sunshine collected polymers, as an example, polyfluorene is employed within the improvement of the optical exhibitions of nanobiotransductors for symptomatic functions. Moreover, associated in nursing analytic check manufactured from a selected oligonucleotides bearing a dyestuff at its 3' finish goes concerning as a nanobiotransducer in deoxyribonucleic acid sexual union. It delivers a lot of grounded visible light signal once hybridized to focus on deoxyribonucleic acid.

4.5.6 Quantum Dots (QDs)

QDs area unit semi-conductor nanoparticles that glow once rested by associate in nursing excitation source of illumination. Besides, QDs area unit inorganic fluorophores introducing vital preferences over standard natural fluorophores used as markers on nucleic acids or proteins for visual detection and observation. The myco-synthesis of semi-conductor nanoparticles was initially careful in living thing yeasts that were incontestable to be equipped for delivering Greenockite DISCS crystallites in lightweight of a metal salt stress. Varied organisms have likewise been used for the biogenesis of compact discs; but scarcely any investigations have targeted its light properties.

Exceptionally light CdSe (cadmium sulphide) were created by the expansion of *Fusarium oxysporum* once hatched with a mix of CdCl2 and SeCl4 at temperature. A capable myco-interceded union of exceptionally fluorescent CdTe QDs was practiced by *Fusarium oxysporum* secludes once responded with a mix of CdCl2 and TeCl2 at encompassing conditions. Portrayal of those biosynthesized CdTe nanoparticles was performed by TEM and selected territory lepton optical phenomenon (SAED).

A compelling biogenesis technique to induce prepared handily reaped biocompatible metal compound (CdTe) QDs with tuned visible light outflow utilizing yeast cells was developed. The confocal pictures of the yeast cells show evident inexperienced emanation from the musical organization CdTe QDs units were unfold clearly within the protoplasm and core of the yeast cells whereas none were found within the cell layer.

Biocompatible semi-conductor precious stones area unit created out of a core and a shell allowing the official of ligands and consequently the association of this fluorescent marker to the target phytopathogen. A hybridization-based biosensor referred to as the quartz–precious stone balance biosensor may be a real model. The quartz gem balance vibrates beneath an electrical upgrade, associate in nursing an adjustment in mass attributable to the association of any compound to its surface is known through a decrease within the return of its vibration. At the purpose once a nucleic corrosive check is joined to the skin of a quartz-gem balance biosensor and after given to correlative PCR, a crossbreed structures and causes the reverberation return of the quartz-gem balance biosensor to diminish considerably. This framework is joined with fast PCR conventions to diminish the best chance for express natural identification of phytopathogens.

4.5.7 Nanobarcodes (A Bio-Barcoded DNA, b-DNA)

Nucleic acids and proteins are located and enhanced by the tag testing known as bio-standardizing. A protein is identified or detected by the use of the gold nanoparticles embedded in the nanobarcodes. From enormous b-DNA to acknowledgement specialist proportion bears a method for significant sign enhancement. It is additionally encouraging by permitting the fast location of various protein focuses at low molecular fixations and nucleotides at high zeptomolar levels under enhanced circumstances. The idea of the biostandardized tag measures is novel and speaks to a potential option in contrast to the PCR strategy.

4.6 DETECTION OF THE MYCOTOXINS USING THE NANOSTRUCTURING TECHNIQUE

Nanostructure technique for mycotoxin location and detoxification area unit auxiliary metabolites created by unhealthful thready microorganisms inflicting plant diseases. Phycotoxicoses area unit toxins caused by the uptake of mycotoxins by the microorganisms, typically through the employment of contamination. This will actuate completely different various organic impacts, as an example growth inflicting, mutagenic, tetrogenic, immunotoxic and toxin effects. Associate in nursing assortment of biosensors are created ad delineate within the on-going writing for phytotoxin investigation, some of that have the aptitude multi-exhibit phytotoxins characteristic proof. Prieto-Simon et al. assessed rising biotechnological ways for phytotoxin investigation and affirmed that engineering has as recently been consolidated into phytotoxin bioassays. The first job of nano-sensors is to diminish the best chance for contagious microorganism detection. Nourishment deterioration is known by nano-sensors, as an example, a range of thousands of nanoparticles meant to be unreal in varied hues in grips with nourishment phytopathogens.

It is set lawfully into the bundling material, wherever they would fill in as associate in nursing device which might determine artificial substances discharged throughout the nourishment decay. Some nano-sensors rely upon microfluidic devices, and they are used to acknowledge phytopathogens competently in a short timeframe with high affectability. Ochratoxin associate in nursing are differentiated by the use of the recombinant Immunoglobulin G (IgG) where the egg whites behave as a BSA that helps in blocking purposes, where the immobilization of the recombinant IgG is pre-curse by ZnO NPs which are in the sol-gel form. Nano-ZnO/TiO stage for discovery of mycotoxins, as an example: aflatoxins (APH), ochratoxin B, citrinin, patulin, ergot alkaloids, fumonisins, trichothecenes and zearalenone (ZEA) continue to be beneath development. Associate in nursing epic ultra-delicate engaging nanoparticles immunochemical assay for phytotoxin location was created to grant continuous quantitative outcomes to recognizing over one phytotoxins. Some of the antibodies are designed in such a way for the purpose of identifying the mycotoxins released from the plant microbes like zearalenone, HT-2 and aflatoxins, for example B1. Lattanzio et al. discussed the transduction of the nano-sensors to identify the mycotoxins, and the limit is 100fg/ml.

An immunosensor strategy combined with a stream infusion framework is used and biased analysis of zearalenol in corn ensilage samples. Peroxidase (HRP) biosensors are applied with no pretreatment to choose OTA in spiked lager tests, and OTA disconnected from simmered roast low. Also, the exhibition of the HRP biosensors has been uncovered yielding traditional restorations of 103% and 99%, one by one within the two instances of OTA as referred earlier. The ozonization and surface assimilation proficiency of adjusted nanodiamonds to diminish the substance of aflatoxin-B1 has been analyzed by Puzyr et al. Silicon dioxide and dirts area unit typically productive in combine with littler estimate water atoms and littler mycotoxins, as an example, aflatoxins and ochartoxins. In any case, dirty or the contaminated area unit less skilled in limiting the larger mycotoxins, as an example, fumonisin and deoxynivalenol (vomitoxin) on the grounds that separation among earth layers isn't capable suit the larger molecules. By utilizing nano-sized mud, the house between the layers of mud has been delayed multiple times. Consequently, the nano-clay will tie the complete cluster of mycotoxins. A fast chemical connected immunosorbent live (ELISA) technique was examined by decreasing the covering, block and competition time needed in traditional enzyme-linked-immunosorbent serologic assay strategy utilizing superparamagnetic nanoparticles. This strategy was compelling for recognizing bioweapon M1 (AFM1) in milk during a direct operating scope of 4–250ng/L. A nanostructured atomic number 58 compound move based mostly immunosensor was to boot made for the identification of nourishment borne mycotoxins. Hare antibody associate in nursing antibodies and BSA are immobilized onto sol-gel-inferred nanostructured atomic number 58 compound film musical organization onto an iodized tin-oxide secured glass plate for the invention of ochratoxin-A. As of late, Paniel et al. engineered associate in nursing chemical science immunosensors for the placement of ultra-follow amounts of AFM1 in staples. These immunosensors were got wind of by utilizing engaging nanoparticles and a significant immunochemical assay utilizing HRP as a mark and had the choice to spot modest quantities of AFM1 (up to 0.1 ppb). Cysteamine functionalized-gold nanoparticles (C-AuNP) aboard covalently joined bioweapon B1 antibodies (aAFB1) were immobilized onto a 4-mercaptobenzoic corrosive place along self-gathered monolayer with relation to a gold anode (MBA/Au) to line up a BSA/aAFB1-C-AuNP/MBA/Au immune-electrode. These cathodes were used to acknowledge AFB1 within the scope of 10–100ng/L. A transferrable machine has recently been engineered that may at the same time distinguish completely different microorganisms, contagious poisons and pathogens in place away nourishment. Clearly biosensors are associated in nursing energizing possibility in distinction to the normal procedures for the popularity of mycotoxins and pathogens in nourishment.

4.7 NANO-SENSORS

4.7.1 Nano-Sensors and Plant Disease Predicting

Plant infection anticipating is an administration technique to gauge the percentages or seriousness of plant maladies and to help ranchers with deciding on financially savvy decisions for dominant diseases. Inquiring concerning is being completed by utilizing nano-sensors to boost infective agent location ways in crop systems. Various electronic organizations are gazing at the use of electrical directional polymers, for instance, polyanaline, polypyrolle and polythiophene. These polymers will likewise be used to manufacture sensors that may acknowledge sub-atomic signs with low force of waste and nourishment-borne pathogens. The nano-biosensors have the aptitude of distended affectability and therefore a basically shrivelled time interval to search out potential infection problems. Such bio-explanatory nano-sensors were accustomed acknowledge and live minute measures of contaminants. For instance, infections microscopic organisms, parasites, toxins and different bio-unsafe substances within the nourishment frameworks. During this manner, these biosensors might have a large result the exactitude cultivating strategies. Nano-sensors are connected to a GPS for on-going checking of illness and sent during the sphere to screen soil conditions and yield resultant. The combination of the biology with the engineering techniques creates a viability in the field of bio-sensing in creating a gear of distend affectability, which allows a reaction towards the natural changes and in the prevailing infections. Nano-sensors can allow the USAQ to acknowledge plant ailments before obvious aspect effects show up and during this manner encourage their management. Exactitude cultivating can allow improved farming creation by giving actual data, serving to cultivators to choose higher decisions.

4.7.2 Nano-Biosensors in Agriculture

The nano-biosensor is the result of a joined methodology of science and engineering science. These sensors hold the aptitude of distended affectability and consequently an altogether shrivelled time interval to discover potential complaint problems in crops, and during this means they will facilitate upgrade creation and improve sanitation in farming. Some of the antibodies are linked with the fibre optical which were later known as the biosensor for the detection and diagnosis of the microbes and the toxins. To assure the presence of the microbe *E. coli*, nano-biosensors were used as the framework. Chemistry estimation of $K_3Fe(CN)_6$, shrivelled by microbic digestion, allowed the quantitative assurance of microscopic organisms and parasites during a short time. Hashimoto et al. designed another biosensor framework for the short analysis of soil-borne pathogens, composed of two biosensors. The framework was designed utilizing equivalent amounts of two distinct organisms, every severally immobilized on a cathode.

A new study revealed the use of the silver element associated with the sulphurazon-ethyl due to its optical density having a convergent activity which shows a change in colour from yellow to orange and later on a reddish colour appears and in the end it shows a purple colour. The effectiveness of this technique is used in the clearing of impurities from water resources and also detoxifies the toxins produced by the microbes.

Fluorescent silicon dioxide nanoparticles (FSNP) joined with response particles effectively distinguished plant pathogens, for instance, bacteria genus *Xanthomonas axonopodis pv. vesicatoria* that causes microorganism spot illness in tomatoes and pepper. Copper oxide (CuO) nanoparticles and nano-layers were combined by sol-gel and shower shift techniques, on an individual basis. For the identification of the pathogen *Aspergillus niger* nanoparticles blended with the copper oxide having the biosensors layered with the nanostructures.

4.8 NANOFABRICATION IMAGING

Indicative imaging alludes to an expansive slew of advancements accustomed glimpse within or outside plant part thus on analyze completely different plant pathogens. The quickly growing utilization of symptomatic imaging advancements keeps on pushing the boundaries of plant medical specialist indicative talents. Utilizing these strategies, plant pathologists will analyze crop ailments previous and additional exactly.

Nanotechnology offers the USA the prospect to accurately tune and manage the compound and physical properties of distinction materials thus on conquer worries with toxicity, valuable imaging

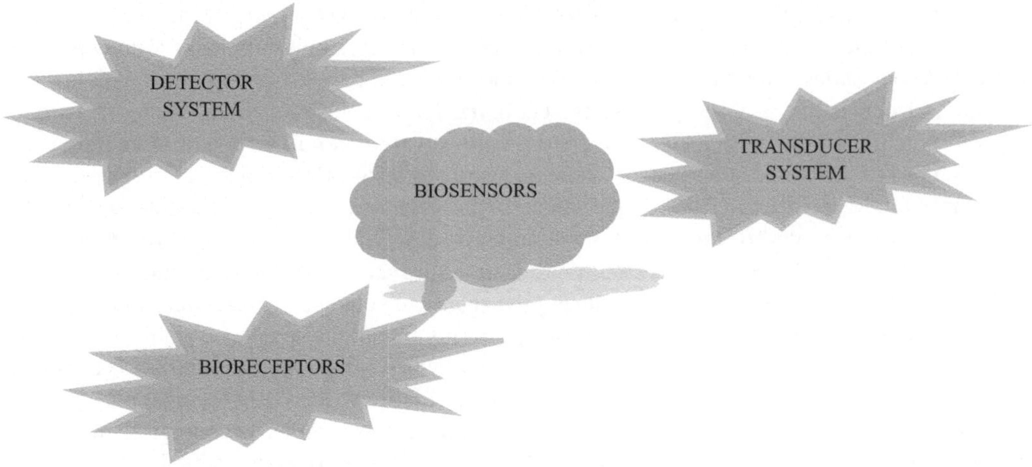

FIGURE 4.3 Components of biosensors.

time, tissue quality and sign quality. Within the mesoscopic size scope of 5–100 nm dimension, nanoparticles to boot have Brobdingnagian surface regions and helpful gatherings for conjugating to varied finding tools. On these lines, progress within the field of nano-scale differentiate operators can assume a key job within the proceeded with improvement of symptomatic imaging capacities within the coming back years.

For instance, negatron bar and lithography ways were used to manufacture geologies that replicate leaf surface highlights even as the inner pipes of the plants, and after nano-imaging innovations were used to concentrate however microorganisms attack and colonize the leaf. Lithography was used to nanofabricate a columned surface on Si wafers. This yard of smaller than expected columns was somewhere within the vary of 1.4 and 20μm wide and spread completely different separations. It absolutely was used to examine the event over the surface by the organisms that imitated a little of the qualities of the host plant.

To create sickness safe cultivars, the contamination procedure and conduct of bacteria infective agent inflicting Penetrate's complaint within grapevine vascular tissue were thought of utilizing nanofabrication technique.

The compact symptomatic gear, nanoparticle-based, bio-barcoded deoxyribonucleic acids device, and therefore the QDs, have the potentiality applied within the completely different identification of plant microbes and toxicity. Up until now, versatile symptomatic tests are created to quickly determine plant maladies and can be used to forestall scourges. These nano-based demonstrative units speed up infective agent discovery also as increment the exactitude of the finding.

Toxins and microbes are easily detected by the use of the nanotechnology by combining the engineering techniques in the field of the plant pathology branch of agriculture. In minutes 40 times the patterns run in the PCR is obtained is the model we got on a small scale. In creating a better rural framework, we should and must be emphasized on using of the gadgets of nano-scale. For instance, these nano-devices will acknowledge plant issues before these become detectable to the cultivator. Such gadgets can be acceptable reacting to extraordinary circumstances, recognizing the problem.

During the subsequent decade, nano-devices may build a large variety of estimations speedily and expeditiously. Future prospects in plant infection indicative can proceed in reducing of silicon chip innovation to the nano-scale run.

Nano-phytopathology is applied as an instrument to grasp plant-pathogen connections, giving new techniques to edit security. Express nano-devices and deoxyribonucleic acids nano-devices may empower actual following, location and determination of plant pathogens within the starting periods of the plant maladies.

4.9 EFFECTS

4.9.1 Effects of Nanoparticles on Pathogens

The impression of the NPs on microbes to saddle the innovation and in plant disease management against the plant microbes using the nano-bounded physio-compounds. Influencing the locomotion of microbes is done by the highly reactive ultra-little sized, even smaller than the infection molecule. *Staphylococcus aureus*, *Pseudomonas aeruginosa*, *Escherichia coli* and *Klebsiella pneumonia* colonization was hindered by the action of the Ag NPs. The antimicrobial activity was observed in the leaf extracts of *Solanum tricobatum* and *Ocimum tenutflorum* which were effective against *S. aureus* and *E. coli*. From the studies and research it has been observed that the NPs have an effective ability on the hinderance of colonization and growth by suppressing the multiplication.

4.9.2 Effect of Nanoparticles on Bacteria

Jayaseelan et al., in 2012 stated that zinc NPs which has antimicrobial activity can be used against the effect and infection caused by *P. aeruginosa*. At 25ng ZnO nanoparticles the highest hinderance in the colonization is recorded with 22+1.8mm sized microorganisms. ZnO nanoparticles is considered a novel antimicrobial compound. Nanoparticles blended with the silver elements and also PVP compounds have the antimicrobial locomotion against three bacterial pathogens like *Staphylococcus aureus* (gram-positive microorganisms), negative microorganisms), *P. aeruginosa* (non-ferment gram-negative microorganisms), against the spores of *E. coli* as concluded by Bryaskova et al., in 2011. Azam et al., in 2012 stated that *S. aureus*, *Bacillus subtilis*, *P. aeruginosa* and *E. coli* were controlled by the action of CuO nanoparticles. The methicillin-safe strains of *E. coli*, *P. aureginosa* and *S. aureus* are controlled by the action of Ag NPs with high bacteria-killing compounds stated by Guzman et al., in 2009. Type of microbial layers cells, permeability and nanoparticle fixation are the properties of the nanoparticle antimicrobial movement.

4.9.3 Effect of Nanoparticles on Fungal Pathogens

Khan (2012c) and Khan et al., in 2012a stated that phytopathogens such as fungi, bacteria, virus etc. have different nutrient requirements for their growth and development and the mechanism of penetration and infection. Khan and Jairajpuri, in 2010a, 2010b and 2012b said that though they used different ways of controlling the phytopathogens from establishing any sort of infections. Two compounds were observed that are CuSO and Na_2BO_4 considered to be the most effective in controlling the rust of peas among 15 other micronutrients combined or blended with the nanoparticles stated by Singh et al., in 2013. Manganese and zinc are the microelements which are effective in controlling the rotting of the seedlings and the charcoal rot of sunflower as mentioned by E-Hai et al., in 2009. *Candida albicans*, *C. kruse I*, *C. tropicalis*, *C. glabrata* and *Aspergillus brasiliensis* were also controlled by the action of the nanoparticles as a fungicide in the form of silver nanoparticles or the PVP. Bryaskova et al., in 2011 said that the tried microorganisms had a greater impact or the effect by the antifungal crossover compounds. *Botrytis cinerea* and *Penicillium expansum* are the postharvest pathogens that cause moulds in the stored grains, and the vegetables or fruits were also reduced by using ZnO NPs. The activities of zinc oxide NPs are also observed by using SEM and Raman spectroscopy which are later used for the consideration of the fungicidal properties of ZnO on the structural and synthesis of the cell of the mycelium. *B. cinerea* and *P. expansum* pathogens which are controlled by the ZnO nanoparticles of the size 70±15nm. The twisting of the mycelium of *Botrytis cinerea* pathogen and also damaging and degrading the conidia and conidiophores of *P. expansum* are observed by the use of the nanoparticles. *Alternaria alternata*, *Sclerotinia sclerotiorum*, *Macrophomina phaseolina*, *Rhizoctonia solani*, *B. cinerea* and *Curvularia lunata* growth and development and colonization were greatly controlled by the fungicide made up of Ag NPs as said by Krishnaraj et al., in 2012 where he concluded that 15–20mg/L has high hinderance on these pathogens. Jayaseelan et al., in 2012 conducted an experiment in which there was restraining of the colony formed by the *Aspergillus flavus*.

Pathogen	Nanoparticles
Pseudomonas syringae	Au NP
Pseudomonas ramorum	Ag NP
Pseudomonas lateralis	Ag NP
Bacillus thuringiensis	Pt NP
Bacillus subtilis	Pt NP
Xanthomonas campestris	Si NP
Ralstonia solanacearum	Au NP
Arabidopsis thaliana	CdSe
Fusarium sp.	CdSe; ZnS
Botrytis cinerea	ZnO
Penicillium expansum	ZnO
Rhizoctonia solani	Ag NP

4.9.4 Effect on Nano-Pesticides

Khan et al., in 2014a, 2014b said that with the help of the nanoparticle component developed in the root zone which provides a protection to the plant from the malady caused by any parasites or any insects etc. and also decreases its economic loss. Khan et al., in 2011a, 2011c stated by saying that any inoculum that is any spores or any part of the infected pathogen does not infect the host plant directly, which are later fixed within the root zone which upgrades the viability. The management should be on the basis of the severity of the attack of the pathogen to reduce the degradation of the quality and the essence of the nature and the environment stated by Khan and Jairajpuri, in 2012b. Using the techniques by combining the nanotechnology tools thus used in the crop protection procedures and fixing the plant malady and the management is achieved. The synthetic compounds are combined with the nanomaterials for fixing of the pathogens by using nanotechnology techniques. Using these chemical it could damage the quality of the environment and also a hazardous impact on the human health and consumption is reduced by combining the pesticides with the nanoparticles which makes them 15 times powerful with the nano-emulsion and also the application dose is smaller as compared to the chemicals stated by Xu et al., in 2010. Bergeson, in 2010b stated that nano-pesticides have the nanoparticles with effective pesticidal properties in fixing the pathogens with an effective scattering and wettability property. The nanoparticles in the nano-pesticides have properties of consistency, crystalline nature, permeability, heat property, easily dissolving and easily biodegradable in nature concluded by Bouwmeester et al., in 2009 and Bordes et al., in 2009. Yan et al., in 2005 stated that there were larger surface areas which are preferred by the pathogen. Lyons and Scrinis, in 2009, Bouwmeester et al., in 2009 and Bergeson, in 2010b demonstrated the plant protection directories programme by the nano-emulsion, nanocontainers etc.

The NPK compost could easily be degraded by the use of chitosan NPs as stated by Corradini et al., in 2010. In 2006, Liu et al. stated that there is natural degradation and the management could be done by the clay mud which are based on the nano-layer technique. The further studies were detailed by Ghormade et al., in 2011. As mentioned earlier, the nano-mud clay particles help in providing a ratio as agrochemicals likewise in compost and other pesticides. There are compounds like PHA (polyhydroxyl alkanates) have an incontestable viability in the pesticides which has perishing microbial action as stated by Pepperman et al., in 1991. In 1992, Cameron et al. said that intelimers have the capacity for the management of the shielding effects by reducing the unwanted filtering. Nano-pesticides are designed in such a way that there are controlled micro-capsules, networks lined with the granules which are then formed into transformed gadgets. A chemical is attained at the centre and encircled by a layer where the energetic compound is released through a micro-capsule. Sotthivirat et al., in 2007 stated that there are spheres of 2,000 metric units which has an operating system which is scattered in a lattice, and this gadget is known as a microsphere. Chuan et al., in

2013 did an investigation by checking the designed nano-pesticides accessory in crop insurance. Basically, the nano-definitions need to corrupt faster within the dirt and bit by bit in plants with build-up levels beneath the executive rules in staples.

Yan et al., in 2005 did a discovery of the SDS element in which there are nanomaterials blended with TiO, Ag compounds with the conjugation DMM (dimetomorph). The traditional pesticides were transformed or modified with the nanoparticles of 96 edge coarseness to reduce the damaging property of the chemicals used in the soil where there is cucumber and cabbage cropping fields as stated by Khot et al., in 2012. DMM is taken up by the altered nanoparticles by utilizing the ODS. This was used in the exemplification of the imidachloprid with nanoparticles for the management of the vegetable crops in the field stated by Guan et al., in 2010. Use of chitosan and algiriate in the imidacloprid combined with the silver or titanium oxide adjusted by the SDS technique which created a nano-pesticide which was experimented on the soybean crop. This was effective in the first 8 days and thereafter in the next 20 days the debris were found out to be negligible. The broadcasting of the nanoparticles on the soil layers with the topping of the SDS helps in the expansion. Mohamed and Khairou, in 2011 used the silver or the titanium oxide in the nano-pesticides with the POL which has the degrading property which reduced the damaging effect of the 2,4-D a systemic herbicide and radiation by the UV rays. Bowman and Hodge, in 2007 stated that nanotechnology techniques must be owed in the animal husbandry field with the permission with the national/international nano consistency structures.

4.10 TOXICITY AND BIOSAFETY OF NANOPARTICLES

Harmfulness or safety of nanoparticles and nano-pesticides could be an important concern of use of technology within the plant malady the board. The application of the pesticides also has an adverse effect on humans, animals and sometimes even on plants and damages the environment. This has a serious effect on the health of the living beings and on the quality of the environment. In 2010, Shi et al. concluded by stating the harmful and toxicity of chlorfenapyr on mice. It was taken into account that the chemical of the range 4–19mg/kg was harmless to mice compared to the heavy dose. Later on, such nano-arranged pesticides could diminish antagonistic natural and human impact once contrasted with ancient pesticides. Therefore, on guarantee eudaemonia and ecological security, additional such examinations on the safety of the NPs. Arrange soundness is to boot a major a part of the safety of nanomaterials. Liu et al., in 2008 effectively planned a stable nano-pesticide (bifenthrin) utilizing compound stabilizers, as an example, poly (acrylic acid)-b-poly (butyl acrylate), poly-vinyl pyrrolidone and poly-vinyl liquor. A glimmer nano-precipitation procedure was used to arrange 60–200nm bifenthrin particles. Whereas utilizing such methods industrially, soundness of the polymers over associate degree panoptic timeframe ought to be thought of. Anjali et al., in 2010 elaborated the meaning of counterfeit while not compound permethrin nanoparticles as a winning larval cidal property that was balanced out by plant freed common surfactants.

Drawback of the use of nanoparticles: The small-sized that build the NPs of large convenience, shockingly an analogous trademark is a necessary reason for many unfavourable impacts which can speak to following perils to condition, creatures, people and flora:

- Use of nanoparticles in pesticides, fertilizers and different formulation once transported by air could land on aerial parts of plants, that will be plugged into pores of the guard cells and building a fine physical and toxic structural phenomenon on disgrace bar germination of the mud and cylinder infiltration. The nanoparticles could likewise penetrate the plant tissue and impede the xylem and phloem mechanisms and also inhibit the mechanism of the photosynthesis products.
- NPs could be inhaled by people and animals profounded into the respiratory systems happening in to totally different sick impacts and clutters. The particles could enter the circulation system and combination in kidney and lungs, Bowman's tubule etc.

- Passage of nanoparticles into the respiratory organs and cardiovascular system, prompting irritation, super-molecule fibrillation, term of enlistment of genotoxicity and so on.
- In the fields and the water resources, the nanoparticles prevent the sedimentation of the pesticides and thus makes them easily transferrable, higher consistency and greatly reactive.

4.11 APPLICATION POSSIBILITIES

Nano-phytopathology can be applied as an instrument to comprehend plant-pathogen connections, giving new techniques to edit security. Explicit nano-devices and DNA nano-devices could empower exact following, location and determination of plant pathogens in the beginning periods of the plant maladies.

- Brisk reaction within coordinated complaint the board framework through outer and embedded nano-sensor frameworks.
- Improvement of quick analysis centre biosensors to spot plant pathogens within the field or post-reap periods.
- Improvement of recovery nano-systems for a selected inspecting from soil, air and plant tests.
- Obtain quick and dependable techniques for discovery of phytotoxins and toxigenic growths.
- Location of chemical deposits within the nutrient medium.

4.12 CONCLUSIONS AND FUTURE PROSPECTS

Specificity of detection of phytopathogens can be done at a higher accuracy level using nanomaterials combined with modern technologies like sensors, quantum, fluorescence microscope etc. The detection using advanced nanotechniques made it easier in the diagnosis of diseases in-vitro as well as in-vivo. Advanced technologies have not only reduced the time but also the cost and the labour pressure. The time taken to identify the pathogen using a conventional technique has reduced greatly to give an accurate result.

There are ill effects of any advanced technology. Thus the challenges faced by the nanodiagnostic tools are the heavy toxicity of nanoparticles, the quick sharing of data as well as the forecasting of the disease and the ability of a nanoparticle to withstand any extreme climatic conditions without tampering the results.

Regardless of the drawbacks, the effect and the impact of using nano-based diagnostic tools have helped scientists as well as farmers to manage the crop disease in an effective manner at an earliest point of time. The techniques can also replace all the heavy pesticides if the nanomaterials are used and prepared judiciously with utmost care and handling of the waste produced by the nanomaterials.

4.13 CONFLICTS OF INTEREST

The authors declare no such conflicts.

4.14 ACKNOWLEDGEMENTS

The authors sincerely thank Lovely Professional University and the Department of Plant Pathology under School of Agriculture for their support and all the guidance, and sincere thanks to Dr. Ajay Tomar, who gave this opportunity to write on nanotechnology.

REFERENCES

Ahmad, A, Mukherjee P, Senapati S, Mandal D, Khan MI, Kumar R, and Sastry M (2003a) Extracellular biosynthesis of silver nanoparticles using the fungus Fusarium oxysporum. Colloids and Surfaces B: Biointerfaces, :28(4):313–318

Ahmad, A, Senapati S, Khan MI, Kumar R, and Sastry M (2003b) Extracellular biosynthesis of monodisperse gold nanoparticles by a novel extremophilic actinomycete. Thermomonospora sp. Langmuir, : 19(8):3550–3553

Amendola, V, Polizzi S, and Meneghetti M (2006) Laser ablation synthesis of gold nanoparticles in organic solvents. The Journal of Physical Chemistry B, :110(14):7232–7237

Anjali, CH, Khan SS, Margulis-Goshen K, Magdassi S, Mukherjee A, and Chandrasekaran N (2010) Formulation of water-dispersible nanopermethrin for larvicidal applications. Ecotoxicology and Environmental Safety, :73(8):1932–1936

Aslam, M, Gopakumar G, Shoba TL, Mulla IS, Vijayamohanan K, Kulkarni SK, Urban J, and Vogel W (2002) Formation of Cu and Cu2O nanoparticles by variation of the surface ligand: Preparation, structure, and insulating-to-metallic transition. Journal of Colloid and Interface Science, :255(1):79–90

Athawale, AA, Katre PP, Kumar M, and Majumdar MB (2005) Synthesis of CTAB – IPA reduced copper nanoparticles. Materials Chemistry and Physics, :91(2–3):507–512

Azam, A, Ahmed AS, Oves M, Khan MS, and Memic A (2012) Size-dependent antimicrobial properties of CuO nanoparticles against Gram-positive and-negative bacterial strains. International Journal of Nanomedicine, :7:3527

Baac, H, Hajós JP, Lee J, Kim D, Kim SJ, and Shuler ML (2006) Antibody-based surface plasmon resonance detection of intact viral pathogen. Biotechnology and Bioengineering, :94(4):815–819

Balaji, DS, Basavaraja S, Deshpande R, Mahesh DB, Prabhakar BK, and Venkataraman A (2009) Extracellular biosynthesis of functionalized silver nanoparticles by strains of Cladosporium cladosporioides fungus. Colloids and Surfaces B: Biointerfaces, :68(1):88–92

Bansal, V, Rautaray D, Ahmad A, and Sastry M (2004) Biosynthesis of zirconia nanoparticles using the fungus Fusarium oxysporum. Journal of Materials Chemistry, :14(22):3303–3305

Basavaraja, S, Balaji SD, Lagashetty A, Rajasab AH, and Venkataraman A (2008) Extracellular biosynthesis of silver nanoparticles using the fungus Fusarium semitectum. Materials Research Bulletin, :43(5): 1164–1170

Bergeson, LL (2010a) Nanosilver pesticide products: What does the future hold? Environmental Quality Management, :19(4):73–82

Bergeson, LL (2010b) Nanosilver: US EPA's pesticide office considers how best to proceed. Environmental Quality Management, :19(3):79–85

Bhainsa, KC, and D'souza SF (2006) Extracellular biosynthesis of silver nanoparticles using the fungus Aspergillus fumigatus. Colloids and Surfaces B: Biointerfaces, :47(2):160–164

Birla, SS, Tiwari VV, Gade AK, Ingle AP, Yadav AP, and Rai MK (2009) Fabrication of silver nanoparticles by Phoma glomerata and its combined effect against Escherichia coli, Pseudomonas aeruginosa and Staphylococcus aureus. Letters in Applied Microbiology, :48(2):173–179

Boonham, N, Glover R, Tomlinson J, and Mumford R (2008) Exploiting generic platform technologies for the detection and identification of plant pathogens. In Sustainable Disease Management in a European Context (pp. 355–363). Springer, Dordrecht

Bordes, P, Pollet E, and Avérous L (2009) Nano-biocomposites: Biodegradable polyester/nanoclay systems. Progress in Polymer Science, :34(2):125–155

Bouwmeester, H, Dekkers S, Noordam MY, Hagens WI, Bulder AS, De Heer C, Ten Voorde SE, Wijnhoven SW, Marvin HJ, and Sips AJ (2009) Review of health safety aspects of nanotechnologies in food production. Regulatory Toxicology and Pharmacology, :53(1):52–62

Brayner, R, Barberousse H, Hemadi M, Djedjat C, Yéprémian C, Coradin T, Livage J, Fiévet F, and Couté A (2007) Cyanobacteria as bioreactors for the synthesis of Au, Ag, Pd, and Pt nanoparticles via an enzyme-mediated route. Journal of Nanoscience and Nanotechnology, :7(8):2696–2708

Bryaskova, R, Pencheva D, Nikolov S, and Kantardjiev T (2011) Synthesis and comparative study on the antimicrobial activity of hybrid materials based on Silver Nanoparticles (AgNps) stabilized by Polyvinylpyrrolidone (PVP). Journal of Chemical Biology, :4(4):185

Buffat, P, and Borel JP (1976) Size effect on the melting temperature of gold particles. Physical Review A, : 13(6):2287

Byrappa, K, and Yoshimura M (2001) Hydrothermal technology – Principles and applications. Handbook of Hydrothermal Technology, :1–52

Cason, JP, Miller ME, Thompson JB, and Roberts CB (2001) Solvent effects on copper nanoparticle growth behavior in AOT reverse micelle systems. The Journal of Physical Chemistry B, :105(12):2297–2302
Chartuprayoon, N, Rheem Y, Chen W, and Myung N (2010) Detection of plant pathogen using LPNE grown single conducting polymer nanoribbon. In Meeting Abstracts (No. 40, pp. 2278). The Electrochemical Society
Chen, H, Lee JH, Kim YH, Shin DW, Park SC, Meng X, and Yoo JB (2010) Metallic copper nanostructures synthesized by a facile hydrothermal method. Journal of Nanoscience and Nanotechnology, :(1):629–636
Chen, L, Zhang D, Chen J, Zhou H, and Wan H (2006) The use of CTAB to control the size of copper nanoparticles and the concentration of alkylthiols on their surfaces. Materials Science and Engineering: A, : 415(1–2):156–161
Chuan, L, He P, Pampolino MF, Johnston AM, Jin J, Xu X, Zhao S, Qiu S, and Zhou W (2013) Establishing a scientific basis for fertilizer recommendations for wheat in China: Yield response and agronomic efficiency. Field Crops Research, :140:1–8
Corradini, E, De Moura MR, and Mattoso LH (2010) A preliminary study of the incorparation of NPK fertilizer into chitosan nanoparticles. Express Polymer Letters, :4(8)
Dameron, CT, Reese RN, Mehra RK, Kortan AR, Carroll PJ, Steigerwald ML, Brus LE, and Winge DR (1989) Biosynthesis of cadmium sulphide quantum semiconductor crystallites. Nature, :338(6216):596–597
Das, R, Vecitis CD, Schulze A, Cao B, Ismail AF, Lu X, . . . and Ramakrishna S (2017). Recent advances in nanomaterials for water protection and monitoring. Chemical Society Reviews, :46(22):6946–7020
Derbalah, AS, El-Mahrouk MS, and El-Sayed AB (2011) Efficacy and safety of some plant extracts against tomato early blight disease caused by Alternaria solani. Plant Pathology Journal, :10(3):115–121
Di Gregorio, S, Lampis S, and Vallini G (2005) Selenite precipitation by a rhizospheric strain of Stenotrophomonas sp. isolated from the root system of Astragalus bisulcatus: A biotechnological perspective. Environment International, :31(2):233–241
Dubchak, S, Ogar A, Mietelski JW, and Turnau K (2010) Influence of silver and titanium nanoparticles on arbuscular mycorrhiza colonization and accumulation of radiocaesium in Helianthus annuus. Spanish Journal of Agricultural Research, :(1):103–108
El-Rafie, MH, Shaheen TI, Mohamed AA, and Hebeish A (2012) Bio-synthesis and applications of silver nanoparticles onto cotton fabrics. Carbohydrate Polymers, :90(2):915–920
El-Sherif, AG, Refaei AR, El-Nagar ME, and Salem HM (2007) Integrated management of meloidogyne incognita infecting eggplant by certain organic amendments, bacillus thuringiensis and oxamyl with reference to N P K and total chlorophyll status. Plant Pathology Journal, :6(2):147–152
Fayaz, AM, Balaji K, Girilal M, Yadav R, Kalaichelvan PT, and Venketesan R (2010) Biogenic synthesis of silver nanoparticles and their synergistic effect with antibiotics: A study against gram-positive and gram-negative bacteria. Nanomedicine: Nanotechnology, Biology and Medicine, :6(1):103–109
Gade, AK, Bonde P, Ingle AP, Marcato PD, Duran N, and Rai MK (2008) Exploitation of Aspergillus niger for synthesis of silver nanoparticles. Journal of Biobased Materials and Bioenergy, :2(3):243–247
Gedanken, A (2004) Using sonochemistry for the fabrication of nanomaterials. Ultrasonics Sonochemistry, : 11(2):47–55
Gericke, M, and Pinches A (2006) Biological synthesis of metal nanoparticles. Hydrometallurgy, : 83(1–4):132–140
Ghormade, V, Deshpande MV, and Paknikar KM (2011) Perspectives for nano-biotechnology enabled protection and nutrition of plants. Biotechnology Advances, :29(6):792–803
Gilbertson, LM, Pourzahedi L, Laughton S, Gao X, Zimmerman JB, Theis TL, . . . and Lowry GV (2020) Guiding the design space for nanotechnology to advance sustainable crop production. Nature Nanotechnology, :15(9):801–810
Guan, H, Chi D, Yu J, and Li H (2010) Dynamics of residues from a novel nano-imidacloprid formulation in soyabean fields. Crop Protection, :29(9):942–946
Guo, Z, Cui K, Zeng G, Wang J, and Guo X (2018) Silver nanomaterials in the natural environment: An overview of their biosynthesis and kinetic behavior. Science of the Total Environment, :643:1325–1336
Gurunathan, S, Kalishwaralal K, Vaidyanathan R, Venkataraman D, Pandian SR, Muniyandi J, Hariharan N, and Eom SH (2009) Biosynthesis, purification and characterization of silver nanoparticles using Escherichia coli. Colloids and Surfaces B: Biointerfaces, :74(1):328–335
Guzmán, MG, Dille J, and Godet S (2009) Synthesis of silver nanoparticles by chemical reduction method and their antibacterial activity. Int J Chem Biomol Eng, :2(3):104–111
Hassan, PA, Raghavan SR, and Kaler EW (2002) Microstructural changes in SDS micelles induced by hydrotropic salt. Langmuir, :18(7):2543–2548

He, S, Guo Z, Zhang Y, Zhang S, Wang J, and Gu N (2007) Biosynthesis of gold nanoparticles using the bacteria Rhodopseudomonas capsulata. Materials Letters, :61(18):3984–3987

Hoar, TP, and Schulman JH (1943) Transparent water-in-oil dispersions: The oleopathic hydro-micelle. Nature, : 152(3847):102–103

Hu, JJ, Cheng YJ, and Zhang XZ (2018) Recent advances in nanomaterials for enhanced photothermal therapy of tumors. Nanoscale, :10(48):22657–22672

Husseiny, MI, El-Aziz MA, Badr Y, and Mahmoud MA (2007) Biosynthesis of gold nanoparticles using Pseudomonas aeruginosa. Spectrochimica Acta Part A: Molecular and Biomolecular Spectroscopy, : 67(3–4):1003–1006

Ikeura, H, Somsak N, Kobayashi F, Kanlayanarat S, and Hayata Y (2011) Application of selected plant extracts to inhibit growth of Penicillium expansum on apple fruits. Plant Pathology Journal, :10(2):79–84

Ingle, A, Rai M, Gade A, and Bawaskar M (2009) Fusarium solani: A novel biological agent for the extracellular synthesis of silver nanoparticles. Journal of Nanoparticle Research, :11(8):2079

Jayaseelan, C, Rahuman AA, Kirthi AV, Marimuthu S, Santhoshkumar T, Bagavan A, Gaurav K, Karthik L, and Rao KB (2012) Novel microbial route to synthesize ZnO nanoparticles using Aeromonas hydrophila and their activity against pathogenic bacteria and fungi. Spectrochimica Acta Part A: Molecular and Biomolecular Spectroscopy, :90:78–84

Jha, AK, Prasad K, and Prasad K (2009) A green low-cost biosynthesis of Sb2O3 nanoparticles. Biochemical Engineering Journal, :43(3):303–306

Kalimuthu, K, Babu RS, Venkataraman D, Bilal M, and Gurunathan S (2008) Biosynthesis of silver nanocrystals by Bacillus licheniformis. Colloids and Surfaces B: Biointerfaces, :65(1):150–153

Kang, TF, Wang F, Lu LP, Zhang Y, and Liu TS (2010) Methyl parathion sensors based on gold nanoparticles and Nafion film modified glassy carbon electrodes. Sensors and Actuators B: Chemical, :145(1):104–109

Kathiresan, K, Manivannan S, Nabeel MA, and Dhivya B (2009) Studies on silver nanoparticles synthesized by a marine fungus, Penicillium fellutanum isolated from coastal mangrove sediment. Colloids and Surfaces B: Biointerfaces, :71(1):133–137

Khan, MM, Khan MR, and Mohiddin FA (2012a) The relative performance of different inoculation methods with Alternaria brassicae and A. brassicicola on Indian Mustard. Plant Pathology Journal, :11(3):93–98

Khan, MR (2007) Prospects of microbial control of root-knot nematodes infecting vegetable crops. Biotechnology: Plant Health Management, :643–665

Khan, MR (2012c) Nematodes, an emerging threat to global forests: Assessment and management. Plant Pathology Journal, :11(4):99–113

Khan, MR, Altaf S, Mohiddin FA, Khan U, and Anwer A (2009) Biological control of plant nematodes with phosphate solubilizing microorganisms. In Phosphate Solubilizing Microbes for Crop Improvement (pp. 395–426) (MS Khan, and A Zaidi, eds.). Nova Science Publishers, New York, NY, USA

Khan, MR, and Anwer MA (2011a) Fungal bioinoculants for plant disease management. In Microbes and Microbial Technology (pp. 447–488). Springer, New York, NY

Khan, MR, Ashraf S, Rasool F, Salati KM, Mohiddin FA, and Haque Z (2014b) Field performance of Trichoderma species against wilt disease complex of chickpea caused by Fusarium oxysporum f. sp. ciceri and Rhizoctonia solani. Turkish Journal of Agriculture and Forestry, :38(4):447–454

Khan, MR, and Haque Z (2013) Morphological and biochemical responses of five tobacco cultivars to simultaneous infection with Pythium aphanidermatum and Meloidogyne incognita. Phytopathologia Mediterranea, :98–109

Khan, MR, Haque Z, and Kausar N (2014a) Management of the root-knot nematode Meloidogyne graminicola infesting rice in the nursery and crop field by integrating seed priming and soil application treatments of pesticides. Crop Protection, :63:15–25

Khan, MR, and Jairajpuri MS (2010a) Nematode infestation in food crops-national scenario. In Nematode Infestations, Part I: Food Crop (pp. 1–6)

Khan, MR, and Jairajpuri MS (2010b) Nematode infestation in industrial crops-national scenario. In Nematode Infestations Part II: Industrial Crops (pp. 1–9). National Academy of Sciences, New Delhi, IN

Khan, MR, and Jairajpuri MS (2012b) Nematode infestation in horticultural crops, national scenario. In Nematode Infestation Part III: Horticultural Crops (pp. 1–30). National Academy of Sciences, IN

Khan, MR, and Khan MM (2011b) Plants response to diseases in sulphur dioxide stressed environment. Plant Pathol J, :10:1–2

Khan, MR, Majid S, Mohidin FA, and Khan N (2011c) A new bioprocess to produce low cost powder formulations of biocontrol bacteria and fungi to control fusarial wilt and root-knot nematode of pulses. Biological Control, :59(2):130–140

Khater, M, De La Escosura-Muñiz A, and Merkoçi A (2017). Biosensors for plant pathogen detection. Biosensors and Bioelectronics, :93:72–86

Khot, LR, Sankaran S, Maja JM, Ehsani R, and Schuster EW (2012) Applications of nanomaterials in agricultural production and crop protection: A review. Crop Protection, :35:64–70

Kitchens, CL, and Roberts CB (2004) Copper nanoparticle synthesis in compressed liquid and supercritical fluid reverse micelle systems. Industrial & Engineering Chemistry Research, :43(19):6070–6081

Komarneni, S (2003) Nanophase materials by hydrothermal, microwave-hydrothermal and microwave-solvothermal methods. Current Science, : 25:1730–1734

Konishi, Y, Ohno K, Saitoh N, Nomura T, and Nagamine S (2004) Microbial synthesis of gold nanoparticles by metal reducing bacterium. Trans Mater Res Soc Jpn, :29:2341–2343

Konishi, Y, Ohno K, Saitoh N, Nomura T, Nagamine S, Hishida H, Takahashi Y, and Uruga T (2007) Bioreductive deposition of platinum nanoparticles on the bacterium Shewanella algae. Journal of Biotechnology, : 128(3):648–653

Kowshik, M, Deshmukh N, Vogel W, Urban J, Kulkarni SK, and Paknikar KM (2002) Microbial synthesis of semiconductor CdS nanoparticles, their characterization, and their use in the fabrication of an ideal diode. Biotechnology and Bioengineering, :78(5):583–588

Krishnaraj, C, Ramachandran R, Mohan K, and Kalaichelvan PT (2012) Optimization for rapid synthesis of silver nanoparticles and its effect on phytopathogenic fungi. Spectrochimica Acta Part A: Molecular and Biomolecular Spectroscopy, :93:95–99

Kumar, D, and Kalita P (2017) Reducing postharvest losses during storage of grain crops to strengthen food security in developing countries. Foods, :6(1):8

Kumaravel, A, and Chandrasekaran M (2011) A biocompatible nano TiO2/nafion composite modified glassy carbon electrode for the detection of fenitrothion. Journal of Electroanalytical Chemistry, :650(2):163–170

Kundu, A, and Saha S (2014) In vitro antifungal activity of novel Picolibnamide 3s against soil borne fungi and structure activity relationship. Plant Pathol J, :13:152–159

Latifa, A, Idriss T, Hassan B, Amine SM, El Hassane B, and Abdellah AB (2011) Effects of organic acids and salts on the development of Penicillium italicum: The causal agent of citrus blue mold. Plant Pathology Journal, :10(3):99–107

Law, N, Ansari S, Livens FR, Renshaw JC, and Lloyd JR (2008) Formation of nanoscale elemental silver particles via enzymatic reduction by Geobacter sulfurreducens. Appl Environ Microbiol, :74(22):7090–7093

Li, Z, Askim JR, and Suslick KS (2018) The optoelectronic nose: Colorimetric and fluorometric sensor arrays. Chemical Reviews, :119(1):231–292

Li, Z, Yu T, Paul R, Fan J, Yang Y, and Wei Q (2020) Agricultural nanodiagnostics for plant diseases: Recent advances and challenges. Nanoscale Advances, :2(8):3083–3094

Lisiecki, I, Filankembo A, Sack-Kongehl H, Weiss K, Pileni MP, and Urban J (2000) Structural investigations of copper nanorods by high-resolution TEM. Physical Review B, :61(7):4968

Liu, S, Yuan L, Yue X, Zheng Z, and Tang Z (2008) Recent advances in nanosensors for organophosphate pesticide detection. Advanced Powder Technology, :19(5):419–441

Liu, Y, Xiao Z, Chen F, Yue L, Zou H, Lyu J, and Wang Z (2021) Metallic oxide nanomaterials act as antioxidant nanozymes in higher plants: Trends, meta-analysis, and prospect. Science of the Total Environment, : 780:146578.

Loo, YY, Chieng BW, Nishibuchi M, and Radu S (2012) Synthesis of silver nanoparticles by using tea leaf extract from Camellia sinensis. International Journal of Nanomedicine, :7:4263

López, MM, Llop P, Olmos A, Marco-Noales E, Cambra M, and Bertolini E (2009) Are molecular tools solving the challenges posed by detection of plant pathogenic bacteria and viruses? Current Issues in Molecular Biology, :11(1):13

Lortie, L, Gould WD, Rajan S, McCready RG, and Cheng KJ (1992) Reduction of selenate and selenite to elemental selenium by a Pseudomonas stutzeri isolate. Appl Environ Microbiol, :58(12):4042–4044

Lovley, DR, Phillips EJ, Gorby YA, and Landa ER (1991) Microbial reduction of uranium. Nature, : 350(6317):413–416

Lovley, DR, Phillips EJ, and Lonergan DJ (1989) Hydrogen and formate oxidation coupled to dissimilatory reduction of iron or manganese by Alteromonas putrefaciens. Appl Environ Microbiol, :55(3):700–706

Lyons, K, and Scrinis G (2009) Under the regulatory radar? Nanotechnologies and their impacts for rural Australia. In Tracking Rural Change: Community, Policy and Technology in Australia, New Zealand and Europe (pp. 151–171). Australian National University E Press, Canberra

Mohamed, MM, and Khairou KS (2011) Preparation and characterization of nano-silver/mesoporous titania photocatalysts for herbicide degradation. Microporous and Mesoporous Materials, : 142(1):130–138

Mourato, A, Gadanho M, Lino AR, and Tenreiro R (2011) Biosynthesis of crystalline silver and gold nanoparticles by extremophilic yeasts. Bioinorganic Chemistry and Applications, :2011

Neethirajan, S, Weng X, Tah A, Cordero JO, and Ragavan KV (2018). Nano-biosensor platforms for detecting food allergens – New trends. Sensing and Bio-Sensing Research, :18:13–30

Park, BK, Jeong S, Kim D, Moon J, Lim S, and Kim JS (2007) Synthesis and size control of monodisperse copper nanoparticles by polyol method. Journal of Colloid and Interface Science, :311(2):417–424

Patil, RS, Kokate MR, and Kolekar SS (2012) Bioinspired synthesis of highly stabilized silver nanoparticles using Ocimum tenuiflorum leaf extract and their antibacterial activity. Spectrochimica Acta Part A: Molecular and Biomolecular Spectroscopy, :91:234–238

Pepperman, AB, Kuan JC, and McCombs C (1991) Alginate controlled release formulations of metribuzin. Journal of Controlled Release, :17(1):105–111

Philip, D (2009) Biosynthesis of Au, Ag and Au – Ag nanoparticles using edible mushroom extract. Spectrochimica Acta Part A: Molecular and Biomolecular Spectroscopy, :73(2):374–381

Pileni, MP, and Lisiecki I (1993) Nanometer metallic copper particle synthesis in reverse micelles. Colloids and Surfaces A: Physicochemical and Engineering Aspects, :80(1):63–68

Pol, VG, Motiei M, Gedanken A, Calderon-Moreno J, and Mastai Y (2003) Sonochemical deposition of air-stable iron nanoparticles on monodispersed carbon spherules. Chemistry of Materials, :15(6):1378–1384

Prasad, K, Jha AK, and Kulkarni AR (2007) Lactobacillus assisted synthesis of titanium nanoparticles. Nanoscale Research Letters, :2(5):248–250

Quesada-González, D, and Merkoçi A (2018) Nanomaterial-based devices for point-of-care diagnostic applications. Chemical Society Reviews, :47(13):4697–4709

Raja, M, Subha J, Ali FB, and Ryu SH (2008) Synthesis of copper nanoparticles by electroreduction process. Materials and Manufacturing Processes, :23(8):782–785

Reverchon, E, and Adami R (2006) Nanomaterials and supercritical fluids. The Journal of Supercritical Fluids, : 37(1):1–22

Sanghi, R, and Verma P (2009) Biomimetic synthesis and characterisation of protein capped silver nanoparticles. Bioresource Technology, :100(1):501–504

Shaligram, NS, Bule M, Bhambure R, Singhal RS, Singh SK, Szakacs G, and Pandey A (2009) Biosynthesis of silver nanoparticles using aqueous extract from the compactin producing fungal strain. Process Biochemistry, :44(8):939–943

Shankar, SS, Ahmad A, Pasricha R, and Sastry M (2003) Bioreduction of chloroaurate ions by geranium leaves and its endophytic fungus yields gold nanoparticles of different shapes. Journal of Materials Chemistry, : 13(7):1822–1826

Shoala, T (2019) Nanodiagnostic techniques in plant pathology. In Nanobiotechnology Applications in Plant Protection (pp. 209–222). Springer, Cham.

Shukla, DD, Lauricella R, and Ward CW (1992) Serology of potyviruses: Current problems and some solutions. In Potyvirus Taxonomy (pp. 57–69). Springer, Vienna

Silverberg, BA, Wong PT, and Chau YK (1976) Localization of selenium in bacterial cells using TEM and energy dispersive X-ray analysis. Archives of Microbiology, :107(1):1–6

Singh, D, Kumar A, Singh AK, and Tripathi HS (2013) Induction of resistance in field pea against rust disease through various chemicals/micronutrients and their impact on growth and yield. Plant Pathology J, : 12(2):36

Singh, S, Singh M, Agrawal VV, and Kumar A (2010) An attempt to develop surface plasmon resonance based immunosensor for Karnal bunt (Tilletia indica) diagnosis based on the experience of nano-gold based lateral flow immuno-dipstick test. Thin Solid Films, :519(3):1156–1159.

Sofi, W, Gowri M, Shruthilaya M, Rayala S, and Venkatraman G (2012) Silver nanoparticles as an antibacterial agent for endodontic infections. BMC Infectious Diseases, :12(1):P60

Song, X, Sun S, Zhang W, and Yin Z (2004) A method for the synthesis of spherical copper nanoparticles in the organic phase. Journal of Colloid and Interface Science, :273(2):463–469

Sotthivirat, S, Haslam JL, and Stella VJ (2007) Evaluation of various properties of alternative salt forms of sulfobutylether-β-cyclodextrin,(SBE) 7M-β-CD. International Journal of Pharmaceutics, :330(1–2):73–81

Suman, TY, Rajasree SR, Kanchana A, and Elizabeth SB (2013) Biosynthesis, characterization and cytotoxic effect of plant mediated silver nanoparticles using Morinda citrifolia root extract. Colloids and Surfaces B: Biointerfaces, :106:74–78

Sun, H, and Fung Y (2006) Piezoelectric quartz crystal sensor for rapid analysis of pirimicarb residues using molecularly imprinted polymers as recognition elements. Analytica Chimica Acta, :576(1):67–76

Suslick, KS, Choe SB, Cichowlas AA, and Grinstaff MW (1991) Sonochemical synthesis of amorphous iron. Nature, :353(6343):414–416

Suslick, KS, Hyeon T, Fang M, and Cichowlas AA (1996) Sonochemical preparation of nanostructured catalysts. In Advanced Catalysts and Nanostructured Materials (pp. 197–212). Academic Press

Talibi, I, Askarne L, Boubaker H, Boudyach EH, and Aoumar AA (2011) In vitro and in vivo antifungal activities of organic and inorganic salts against citrus sour rot agent Geotrichum candidum. Plant Pathol J, :10:138–145

Taniguchi, N (1974) On the basic concept of nanotechnology. In Proceeding of the ICPE

Tomei, FA, Barton LL, Lemanski CL, Zocco TG, Fink NH, and Sillerud LO (1995) Transformation of selenate and selenite to elemental selenium by Desulfovibrio desulfuricans. Journal of Industrial Microbiology, : 14(3–4):329–336

Umer, A, Naveed S, Ramzan N, and Rafique MS (2012) Selection of a suitable method for the synthesis of copper nanoparticles. Nano, :7(05):1230005

Vamvakaki, V, and Chaniotakis NA (2007) Pesticide detection with a liposome-based nano-biosensor. Biosensors and Bioelectronics, :22(12):2848–2853

Van Dyk, JS, and Pletschke B (2011) Review on the use of enzymes for the detection of organochlorine, organophosphate and carbamate pesticides in the environment. Chemosphere, :82(3):291–307

Vigneshwaran, N, Kathe AA, Varadarajan PV, Nachane RP, and Balasubramanya RH (2006) Biomimetics of silver nanoparticles by white rot fungus, Phaenerochaete chrysosporium. Colloids and Surfaces B: Biointerfaces, :53(1):55–59

Vijayakumar, M, Priya K, Nancy FT, Noorlidah A, and Ahmed AB (2013) Biosynthesis, characterisation and anti-bacterial effect of plant-mediated silver nanoparticles using Artemisia nilagirica. Industrial Crops and Products, :41:235–240

Vijayakumar, R, Koltypin Y, Felner I, and Gedanken A (2000) Sonochemical synthesis and characterization of pure nanometer-sized Fe3O4 particles. Materials Science and Engineering: A, :286(1):101–105

Wang, C, Liu X, Li J, Yue L, Yang H, Zou H, . . . and Xing B (2021) Copper nanoclusters promote tomato (Solanum lycopersicum L.) yield and quality through improving photosynthesis and roots growth. Environmental Pollution, :289:117912

Wang, M, and Li Z (2008) Nano-composite ZrO2/Au film electrode for voltammetric detection of parathion. Sensors and Actuators B: Chemical, :133(2):607–612

Wang, S, Lawson R, Ray PC, and Yu H (2011) Toxic effects of gold nanoparticles on Salmonella typhimurium bacteria. Toxicology and Industrial Health, :27(6):547–554

Wang, Y, Chen P, and Liu M (2006) Synthesis of well-defined copper nanocubes by a one-pot solution process. Nanotechnology, :17(24):6000

Wang, Y, Zhang P, Li M, Guo Z, Ullah S, Rui Y, and Lynch I (2020) Alleviation of nitrogen stress in rice (Oryza sativa) by ceria nanoparticles. Environmental Science: Nano, :7(10):2930–2940

Wang, Z, Wei F, Liu SY, Xu Q, Huang JY, Dong XY, Yu JH, Yang Q, Zhao YD, and Chen H (2010) Electrocatalytic oxidation of phytohormone salicylic acid at copper nanoparticles-modified gold electrode and its detection in oilseed rape infected with fungal pathogen Sclerotinia sclerotiorum. Talanta, :80(3):1277–1281

Woolfolk, CA, and Whiteley HR (1962) Reduction of inorganic compounds with molecular hydrogen by micrococcus lactilyticus I.: Stoichiometry with compounds of arsenic, selenium, tellurium, transition and other elements. Journal of Bacteriology, :84(4):647–658

Xu, L, Liu Y, Bai R, and Chen C (2010) Applications and toxicological issues surrounding nanotechnology in the food industry. Pure and Applied Chemistry, :82(2):349–372

Yadav, V, Sharma N, Prakash R, Raina KK, Bharadwaj LM, and Prakash NT (2008) Generation of selenium containing nano-structures by soil bacterium, Pseudomonas aeruginosa. Biotechnology, :7(2):299–304

Yan, J, Huang K, Wang Y, and Liu S (2005) Study on anti-pollution nano-preparation of dimethomorph and its performance. Chinese Science Bulletin, :50(2):108–112

Yang, J, Cheng GH, Zeng JH, Yu SH, Liu XM, and Qian YT (2001) Shape control and characterization of transition metal diselenides MSe2 (M= Ni, Co, Fe) prepared by a solvothermal-reduction process. Chemistry of Materials, :13(3):848–853

Yao, KS, Li SJ, Tzeng KC, Cheng TC, Chang CY, Chiu CY, Liao CY, Hsu JJ, and Lin ZP (2009) Fluorescence silica nanoprobe as a biomarker for rapid detection of plant pathogens. In Advanced Materials Research (Vol. 79, pp. 513–516). Trans Tech Publications Ltd

Yin, J, Wang Y, and Gilbertson LM (2018) Opportunities to advance sustainable design of nano-enabled agriculture identified through a literature review. Environmental Science: Nano, :5(1):11–26.

Yong, P, Rowson NA, Farr JP, Harris IR, and Macaskie LE (2002) Bioreduction and biocrystallization of palladium by Desulfovibrio desulfuricans NCIMB 8307. Biotechnology and Bioengineering, :80(4):369–379

Zhu, HT, Zhang CY, and Yin YS (2004) Rapid synthesis of copper nanoparticles by sodium hypophosphite reduction in ethylene glycol under microwave irradiation. Journal of Crystal Growth, :270(3–4):722–728

5 Antimicrobial Resistance in Agrarian Distress

Neelam Chaudhary, Chandan Singh, and Deepak Vyas

5.1 INTRODUCTION

Distress is a joint function of factors like economic conditions, political dilemmas, social structure, and inner psychological conflicts, and these stressors change the mind balance and internal functioning. Peasants face prolonged, uninterrupted, and unmanageable stress, the most critical factors leading to agrarian distress. Among these stress/or aspects, the input of chemical fertilizer and antibiotic-resistant microbes escalated the distress manifold with several spheres of peasant life as well as the other stakeholders, including environmental degradation and day-by-day increasing resistant phytopathogens and livestock zoonotic pathogens.

Antibiotics are a class of secondary metabolites produced by microorganisms, in addition to chemically synthesized or semi-synthesized analogous compounds, that may prevent the growth and survival of other microorganisms (Demain and Sanchez, 2009). These substances are primarily used to treat human illnesses, but they are also widely used in agriculture to increase crop productivity, in animal husbandry to treat illnesses and prevent infections in live animals, as growth promoters in animal feed at specific concentrations, and in other fields of human and animal care. When extensive global use of antibiotics was examined, it was found that China consumes around half of the antibiotics used in all animal husbandry, followed by the US, Brazil, India, and Germany (Manyi-Loh et al., 2018; Mann et al., 2021).

According to Ashbolt et al. (2013), there are many avenues for antibiotics to infiltrate the water and land environment, including runoff from metropolitan areas, aquaculture ponds, and agricultural fields holding cattle dung. Antibiotic residues in the environment can be considered a "persistent" organic contaminant due to the frequent and extensive use of antibiotics and continuous emissions (Ji et al., 2010). In particular, irrigation with treated wastewater and fertilization with livestock manure are high-potential pathways for introducing antibiotics in the agroecosystem (Ben et al., 2019; Ji et al., 2010). Antibiotic use on a large scale could exert enormous selection pressure on natural and human microbial systems (Figure 5.1) (National Academies of Sciences, Engineering, and Medicine, 2018; Qiao et al., 2017). The presence of antibiotics in a microbial system may cause usually sensitive bacteria to undergo genetic or mutational alterations that enable the bacteria to survive and spread as antibiotic resistance bacteria (ARB) that carry antibiotic resistance genes (ARGs) (Ben et al., 2019). The dispersion and encouragement of genetic recombination of ARGs through transformation, conjugation, or transduction, collectively known as horizontal gene transfer (HGT), is further aided by mobile genetic elements (MGE), such as plasmids, integrins, and prophages (Martinez et al., 2015). ARGs started to be recognized as a new type of environmental contaminant. It has also been predicted that by 2030, the usage of antibiotics will increase by 67% in a number of the world's most populous nations, including India. Although the widespread use of antibiotics has led to a rise in the demand for a variety of agricultural and animal products, their prolonged use, even at subtherapeutic concentrations, either directly on the fields or indirectly through animal manure, is negatively affecting the soil microbiota in agricultural areas (Ben et al., 2019).

The present century has seen a significant increase in the emergence of microbes due to this overuse of antibiotics. These microbes have been observed to modify their genes more quickly

DOI: 10.1201/9781003269380-5

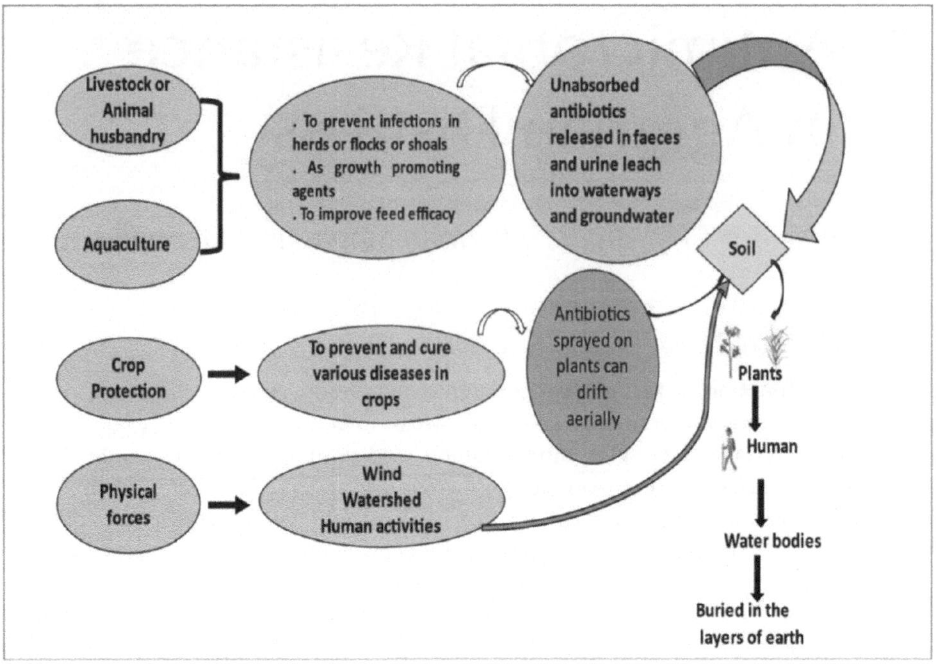

FIGURE 5.1 Antibiotics and ARG pathway through livestock and agriculture (modified from **Mann et al., 2021**).

and effectively, aiding them in developing resistance against various antibiotic groups, particularly broad-spectrum antibiotics. This has led to a significant problem for people who live in rural areas because even the latest generation of antibiotics is reportedly ineffective against such germs.

The goal of the current chapter is to shed light on the various issues caused by the excessive use of various types of antibiotics in agriculture, the mechanisms underlying the rapid emergence of bacterial resistance to the most widely used antibiotics in agriculture, and some of the approaches under investigation as potential antibiotic alternatives.

5.2 INPUT OF ANTIBIOTICS IN AGRICULTURE

The input of antibiotics in agriculture increased to increase the food yield to meet the demand of growing populations. The defined doses for each antibiotic were manipulated knowingly or unknowingly by peasants to improve the products without knowing the serious effects of the antibiotics. Some of the mechanisms of antibiotic application are described in the following.

5.2.1 Direct Application of Antibiotic for Pathogen Management

The magnitude of bacterial pathogen populations on sensitive plant hosts and the scarcity of bactericides makes it challenging to manage bacterial plant diseases effectively. In many pathosystems, antibiotics have served as the greatest alternative for controlling bacterial disease because they are the most effective means of reducing bacterial population size and preventing disease outbreaks in the absence of long-lasting and powerful host disease resistance. The use of antibiotics for plant disease control was restrained by a number of factors, including ineffectiveness at lower doses, phytotoxicity issues at higher doses, and cost compared to other available methods of disease control, despite the fact that many new types of antibiotics were quickly tested and then deployed in agriculture beginning in the 1950s. Therefore, only streptomycin and oxytetracycline were finally used in plant agriculture and only in particular disease pathosystems, despite the fact that penicillin, streptomycin, aureomycin, chloramphenicol, and oxytetracycline were studied for the control of plant diseases in the late 1940s (Raman et al., 2019). The primary antibiotic used to treat plant diseases today is streptomycin, which fights pathogens like *Xanthomonas campestris*, which causes tomato and pepper bacterial spot; *Erwinia amylovora*, which causes fire blight on apple and pear trees; *Pseudomonas syringae*, which infects apple and pear tree flowers and fruits; and *Erwinia amylovora*, which causes fire blight on pome fruit trees (McManus et al., 2002). In some circumstances, such as those involving the treatment of *Xanthomonas arboricola* pv. *pruni*, the cause of the bacterial spot of peach and nectarine, oxytetracycline has been utilized as the main antibiotic (McManus et al., 2002). Additionally, oxytetracycline has been employed in the management of fire blight in the US, most notably where streptomycin resistance has become an issue (McManus and Jones, 1994; Raman et al., 2019).

5.2.2 Application of Manure in Farmland Soils (Indirect Entry of Antibiotic)

The production of animal food and the control of disease are greatly aided by the use of antibiotics in livestock. Higher doses of antibiotics do not fully metabolize before being released and excreted into the environment, where they interact with multiple microbes and cause mutations or multidrug resistance in those microbes. Additionally, the organic waste left behind by livestock used as manure in agricultural soil indirectly contributes antibiotics to farm soil.

And this external release of animal manure into the environment is a significant route for the spread of antibiotic-resistant organisms in agroecosystems, which may lead to the development of ARGs and ARBs in human communities (Heuer et al., 2011). Manure, a fertilizer rich in nutrients and organic matter, is frequently utilized in agriculture across the globe to improve soil characteristics and promote crop growth. However, manure has developed into a significant source of antibiotics, ARBs, ARGs, and MGEs (Aggarwal et al., 2012; Han et al., 2018; Karkman et al., 2019; Thanner et al., 2016; Zhu et al., 2013). Research correlated that the ARGs and ARB population significantly increased when manure was applied in the soil (Abraham, 1963; Han et al., 2018; Heuer et al., 2011; Muurinen et al., 2017; Xie et al., 2018). Due to the varied resident bacteria that produce different antibiotics and the complicated microbial community, soil typically carries a variety of intrinsic ARGs (Zhu et al., 2019). As a result, the increased level of ARGs following manure application is attributable to the introduction of novel ARGs from faecal contamination and the increased abundance of resident ARGs chosen by applied manure (Nikolina, 2014).

Furthermore, under regular manure treatments, ARGs may become more persistent and well established in the soil microbial population (Heuer and Smalla, 2007). ARGs in manure-impacted soils have the ability to penetrate the food chain, and eating organically grown produce from manured soils is probably another way for ARGs to spread to humans (Berger et al., 2010; Mann et al., 2021; Zhang et al., 2019). Given the significant functional overlap between bacteria produced from leaves and roots, soil microorganisms significantly influence the formation of bacterial communities in both above- and below-ground sections of plants (Bai et al., 2015). Numerous facultative endophytic bacteria can also survive in the rhizosphere, and there is a strong correlation between endophytic and rhizosphere microorganisms (Van der Lelie et al., 2009). Additionally, the rhizosphere is well known for being a hub for horizontal gene transfer. An earlier study showed that soil bacteria might persist inside the root and develop into root endophytes (Bulgarelli et al., 2012). As a result, the bacterial populations in the soil can impact the bacterial community of plants growing in manure-amended soil. The horizontal transfer of ARGs in soils and from soils to vegetable microbiomes has been accelerated by manure (Zhang et al., 2019). Previous research has found that after growing in soil that has been treated with manure, several ARGs, such as root endophytes, leaf endophytes, have been found in vegetables and fruits (Marti et al., 2013; Ruimy et al., 2010; Zhu et al., 2017) carried out a high-throughput quantitative polymerase chain reaction (HTqPCR) to identify a total of 134 ARGs in commercially and organically grown lettuce, where they found that the phyllosphere of organic product contained more ARGs than the phyllosphere of conventionally grown lettuce.

Therefore, manure-contaminated produce, especially eaten raw, may significantly subsidize the growth of antibiotic resistance in humans (Berger et al., 2010). In this regard, deep knowledge of ARGs is required to understand their mechanism in the soil-plant system under the application of manure. Agricultural techniques impact how manure application affects soil ARGs (Muurinen et al., 2017).

5.2.3 Through Aquatic Environments

The aquatic environment is thought to be the most suitable medium for the acquisition and spread of ARGs in order to influence the evolution of aquatic microbes, and genetic recombination regularly happens in the aquatic environment (Watts, 2017). Aquatic bacterial populations and other communities quickly shift ARGs to adjust by themselves to manage antibiotic contaminants and boost competitiveness. This all happens in bacterial communities due to the genetic plasticity of the aquatic microbes. Once the microbes acquire ARGs, the ARGs have more tendency to be more durable in aquatic conditions than in terrestrial conditions, holding even in the absence of selection pressure (Manu, 2010). Considering this fact, aquaculture opens the door to the growth and internationalization of AMR in aquatic environments (Cabello et al., 2016). Large quantities of selective agents and faecal contaminations containing ARGs/ARBs from aquaculture get into our bodies of water, causing bacterial mutation, recombination, and horizontal gene transfer. As a result, the level of ARGs in natural aquatic environments rises, as does the danger of ARGs from aquaculture getting into humans. The pattern of ARGs dissemination from aquaculture greatly varies from type of aquaculture system, i.e., in an open system of aquaculture large amount of antibiotic, nearly 70%–80%, remained as the residue and persists long enough in the system that it transforms the normal bacteria into ARB, which, along with the ARG that excreted from fish and other aquatic animals, imprints the effects of antibiotics on aquatic microbial communities (Watts, 2017). At the same time, closed farming systems are commonly separate from the general environments. However, very limited exchange between aquaculture and the surrounding environment took place. However, it was seen that the accumulation of antibiotics in a closed recirculating aquaculture system biofilter promoted the ARBs (Okeke et al., 2022; Zhang et al., 2017). It has been seen that in the integrated farming system, components of integrated farming are designed such that the waste of one component is utilized by the other components of the integrated farming (Singh et al., 2020) – for instance, aquaculture in a fish farming system is connected with livestock farming or/and crops for good produce and growth to efficiently use the resource, and by doing so, the antibiotic used in one component moves to other components, which represents the most degree of ARG pollution and gene transfer in the soil-water-plant system, providing the high risk of resistome transmission to humans and also adversely affecting the phytoplanktons and zooplanktons of the aquatic habitats (Klase et al., 2019). Hence untreated aquaculture wastewater may also directly flow into the nearby water bodies and contaminate them, which promotes the spreading risk of ARGs from aquaculture to humans and other animals through an aquatic environment, either an open or closed system, and then to the soil ecosystem (US Centers for Disease Control and Prevention, 2018; Zhao et al., 2021).

5.3 IMPACT OF ANTIBIOTICS ON AGROECOLOGY

Agroecology is an integrated approach which applies ecological and social concepts and principles to the design and management of food and agricultural systems. It optimizes the interaction among plants, animals, humans, and the environment while taking into consideration the social aspect that needs to be addressed for a sustainable and fair food system (**FAO, 2018**). The application of antibiotics and contaminants of antibiotics from different sources have raised concern over the stability of agroecology (Durso and Cook, 2019).

Numerous findings demonstrate how antibiotics affect the physiological integrity and composition of soil microorganisms (Gutierrez et al., 2010; Hammer et al., 2016; Toth et al., 2011), soil serves as the reservoir of the antibiotic and intrinsically or acquired through transfer elements, bacteria developed into the antibiotic resistance bacteria, thus the human and the natural biota exposed to these increased level of ARB. There are no known pathways to trace the complete fate of antibiotics in soil; however, the input of the antibiotic through agriculture ultimately to agroecology is gradually increasing due to high consumption in animal health management worldwide by weight (Williams-Nguyen et al., 2016). Besides the risk of developing antibiotic resistance bacteria, the residue of antibiotics absorbed by plants interferes with the different physiological process of the plants and causes potential ecotoxicology effects (Williams-Nguyen et al., 2016). Apart from this, a number of adverse effects of antibiotic residue were tested, which revealed the acute and chronic effects of antibiotic residue on photosynthesis and mitochondria, probably due to the bacterial origin of the chloroplast and mitochondrial genome (Williams-Nguyen et al., 2016; Durso and Cook, 2019). The antibiotic residue in the agricultural soil delay germination, reduce biomass, and consequently may negatively affect yield in farmland (Williams-Nguyen et al., 2016). The extensive use of antibiotics decreased microbial efficiency, which can be indicated by the increased mass-specific respiration that could subsequently decline soil carbon (C) retention. This is similar to the response of the soil microbial communities in drought stress, and therefore microbes shift their allocation of carbon and the nutrient from growth to the synthesis and upkeep of molecules (e.g., osmolytes) in order to survive (Schimel et al., 2007; Wepking et al., 2017). Therefore, substantial treatment with antibiotics also results in the reduction of beneficial microorganism populations. Loss of beneficial bacteria may cause soil degradation and infertility. The overuse of antibiotics in the agricultural system will result in deleterious effects on agroecology and sustainability.

5.4 FACTORS CAUSING RESISTANCE IN MICROBES AGAINST ANTIBIOTICS

The primary resistance in the bacteria against the antibiotic is inferred from the selection pressure placed by the therapeutical agents over the susceptible bacteria (Polianciuc et al., 2020). Besides the selection pressure, social and administrative factors are also responsible for the spreading and development of antibiotic resistance bacteria. The misuse and overuse of antibiotics are the key drivers of the resistant development of bacteria and pathogens. Lack of sanitization and inadequate infectious disease prevention promote the spread of microbes which are resistant to antibiotic treatments. Such resistant microbes developed into multi-drug resistant (MDR), which is a very serious concern for microbes like contagious microbes for infectious diseases. Bacteria have a high rate of mutation ability, and contact with a low dose of the antibiotic result in mutation in their genome, which allow the bacteria to survive these antibiotic and even pass these genes to the next generations (Chandler, 2019).

5.5 MECHANISMS OF ANTIBIOTIC RESISTANCE ADOPTED BY DIFFERENT MICROORGANISMS

Bacteria have evolved five main mechanisms to counteract the activity of antibiotics through resistance. Resistance genes encode for enzymes which degrade or modify the target antibiotic, rendering it inactive. Alternatively, antibiotic-efflux pumps can pump the antibiotic out of the bacterial cell before it is able to cause damage. Other adaptations include modification of the bacterial cell surface to reduce uptake of the antibiotic through a reduction in the cell wall or cell membrane permeability, the production of an alternative metabolic pathway bypassing the pathogenic bacteria, or other commensals facilitating the spread of resistance genes within the bacterial community (Shaban et al., 2014).

The antibiotic resistance mechanisms adopted by different microorganisms can be broadly classified into (i) mechanisms having a genetic basis of microbial resistance and (ii) mechanisms with a mechanistic basis of antimicrobial resistance. Besides these major mechanisms, more recently,

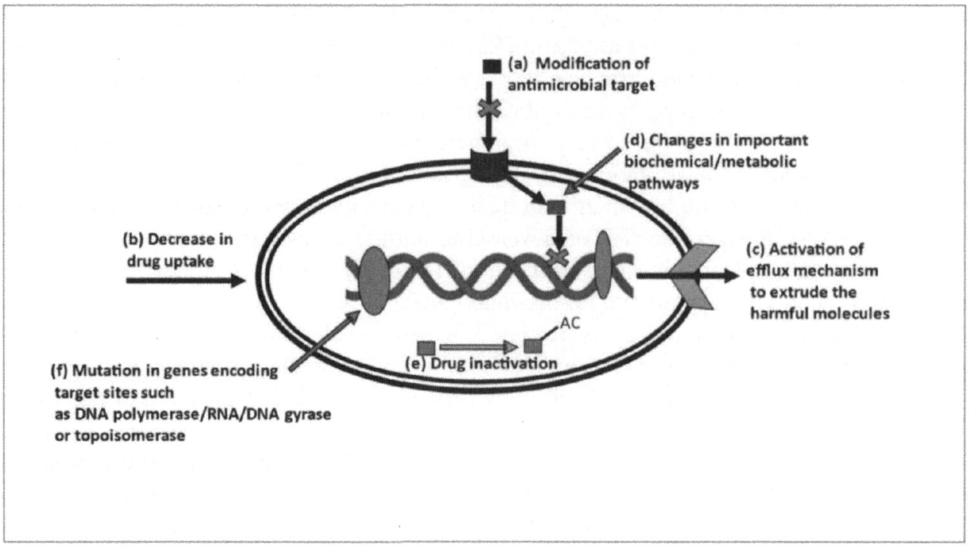

FIGURE 5.2 Mechanisms of antibiotic resistance adopted by bacteria: (a) **Modification of antimicrobial targets** – some bacterial membranes lack the target molecule for a particular antibiotic and thus become impermeable to the passage of such antibiotics across their cell membranes; (b) **decrease in drug uptake** – a few bacteria modify their cell membranes in order to reduce or diminish the drug uptake; (c) **activation of efflux mechanisms to extrude the harmful molecules** – these mechanisms increase the efflux of antibiotics at a faster pace such that they are not able to produce much effect on the cell; (d) **changes in important metabolic pathways –** effect of antibiotics can be produced through many biochemical pathways, and hence by incorporating changes in these pathways, bacteria rescue themselves from the effect of antibiotics; (e) **drug inactivation –** drugs are inactivated by processes such as acetylation or methylation once they enter in the cell, thus making them ineffective; (f) **mutations in genes encoding target sites** – drugs have targets such as DNA polymerase/RNA/DNA gyrase and topoisomerase etc., so the genes encoding such target molecules are mutated/suspended by some bacteria to protect them from the effect of antibiotics (Mann et al., 2021).

a phenomenon called protein promiscuity has also been reported to be responsible for antibiotic resistance. According to classical biochemistry, proteins bind to their ligands/substrates in a specific manner via their pre-formed binding sites. However, several types of variations from this form, known as promiscuous behaviour/protein promiscuity, have been observed in recent years (Mann et al., 2021; Gupta et al., 2020).

5.6 AGRARIAN DISTRESS ASSOCIATED WITH ANTIMICROBIAL RESISTANCE (AMR)

Agrarian distress is the economical anguish that the farmers faces due to farming and farmed land, as well as the high input costs and low output profit margin. The underlying reasons for agrarian distress are many, like unviable agriculture, adverse economic trade, insufficient value chain in agriculture, and fluctuating climate, and to this, antibiotic resistance microbes add up another challenge to the peasant.

Antibiotics play an important role in the treatment of terrestrial and aquatic food-producing animals and plants to assure food safety and quality, animal health, and the welfare of humans; basically, in food-producing animals, antibacterials are used to treat the sick animal to prevent the spread of disease in animals, or antibiotics are used to promote the fast growth of animals by applying a low concentration of antibiotics; this practice has put pressure for antimicrobial resistant

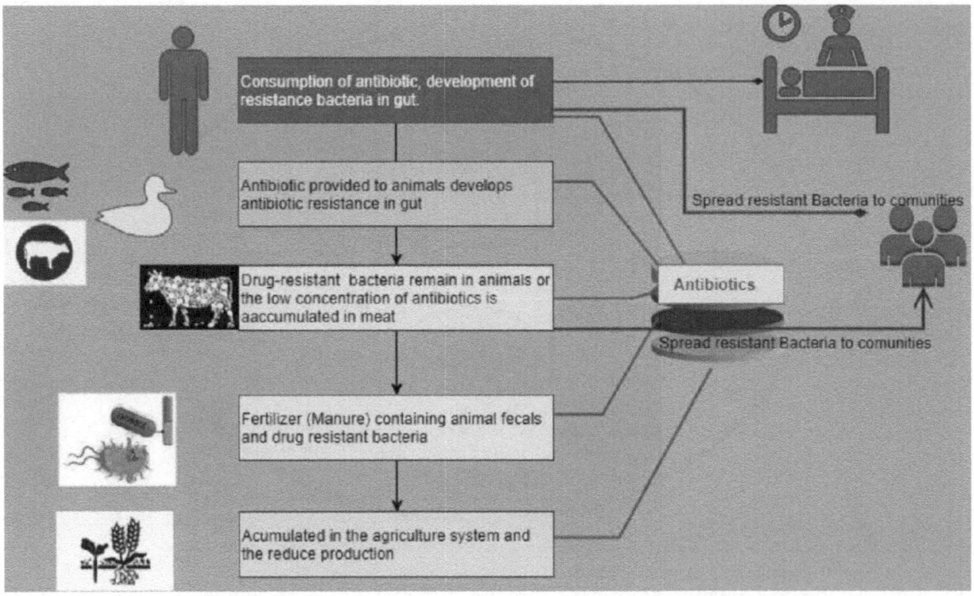

FIGURE 5.3 Mechanism of antibiotic transfer to agriculture and the public health.

microbes (FAO, 2017). As a result, the antibiotic that was once an effective treatment for diseases in animals or humans becomes less effective or ineffective, leading to a reduced ability to successfully treat the infection. This, in turn, leads to more severe or prolonged illnesses, increased mortality, production losses in agriculture, and reduced livelihoods and food security, causing agrarian distress (Figure 5.3) (Taylor and Reeder, 2020). It is difficult to estimate global antimicrobial use in agriculture sectors because of insufficient and poor data collection in many countries; however, it is estimated to be 60,000 tons (FAO, 2017). However, the total volume is expected to rise over time with the increasing demand for food and food products. Hence, the chance of antimicrobial resistance microorganisms will also be increased if it is not properly administered and addressed, leading to a global food crisis and adversely worsening agrarian distress.

Since there is a strong connection among soil, plants, and animals, the spread of antibiotics and antibiotic resistance in our environment is not limited to any one ecosystem. Antibiotics travel among all the routes (Karesh et al., 2012) and create a network where ARGs also travel along with them (animals-plant-animals), as shown in Figure 5.4. At the same time, antibiotics are provided in animal feed, agriculture, and aquaculture to improve growth and in crop protection to prevent and cure several diseases. In the case of animals, however, complete absorption of these antibiotics does not take place inside the gut, and they are excreted out of the body in the form of faeces and urine, which ultimately leads to the formation of manure. This manure is a rich source of organic matter and nutrients; although it helps in improving the fertility of the soil, additionally, it also becomes responsible for the transfer of traces of antibiotics from the animal kingdom to the soil ecosystem and ultimately to the plants and again from plants to the animal kingdom (Mann et al., 2021; Chambers and Deleo, 2009).

5.7 REDUCTION OF ANTIBIOTIC RESISTANCE IN AGRICULTURE

The unregulated application and use of antibiotics on the farm have compounded the problem of antibiotic resistance. Considering that the implementation of control measures will help reduce the abusive use of antibiotics in livestock farming, the following strategy is set forth:

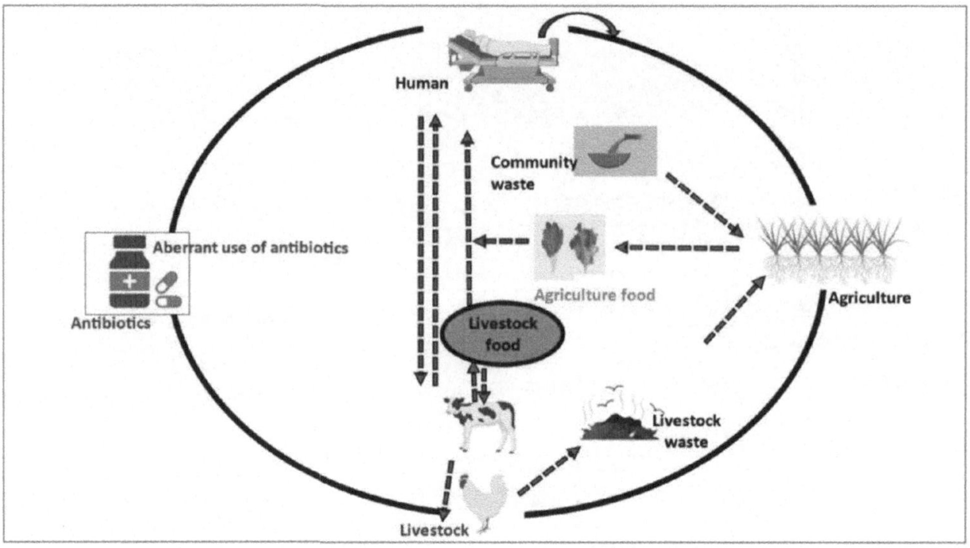

FIGURE 5.4 Antibiotic contamination cycle through different kinds of sources to ecosystems (**Aslam et al., 2021**).

- Farmers generally lack understanding about the proper use and storage of antibiotics, their misuse, and the resulting risks to the environment and public health (Islam et al., 2016). Therefore, a campaign to raise awareness about the risks of using antibiotics carelessly, storing them improperly, and the resulting antibiotic resistance, with its effects on public health, is likely to target both farmers and the general public.
- A qualified veterinarian should perform activities or therapy in industrial and semi-industrial production systems rather than the farm owner (Kim et al., 2013).
- Veterinarian services must be available at any location of the farm with all required facilities. Despite the fact that raising livestock contributes to the nation's economy, we hereby propose that the government and reputable private businesses that are concerned with the growth of the economy of the nation support the use of regular, proper, and effective veterinary services by providing subsidies to farmers, particularly to lower-income farmers in rural and remote areas of the nation. Farmers should avoid the indiscriminate use of antibiotics (that is, without prescription, supervision, and guidance from veterinarians) in animal farming.
- Policies governing the prescription of antibiotics used in animal farming should be properly followed by practitioners. They should also make sure the farmers adhere to the dosage, length, and administration of the treatment and withdrawal periods of the antibiotics prescribed for a purpose, as well as the diagnosis of the disease's causative agent and antibiotic susceptibility testing against it, before the prescription and employment of the antibiotics (Kim et al., 2013; Sekyere, 2014).
- Before food is distributed or delivered for human consumption, the presence of antibiotics must be tested. Due to the variances in antibiotic consumption patterns and animal production systems among nations, it is advised that strict controls and/or restrictions be put in place by policymakers to ensure that antibiotics are only used and purchased legally in animal husbandry (Van Boeckel et al., 2015; Kim et al., 2013).
- The use of antibiotics in food-producing animals must be minimized by substituting them with possible alternative substances, like probiotics, prebiotics, and plant-derived or crude plant extracts, for the treatment and prevention of diseases (Moyane et al., 2013; Guetiya-Wadoum et al., 2016).

- Governments of respective countries should conduct collaboration with the public health department and authorities to monitor antibiotic consumption by animals on the farm; moreover, the government and the public health sector should make available services for research and analysis of data to track the drug and its residue level in most environmental samples (Manyi-Loh et al., 2018).

5.8 CONCLUSIONS

Antibiotics (agricultural, veterinary, and clinical) are essential in all settings (whether they are produced naturally, semi-synthetically, or synthetically). They prevent diseases caused by bacteria and microbes; despite these positive results, using antibiotics has negative side effects. The application of subtherapeutic doses of antibiotics as growth promoters and water that animals eat over a lengthy period would result in antibiotic contamination, leaving antibiotic residues in animal-derived goods like meat, milk, eggs, and edible tissues. When the products reach humans for consumption, the contaminated animal-derived goods cause accumulation in the human body and provide immunity to lethal human pathogens. This chapter gives the route for antibiotic entry to the agroecosystem and its consequential role in agrarian distress. This chapters has also discussed the pathways of ARB and ARG transmission within various agroecosystem components. More research is required to quantify human exposure to antibiotics, ARBs, and ARGs as well as the number of antibiotics released into the agricultural ecosystem. Further research is needed to fill in information gaps and uncover possible threats that environmental antibiotics, ARBs, and ARGs pose to people's health and the health of the larger ecosystem. Furthermore, it is essential that researchers, farmers, veterinarians, medical professionals, and nongovernmental organizations work together to break the spread of agricultural stress. Food systems affect nutritional and other health outcomes. There is a need to address policy aspects of nutritional implications of specific foods (e.g., fruits, vegetables, and trans fats) and identified opportunities to tackle the double burden of malnutrition. This chapter attempts to deepen the understanding on how health concerns and the role of the health sector are addressed across food systems policies across the world community.

5.9 ACKNOWLEDGEMENT

The author thanks the University Grant Commission for providing fellowship and the Department of Botany, Dr. Harisingh Gour Vishwavidyalaya, Sagar (M.P.), 470003, India, for sharing their facilities and those who made their literature available for the compilation of this manuscript.

REFERENCES

Abraham, E. P., 1963. "The antibiotics." *Comprehensive Biochemistry*, 11(4), 181–224. https://doi.org/10.1016/B978-1-4831-9711-1.50022-3.

Aggarwal, A., Mehta, S., Gupta, D., Sheikh, S., Pallagatti, S., Singh, R., and Singla, I., 2012. "Clinical and immunological erythematosus patients characteristics in systemic lupus maryam." *Journal of Dental Education*, 76(11), 1532–1539. https://doi.org/10.4103/ijmr.IJMR.

Ashbolt, N. J., Amézquita, A., Backhaus, T., Borriello, P., Brandt, K. K., Collignon, P., Coors, A., Finley, R., Gaze, W. H., and Heberer, T., 2013. "Human Health Risk Assessment (HHRA) for environmental development and transfer of antibiotic resistance." *Environ. Health Perspect.*, 121, 993.

Aslam, B., Khurshid, M., Arshad, M. I., Muzammil, S., Rasool, M., Yasmeen, N., Shah, T., Chaudhry, T. H., Rasool, M. H., Shahid, A., Xueshan, X., and Baloch, Z., 2021. "Antibiotic resistance: One health one world outlook." *Front. Cell. Infect. Microbiol.*, 11, 771510. doi: 10.3389/fcimb.2021.771510.

Bai, Y., Müller, D. B., Srinivas, G., Garrido-Oter, R., Potthoff, E., Rott, M., Dombrowski, N., Münch, P. C., Spaepen, S., Remus-Emsermann, M., Hüttel, B., McHardy, A. C., Vorholt, J. A., and Schulze-Lefert, P., 2015. "Functional overlap of the Arabidopsis leaf and root microbiota." *Nature*, 528(7582), 364–369. https://doi.org/10.1038/nature16192.

Ben, Y., Fu, C., Hu, M., Liu, L., Wong, M. H., and Zheng, C., 2019. "Human health risk assessment of antibiotic resistance associated with antibiotic residues in the environment: A review." *Environmental Research*, 169, 483–493. https://doi.org/10.1016/j.envres.2018.11.040.

Berger, C. N., Sodha, S. V., Shaw, R. K., Griffin, P. M., Pink, D., Hand, P., and Frankel, G., 2010. "Fresh fruit and vegetables as vehicles for the transmission of human pathogens." *Environmental Microbiology*, 12(9), 2385–2397. https://doi.org/10.1111/j.1462-2920.2010.02297.x.

Bulgarelli, D., Rott, M., Schlaeppi, K., Ver Loren van Themaat, E., Ahmadinejad, N., Assenza, F., Rauf, P., Huettel, B., Reinhardt, R., Schmelzer, E., Peplies, J., Gloeckner, F. O., Amann, R., Eickhorst, T., and Schulze-Lefert, P., 2012. "Revealing structure and assembly cues for Arabidopsis root-inhabiting bacterial microbiota." *Nature*, 488(7409), 91–95. https://doi.org/10.1038/nature11336.

Cabello, F. C., Godfrey, H. P., Buschmann, A. H., and Dölz, H. J., 2016. "Aquaculture as yet another environmental gateway to the development and globalization of antimicrobial resistance." *The Lancet Infectious Diseases*, 16(7), e127–e133. https://doi.org/10.1016/S1473-3099(16)00100-6.

Chambers, H. F., and Deleo, F. R., 2009. "Waves of resistance: Staphylococcus aureus in the antibiotic era." *Nat. Rev. Microbiol.*, 7, 629–641.

Chandler, C. I. R., 2019. "Current accounts of antimicrobial resistance: Stabilization, individualization and antibiotics as infrastructure." *Palgrave Communications*, 5(1), 15–17. https://doi.org/10.1057/s41599-019-0263-4.

Demain, A. L., and Sanchez, S., 2009. "Microbial drug discovery: 80 years of progress." *J. Antibiot.*, 62, 5–16.

Durso, L. M., and Cook, K. L., 2019. "One health and antibiotic resistance in agroecosystems." *EcoHealth*, 16(3), 414–419. https://doi.org/10.1007/s10393-018-1324-7.

FAO., 2017. "Antimicrobial resistance in food and agriculture amr in the food and agriculture sectors food chain crisis | emergency prevention system | information sheets." i(April). www.fao.org/antimicrobial-resistance.

FAO., 2018. "The 10 elements of agroecology guiding the transition to sustainable food and agricultural systems." *Fao*, 15. www.fao.org/3/I9037EN/i9037en.pdf.

Guetiya-Wadoum, R. E., Zambou, N. F., Anyangwe, F. F., Njimou, J. R., Coman, M. M., Verdenelli, M. C., Cecchini, C., Silvi, S., Orpianesi, C., and Cresci, A., 2016. "Abusive use of antibiotics in poultry farming in Cameroon and the public health implications." *Br. Poult. Sci.*, 57, 483–493.

Gupta, M. N., Alam, A., and Hasnain, S. E., 2020. "Protein promiscuity in drug discovery, drug-repurposing and antibiotic resistance." *Biochimie*, 175, 50–57. doi: 10.1016/j.biochi.2020.05.004.

Gutierrez, I. R., Watanabe, N., Harter, T., Glaser, B., and Radke, M., 2010. "Effect of sulfonamide antibiotics on microbial diversity and activity in a Californian Mollic Haploxeralf." *J. Soils Sediments*, 10, 537–544. doi: 10.1007/s11368-009-0168-8.

Hammer, T. J., Fierer, N., Hardwick, B., Simojoki, A., Slade, E., Taponen, J., Viljanen, H., and Roslin, T., 2016. "Treating cattle with antibiotics affects greenhouse gas emissions, and microbiota in dung and dung beetles." *Proc. R. Soc. B.*, 283, 20160150. doi: 10.1098/rspb.2016.0150.

Han, X. M., Hu, H. W., Chen, Q. L., Yang, L. Y., Li, H. L., Zhu, Y. G., Li, X. Z., and Ma, Y. B., 2018. "Antibiotic resistance genes and associated bacterial communities in agricultural soils amended with different sources of animal manures." *Soil Biology and Biochemistry*, 126, 91–102. https://doi.org/10.1016/j.soilbio.2018.08.018.

Heuer, H., Schmitt, H., and Smalla, K., 2011. "Antibiotic resistance gene spread due to manure application on agricultural fields." *Current Opinion in Microbiology*, 14(3), 236–243. https://doi.org/10.1016/j.mib.2011.04.009.

Heuer, H., and Smalla, K., 2007. "Manure and sulfadiazine synergistically increased bacterial antibiotic resistance in soil over at least two months." *Environmental Microbiology*, 9(3), 657–666. https://doi.org/10.1111/j.1462-2920.2006.01185.x.

Ji, K., Kho, Y., Park, C., Paek, D., Ryu, P., Paek, D., Kim, M., Kim, P., and Choi, K., 2010. "Influence of water and food consumption on inadvertent antibiotics intake among general population." *Environ. Res.*, 110, 641–649.

Karesh, W. B., Dobson, A., Lloyd-Smith, J. O., Lubroth, J., and Dixon, M. A., 2012. "Ecology of zoonoses: Natural and unnatural histories." *Lancet*, 380, 1936–1945. doi: 10.1016/S0140-6736(12)61678-X.

Karkman, A., Parnanen, K., and Larsson, D. G. J., 2019. "Fecal pollution can explain antibiotic resistance gene abundances in anthropogenically impacted environments." *Nature Communications*, 10(1), 80. https://doi.org/10.1038/s41467-018-07992-3.

Kim, D. P., Rman, C. S., Douny, C., Dinh, T. V., Xuam, B. H., Vu, B. D., Hong, N. P., Scippo, M. L., 2013. "First survey on the use of antibiotics in pig and poultry production in the red river Delta region of Vietnam." *Food Public Health*, 3, 247–256.

Klase, G., Lee, S., Liang, S., Kim, J., Zo, Y. G., and Lee, J., 2019. "The microbiome and antibiotic resistance in integrated fishfarm water: Implications of environmental public health." *Science of the Total Environment*, 649, 1491–1501. https://doi.org/10.1016/j.scitotenv.2018.08.288.

Mann, A., Nehra, K., Rana, J. S., and Dahiya, T., 2021. "Antibiotic resistance in agriculture: Perspectives on upcoming strategies to overcome upsurge in resistance." *Current Research in Microbial Sciences*, 2. https://doi.org/10.1016/j.crmicr.2021.100030.

Manu, T., 2010. "Tetracycline resistance genes persist at aquaculture farms in the absence of selection pressure." *Environmental Science & Technology*, 45(2), 386.

Manyi-Loh, C., Mamphweli, S., Meyer, E., and Okoh, A., 2018. "Antibiotic use in agriculture and its consequential resistance in environmental sources: Potential public health implications. *Molecules*, 23. https://doi.org/10.3390/molecules23040795.

Marti, R., Scott, A., Tien, Y.- C., Murray, R., Sabourin, L., Zhang, Y., and Topp, E., 2013. "Impact of manure fertilization on the abundance of antibiotic-resistant bacteria and frequency of detection of antibiotic resistance genes in soil and on vegetables at harvest." *Applied and Environmental Microbiology*, 79(18), 5701–5709. https://doi.org/10.1128/AEM.01682-13.

Martinez, J. L., 2009. "The role of natural environments in the evolution of resistance traits in pathogenic bacteria." *Proc. Biol. Sci.*, 276, 2521–2530.

Martínez, J. L., Coque, T. M., and Baquero, F., 2015. "What is a resistance gene? Ranking risk in resistomes." *Nat. Rev. Microbiol.*, 13, 116.

McManus, P. S., and Jones, A. L., 1994. "Epidemiology and genetic analysis of streptomycin-resistant *Erwinia amylovora* from Michigan and evaluation of oxytetracycline for control." *Phytopathology (USA).*

McManus, P. S., Stockwell, V. O., Sundin, G. W., and Jones, A. L., 2002. "Antibiotic use in plant agriculture." *Annual Review of Phytopathology*, 40(1), 443–465.

Moyane, J. N., Jideani, A. I. O., and Aiyegoro, O. A., 2013. "Antibiotics usage in food-producing animals in South Africa and impact on human: Antibiotic resistance." *Afr. J. Microbiol. Res.*, 7, 2990–2997.

Muurinen, J., Stedtfeld, R., Karkman, A., Pärnänen, K., Tiedje, J., and Virta, M., 2017. "Influence of manure application on the environmental resistome under Finnish agricultural practice with restricted antibiotic use." *Environmental Science & Technology*, 51(11), 5989–5999. https://doi.org/10.1021/acs.est.7b00551.

National Academies of Sciences, Engineering, and Medicine (2018). The National Academies Press. doi: 10.17226/21896.

Nikolina, U. K., 2014. "Bloom of resident antibiotic-resistant bacteria in soil following manure fertilization." *Proceedings of the National Academy of Sciences of the United States of America*, 111(42), 15202–15207.

Okeke, E. S., Kingsley, R. D., Okoye, C. O., Road, K., and Nairobi, K., 2022. "Antibiotic resistance in aquaculture and aquatic organisms: A review of current nanotechnology applications for sustainable management Organisation of African Academic Doctors (OAAD), off." *Environmental Science and Pollution Research*, 29, 69241–69274. https://doi.org/10.1007/s11356-022-22319-y.

Polianciuc, S. I., Gurzău, A. E., Kiss, B., Ştefan, M. G., and Loghin, F., 2020. "Antibiotics in the environment: Causes and consequences." *Medicine and Pharmacy Reports*, 93(3), 231–240. https://doi.org/10.15386/mpr-1742.

Qiao, M., Ying, G.- G., Singer, A. C., and Zhu, Y.- G., 2017. "Review of antibiotic resistance in China and its environment." *Environ. Int.*, 110, 160–172.

Raman, N. M., Easwaran, M., Kaul, R., Bharti, J., Motelb, K. F. A., and Kaul, T., 2019. "Antimicrobial resistance with special emphasis on pathogens in agriculture." *Intechopen*, 32, 137–144.

Ruimy, R., Brisabois, A., Bernede, C., Skurnik, D., Barnat, S., Arlet, G., Momcilovic, S., Elbaz, S., Moury, F., Vibet, M.- A., Courvalin, P., Guillemot, D., and Andremont, A., 2010. "Organic and conventional fruits and vegetables contain equivalent counts of Gram-negative bacteria expressing resistance to antibacterial agents." *Environmental Microbiology*, 12(3), 608–615. https://doi.org/10.1111/j.1462-2920.2009.02100.x.

Islam, K. B. M. S., Shiraj-Um-Mahmuda, S., and Hazzaz-Bin-Kabir, M., 2016. "Antibiotic usage patterns in selected broiler farms of Bangladesh and their public health implications." *J. Public Health Dev. Ctries.*, 2, 276–284.

Schimel, J., Balser, T. C., and Wallenstein, M., 2007. "Microbial stress-response physiology and its implications for ecosystem function." *Ecology*, 88, 1386–1394. doi: 10.1890/06-0219.

Sekyere, J. O., 2014. "Antibiotic types and handling practices in disease management among pig farms in Ashanti region, Ghana." *J. Vet. Med.*, 531952.

Shaban, R. Z., Simon, G. I., Trott, D. J., Turnidge, J., and Jordan, D., 2014. *Surveillance and reporting of antimicrobial resistance and antibiotic usage in animals and agriculture in Australia*. Canberra, AU: Department of Agriculture, The Australian Government, 1–208.

Singh, C., Pathak, P., Chaudhary, N., Rathi, A., Dehariya, P., and Vyas, D., 2020. "Mushrooms and mushroom Composts in integrated farm management." *Research Journal of Agricultural Sciences*, 11(6), 1436–1443.

Taylor, P., and Reeder, R., 2020. "Antibiotic use on crops in low and middle-income countries based on recommendations made by agricultural advisors." *CABI Agriculture and Bioscience*, 1(1), 1–14. https://doi.org/10.1186/s43170-020-00001-y.

Thanner, S., Drissner, D., and Walsh, F., 2016. "Antimicrobial resistance in agriculture." *MBio*, 7(2), 1–7. https://doi.org/10.1128/mBio.02227-15.

Toth, J. D., Feng, Y. C., and Dou, Z. X., 2011. "Veterinary antibiotics at environmentally relevant concentrations inhibit soil iron reduction and nitrification." *Soil Biol. Biochem.*, 43, 2470–2472. doi: 10.1016/j.soilbio.2011.09.004.

US Centers for Disease Control and Prevention., 2018. "Initiatives for addressing antimicrobial resistance in the environment: Current situation and challenges." *Wellcome Trust.*

Van Boeckel, T. P., Brower, C., Gilbert, M., Grenfell, B. T., Levin, S. A., Robinson, T. P., Teillant, A., and Laxminarayan, R., 2015. "Global trends in antimicrobial use in food animals." *Proc. Natl. Acad. Sci. USA.*, 112, 5649–5654.

Van Der Lelie, D., Taghavi, S., Monchy, S., Schwender, J., Miller, L., Ferrieri, R., Rogers, A., Wu, X., Zhu, W., Weyens, N., Vangronsveld, J., and Newman, L. (2009). "Poplar and its bacterial endophytes: Coexistence and harmony." *Critical Reviews in Plant Sciences*, 28, 346–358.

Watts, J. E. M., 2017. "The rising tide of antimicrobial resistance in aquaculture: Sources, sinks and solutions." *Marine Drugs*, 15(6), 158.

Wepking, C., Avera, B., Badgley, B., Barrett, J. E., Franklin, J., Knowlton, K. F., Ray, P. P., Smitherman, C., and Strickland, M. S., 2017. "Exposure to dairy manure leads to greater antibiotic resistance and increased mass-specific respiration in soil microbial communities." *Proceedings of the Royal Society B: Biological Sciences*, 284(1851). https://doi.org/10.1098/rspb.2016.2233.

Williams-Nguyen, J., Sallach, J. B., Bartelt-Hunt, S., Boxall, A. B., Durso, L. M., McLain, J. E., Singer, R. S., Snow, D. D., and Zilles, J. L., 2016. "Antibiotics and antibiotic resistance in agroecosystems: State of the science." *Journal of Environmental Quality*, 45(2), 394–406. https://doi.org/10.2134/jeq2015.07.0336.

Xie, W. Y., Shen, Q., and Zhao, F. J., 2018. "Antibiotics and antibiotic resistance from animal manures to soil: A review." *European Journal of Soil Science*, 69(1), 181–195. https://doi.org/10.1111/ejss.12494.

Zhang, J., Lu, T., Chai, Y., Sui, Q., Shen, P., and Wei, Y., 2019. "Which animal type contributes the most to the emission of antibiotic resistance genes in large-scale swine farms in China?" *Science of the Total Environment*, 658, 152–159. https://doi.org/10.1016/j.scitotenv.2018.12.175.

Zhang, J., Wang, Z., Wang, Y., Zhong, H., Sui, Q., Zhang, C., and Wei, Y., 2017. "Effects of graphene oxide on the performance, microbial community dynamics and antibiotic resistance genes reduction during anaerobic digestion of swine manure." *Bioresource Technology*, 245(Pt A), 850–859. https://doi.org/10.1016/j.biortech.2017.08.217.

Zhao, Y., Yang, Q. E., Zhou, X., Wang, F. H., Muurinen, J., Virta, M. P., Brandt, K. K., and Zhu, Y. G., 2021. "Antibiotic resistome in the livestock and aquaculture industries: Status and solutions." *Critical Reviews in Environmental Science and Technology*, 51(19), 2159–96. https://doi.org/10.1080/10643389.2020.1777815.

Zhu, Y. G., Johnson, T. A., Su, J. Q., Qiao, M., Guo, G.- X., Stedtfeld, R. D., Hashsham, S. A., and Tiedje, J. M., 2013. "Diverse and abundant antibiotic resistance genes in critical reviews in environmental science and technology 2195 Chinese swine farms." *Proceedings of the National Academy of Sciences of the United States of America*, 110(9), 3435–3440. https://doi.org/10.1073/pnas.1222743110.

Zhu, Y. G., Zhao, Y., Li, B., Huang, C. L., Zhang, S. Y., Yu, S., Chen, Y. S., Zhang, T., Gillings, M. R., and Su, J. Q., 2017. "Continental-scale pollution of estuaries with antibiotic resistance genes." *Nature Microbiology*, 2, 16270. https://doi.org/10.1038/nmicrobiol.2016.270.

Zhu, Y. G., Zhao, Y., Zhu, D., Gillings, M., Penuelas, J., Ok, Y. S., Capon, A., and Banwart, S., 2019. "Soil biota, antimicrobial resistance and planetary health." *Environment International*, 131, 105059. https://doi.org/10.1016/j.envint.2019.105059.

6 Host Plant Resistance and Sustainable Management of Insect Pests

Sunilkumar Anilkumar Dwivedi and Trupti Rajesh Pandit

6.1 INTRODUCTION

Even in the simplified environments typical of agriculture, interactions between herbivorous arthropods and their plant hosts are complicated and diverse (Rasmann *et al.*, 2009). The entire process by which an herbivore uses a plant generally comprises many phases: a seeking phase in which the herbivore travels, typically in response to sight and smell signals, and a feeding phase in which the herbivore eats the plant. The herbivore's performance is influenced by interacting suites of nutrients, toxins, digestibility reducers, and other factors in the plant during the contact evaluation phase, which progresses from the absence of a host plant to the possibility of a host; and the herbivore's performance is influenced by an expanded set of visual, physical, and chemical cues from the plant during the host utilization phase (Schoonhoven *et al.*, 1998). The herbivore interacts with other species at the same trophic level, such as competing herbivores, as well as species at different trophic levels, such as predators and parasitoids and microbes, at each stage of this process. Plant resistance is caused by the expression of resistance-related plant characteristics by the host plant that influences one or more elements of the herbivore's interaction with the host plant and other plant-associated organisms. Plant resistance may be described as the "total of genetically inherited characteristics" that influences the extent of herbivore damage (yield loss) to the plant (Smith and Smith, 2012).

Organic substances that do not fall into the categories of carbohydrates, protein, lipid, vitamin, steroid, or mineral are referred to as xenobiotics. The majorities of them are harmful and must be expelled. When a pesticide is administered to a live creature (plant or animal), chemical changes occur, resulting in the production of new products known as "metabolites." The liver and kidney are the major organs for pesticide metabolism in vertebrates, whereas the fat body, Malpighian tubules, alimentary canal, and tracheal system are the major organs for volatile chemical metabolism in insects. Metabolic activation (where the product is more active than the parent component) and detoxification (when the product is less or non-toxic) are two examples of metabolic processes. The former is the instance of synergism while the latter is the instance of insecticide resistance. "Plant resistance" is a term used to describe an integrated management strategy in which a resistant plant genotype is used to decrease the influence of herbivorous arthropods on crop production or quality, either alone or in conjunction with other strategies. Plant resistance offers several advantages over conventional pest control methods. Plant resistance has a consistent and cumulative effect on the target pest, and implementing plant resistance is typically straightforward and affordable for farmers once a resistant variety has been established. Plant resistance is also typically compatible with other approaches, such as pesticide use and biological control. This chapter provides a summary of the traditional approach to an exploitation of plant resistance in agricultural protection and shortly summarizes the current use and importance of plant resistance, including case studies from rice (*Oryza sativa*).

DOI: 10.1201/9781003269380-6

It conjointly highlights recent advances in our understanding of an inducible plant resistance and discusses how these recent advances might engender novel approaches to a victimization plant resistance in tormentor management. It mostly excludes the current commercial usage of resistant transgenic plants since they generally entail the introduction of foreign genes into agricultural plants and so give minimal insight into naturally occurring plant resistance. In the last decade, several more monographs and reviews on various areas of plant resistance have been published.

6.2 CATEGORIZATION OF RESISTANCE

The discovery of resistant genotypes is frequently followed by research aimed at assigning resistant genotypes to one or more of Painter's three categories of resistance, which were first described in 1951. "Antibiosis" is the term used to explain the negative impacts of resistant plants on herbivore physiology and life history, such as decreased growth, survival, and fertility. The second category, initially called "non-preference" by Painter, was renamed "antixenosis" by Kogan in 1978. This refers to plant characteristics that impact herbivore behaviors in ways that diminish an herbivore's affinity for, or acceptance of, a plant as a host. Finally, "tolerance" refers to a plant's ability to resist herbivore harm to the point that agronomic yields or quality are lowered less than in a less tolerant plant subjected to similar harm. Even though antibiosis, antixenosis, and tolerance are best conceived of as kinds or "functional categories" of effects, antibiosis, antixenosis, and tolerance are nevertheless commonly characterized as processes in the literature. There are also certain plant resistances mechanisms that can't be categorized into one of these three groups. These categories, such as herbivore-induced emission of volatiles that attract parasitoids, may not be accurate enough to indicate their potential significance in a management program in some situations. Plant resistance that causes significant levels of death in the early phases of insect development, as well as plant resistance that just delays an insect's growth, are both considered antibiotic. However, the two categories may have quite distinct values in a management program and may be employed in very different ways. Nonetheless, these words are still extensively employed in the literature on applied plant resistance.

6.3 HOST PLANT RESISTANCE AND ECONOMIC INJURY LEVEL

Economic constraints or thresholds of action in relation to hosted crops are one of the first and most important changes in crop management proposals to be made (Sharma, 1993). There are many types of plants with different levels of resistance in some cases. To test the cultivar resistance level, test and empirical data should be collected. This information is important in determining the type and timing of the intervention (pesticide treatment or release of natural enemies) needed to control the growing number of pests. The use of economic injury levels (EILs) to define the number of herbicides in the newly released herbicar is important for pest control as the term "resistance" raises different expectations for different people and farmers do not want to risk it. Insect-resistant plants not only reduce pest congestion but also increase the time it takes for insects to reach EILs, depending on the level of resistance. If the EIL is based on insect damage (e.g., the percentage of dead heart stem and stem borer, the number of leaves infested with lice, or the part of the leaves eaten by armyworm worms) and the number of insects growing throughout the season, that could be dangerous. The cultivar will lose money in July, the average resistant cultivar will lose money in August, and the resistant cultivar will be able to withstand pests throughout the season (Sharma, 1993).

The cultivar can be planted up to the time when pest infestation is predicted to be below EIL if harmful pests are limited to a certain period and a short period of time (e.g., dead heart growth due to flies and stalks boredom). Adults, which are a non-invasive class of insects (e.g., sorghum or the number of larvae trapped in pheromone), are used to calculate EIL. As bedbug resistance levels increase, the EIL will also increase. The link between insect infestation and grain damage to the midge of the breast varies depending on the resistance level. The EIL of sorghum midge is one adult period in the affected cultivar (CSH 1), 25 adults per panicle per medium-resistant cultivar (ICSV 745), and > 50 adults per panicle in a cultivar that is very resistant (ICSV 745). If the EIL is based on adult allergies (e.g., bedbugs), and resistance is based on allergies and antibiosis (which reduces initial doses and the rate of population growth), then resistant plants will increase the duration it takes an insect to reach the EIL (Sharma, 1993). A list of promising cultivars of pulses showing resistance against insect pest due to morphological characters is provided in Table 6.1.

6.4 HOST PLANT RESISTANCE IN INTEGRATED PEST MANAGEMENT

Plant resistance is only achieved at high levels against a few species of insects. High levels of resistance, on the other hand, are not necessarily due to the use of host plant resistance (HPR), which refers to the intrinsic capability exhibited by specific plant cultivars or species to protect themselves from herbivorous pathogens, pathogens, or parasites. Various mechanisms that plants have developed over time to deter or limit the damage caused by these organisms frequently contribute to this natural resistance. In conjunction with other pest control components, species with low or moderate levels of resistance or that can prevent pest damage can be used to control pests. The goal of planting pest-resistant crops should be to conserve natural enemies while reducing the number of pesticides. Other pest control methods, such as synthetic pesticides, work best when using insect-resistant plants, in addition to cultural, biological, cultural, and chemical management. The utilization of resistant crop varieties can serve as a crucial element in pest management strategies, effectively mitigating the need for the introduction of endangered plant species into the environment.

6.5 HPR AS A PRINCIPAL METHOD OF INSECT CONTROL

In the context of integrated pest management (IPM), HPR as an insect control approach offers higher promise than any other form of pest suppression. Unlike chemical and biological control techniques, the usage of pest-resistant cultivars is not susceptible to the whims of nature. Insect-resistant rice cultivars have made a significant contribution to Asia's long-term rice production. HPR is a critical component in reducing insect pest losses in sorghum (Sharma, 1993). Plant resistance as a pest control approach has several advantages, and in certain situations, it is the only feasible and efficient method available. If we rely solely on plant resistance for pest management, however, there may be issues, such as high levels of resistance being linked with poor yield potential or unacceptable quality features and resistance not being exhibited in every location where a variety is cultivated. As a result, pest-resistant cultivars must be properly integrated into pest control programs in various agricultural settings. Prior to the invention of pesticides, plant resistance to insects was a common technique of insect control. The adoption of insect-resistant cultivars has helped keep several insect pests in check. The advantages of HPR are dependent on the pattern of insect invasion; for example, numerous insects such as aphids, whiteflies, and mites attack the crop in small quantities. Before reaching economic threshold levels, their abundance rises over numerous generations. Even minor levels of antixenosis and antibiotic resistance might help these insects postpone the time it takes to reach harmful levels. Antixerotic resistance is expected to slow the initial and subsequent growth of insect populations. Antibiotic resistance reduces insect survival and reproduction while also lengthening generation time, but recovery resistance can help a host plant tolerate damage from an insect population that would harm other cultivars without a tolerance mechanism of resistance. Antixerotic resistance may cause the insect population to migrate to other fields of the same crop or to other crops, resulting in a change in the insect population (Sharma, 1993).

6.6 ASSOCIATION OF BIOLOGICAL CONTROL FOR HPR

Insect-resistant plants are often associated with natural enemies and are ready to control pests. Pest control in conjunction with natural enemies is best suited to moderately resistant species that allow the concentration of pests to remain below economic levels. Enemies of nature not only help control targeted pests but also help reduce overcrowding of other insects in the range of these pests (Maxwell and Davis, 1972). Due to the good balance between targeted pest and its natural enemies, insect-resistant species also improve the efficiency of natural enemies. In plants with a tolerance mechanism, such a combination is very effective (Kogan and Kuhlman, 1982). The insect is vulnerable to predators due to its persistent practise of developing resistant species (Pathak, 1991). Pest sensitivity time targeted to natural enemies is extended by the long growing period of immature stages, leading to the harmonization of insect growth stages with higher activity and more natural enemies, key components of integrated pest control and biodiversity management (Starks *et al.*, 1972). Medium levels of resistance to the chickpea plant combined with natural enemies (e.g., *Campoletis chloride* Uchida) can have a significant effect on *H. armigera* dynamics (Table 6.2). Assuming there are 10 female hens per hectare at the beginning of the season, each female *H. armigera* produces 500 eggs on average, there are three generations at the time of planting, the larval parasite *C. chlorideae* causes 20% of worm deaths per generation (based on Romeis *et al.*, 1997), and worm mortality is 15% in –ICCV 37, 35% in ICCV 2, and 40% in ICC 506. The combination of HPR and biological management reduces the genetic response to several stress insects in choosing to resist plants or natural enemies by combining unrelated causes of death. When they work together, they produce independent deaths when the insect population is low, and the death depends on the density when the insect density is high (Bergman and Tingey, 1979). Selective pressure generated by natural enemies can increase the impact of plant resistance on insect repellent, in addition to the direct and indirect effects of plant resistance on insects (Van Emden, 1990). In general, where pest resistance is achieved with a combination of plant resistance and innate adversaries, only plants exhibit elevated levels of resistance and the rate at which pests adapt to plant-resistant species is minimal (Gould *et al.*, 1991). Tritrophic relationships between plants, target insects, and natural enemies are involved in biological control mechanisms. These interactions include the direct effects of plants on natural enemies, as well as the effect of plant compounds on the target pest and the effect of insect repellent on natural enemies. Plants have probably developed ways to attract natural enemies to reduce insect damage. For example, the female parasitoid female *Campoletis sonorensis* (Cam.) reacts to cotton buds' short-term flexibility when looking for food. It is easy for a wasp to find a place to handle first, followed by a deer at a short distance from the host plant. In response to damage from *H. virescens* and *H. zea*, tobacco, cotton, and corn plants produce a wide variety of variables. *Cardiochiles nigriceps* (Vier.) uses these features to distinguish between *H. virescens* and *H. zea* plant infestations and prefers damaged plants. As a result, the type of host insect interaction is important in defining the parasite number of natural enemies. Biological control is in line with this process. Conversely, the inclusion of secondary plant compounds, such as tomatine, in the insect diet could potentially promote the growth of the parasite. In some cases, changes in host fitness due to insect intake may affect parasitoid growth rate, size, percentage appearance, parasitization success, gender ratio, fecundity, and longevity (Vinson *et al.*, 2015).

6.7 COMPATIBILITY OF PLANT RESISTANCE WITH BIO AGENT'S EFFICACY

Insect-resistant pests have a slow growth rate and a long growing season. Untreated pests are highly vulnerable to natural enemies for a long time, which increases the risk of biotic and abiotic causes of death. Slow-growing insects on the resistant species are better treated by predators than endangered species. In some cases, second-hand plant chemicals that provide insect resistance are associated with the action of natural enemies; for example, *the Cotesia congregate* (Say), a monophagous parasitoid of the *Manduca sexta*, has no adverse effects on nicotine exposure to nicotine *H. virescens* import gossypol from cotton pigment glands, while *C. sonorensis* remains unaffected. When insect larvae are exposed to a restricted quantity of *C. gossypol*, they develop into adults (Williams and Green, 1988). In the case of green pests, plant resistance and biological control are equally beneficial (Starks *et al.*, 1972). The parasitization of cultivar-resistant sorghum cultivars increases with the migration of bugs (Starks and Burton, 1977) because pesticides are not commonly used. The natural enemies of the green bugs are also protected from resistant cultivars. In the presence of natural enemies, lice with low fertility in moderately resistant plants can increase plant resistance (Van Emden, 1965). In the affected field, the greed of a particular hunting animal fails to control daily repetition by an aphid value of 1.2 but is best controlled if the aphid growth rate drops to 1.15 in the resistant cultivar (Bombosch, 1963). Plants with low levels of resistance work well with predators of *B. brassicae* on Brussels sprouts. Low levels of resistance to rice plants IR 46 and Utri Rajapan successfully increase the risk of being eaten by brown spider mites (*Lycosa* spp.). The first instars of *H. virescens* are attacked by a variety of predatory animals and parasitoids, including *C. sonorensis* and *C. nigriceps*, but large worms are rarely attacked. Due to the plants' moderate resistance to low concentrations of antibiosis and nematodes, perhaps parasites, *H. Virescens* is a relatively recent species. Due to overcrowding and/or frequency dependence, the antixenotic effect may reduce or improve the suitability of the insect. Smaller caterpillars are associated with a limited number of food sources due to the avoidance of plants or parts of plants that are anti-antixenotic, and this condition is particularly beneficial to natural enemies when exploring prey (Gould 1991). In the case of sorghum, the correlation between low plant resistance levels and natural enemies has also been documented (Sharma and Davis, 1994). The interaction between plant resistance and biological control is expected to be beneficial in creating effective pest control strategies. *Aphidius matricariae Hal.* is effective in suppressing *M. persicae* in aphid-resistant chrysanthemums. Plants resistant to barley and sorghum are linked to the parasite *Lysiphlebus testaceips* (Cresson), which aids in mitigating the detrimental effects of *Schizaphis graminum* (Rondani), commonly known as green bugs. Genotype IS 10132 and PM 9760 are moderately resistant to pick level of sorghum midge (*S. sorghicola*) parasitization by *Neotrichoporoides* (Trtrastichus) *diplosidis* Crawford than genotypes that are more resistant to IS 3461, IS 7005, IS2 98071, IS2 980AF and 980AF (with colored grain and high tannin content). Nevertheless, there were no significant alterations observed in the parasitization of sorghum midge larvae. On *spotted* stem borer larvae, five parasitoid species have been discovered: *Glyptomorpha deesae* (Cam.), *Cotesia* spp., *Sturmiopsis inferens* (Thn.), *Temelucha* spp., and *Sturmiopsis inferens* (Thn.). Parasitoid activity rose with crop age in general, peaking at 40 days following emergence. Regardless of the time an individual was parasitized, stem borer–resistant genotypes had higher levels of parasitization than susceptible genotypes (Table 6.1).

6.8 INCOMPATIBILITY OF PLANT RESISTANCE WITH BIO AGENT'S EFFICACY

Plant external characteristics sometimes are responsible for pest attack and plant damage. The leaf hairs and trichomes provide an effective plant defense against arthropods, although they also inhibit the movement of natural enemies (Stipanovic, 1983). The genotype of tobacco with hooked trichomes or sparse severely limit *Telenomus* sp. and *Trichogramma minutum* Riley parasitization of tobacco hornworm eggs (Rabb and Bradley, 1968). The presence of trichomes and exudates in pigeon pea and chickpea prevents the egg parasitoid *Trichogramma chilonis* Ishii from parasitizing *H. armigera* eggs (Romeis *et al.*, 1997). The parasitoids *T. pretiosum* Riley and *Chrysopa rufllabris* (Burm.) are less able to detect and parasitize com earworm eggs when the trichome density on cotton leaves is increased. In parasitoids, the pace of development is slowed. Corn earworm larvae were fed fake diets including nicotine, tomatine, and gossypol while being raised (Vinson and Barbosa, 1987). Midge-resistant cereals with colored lettering and high tannin content indicate low N levels. Diplosidis parasitization consists of termite or affected antibodies with white grain and little or no tannins. When *H. zea* eats wild tomato leaves PI 134417, tomatine has adverse effects on egg-eating animals *Coleomegilla maculata* (DeGeer) and *Geocoris puctipes* (Say). Tomatine is absorbed by the endoparasitoid *Hyposoter exiguae* (Vier.). Tomato is one of the host of, *Hyposoter zea* expands the parasitoid larvae stage, reduces jaw closure and adult size, and shortens parasitoid life and fecundity. Composition of secondary plants may extend postembryonic growth or limit the

TABLE 6.1
List of Promising Cultivars Reported as Less Susceptible to Various Pest Species under All India Co-Ordinated Pulses Improvement Project (Lal, 1990)

Crop	Location	Pest species	Cultivars found promising
Mungbean	Durgapur	**Pod borer**	J1, LM 11, P 526, P336
	Badnapur	**Galerucid beetle**	S 9, ML 4, ML 6, B1, Jawahar 45, ML 24, ML 73, Ml 12, Ml 69, PIMS 3, PIMS 4, Gujarat 1
	Kanpur	**Galerucid beetle**	Jawahar 45, 122 S 8, T44, ML 8, ML 5
	Coimbatore	**Thrips, stem borer, pod borer**	Co 3
Urd bean	Badnapur	**Galerucid beetle**	Pusa 1, KG 3, Krishna, T 9, T 27 Pod borer Kalai, 338–3 Jassid Sinkheda 1, Krishna, H 70–3, No. 55, UPB 1
	Kanpur	**Galerucid beetle**	JU 78–3, G 1, H 11
		Pod borer	Krishna
	Coimbatore	**Stem fly**	Karaikal, Killikulum, 338/3, P 58
		Pod borer	No.55
		Thrips, stem fly, pod borer	Co 3, Co 4, Co 5
Cowpea	Durgapur	**Pod borer**	Banswara, G 20, C 55, CR 2–55, P 1461
	Badnapur	**Jassids**	JG 10–72, NS 19–4–1, C 152, 3–779
		Aphid	P 1473, P 1476
	Kanpur	**Galerucid beetle**	5269
		Pod borer	G 7
	Coimbatore	**Aphid**	MS 9369

number of adult parasitoids reaching the host. Plant features can also be modified to improve natural enemy efficacy. Cucumber hairiness, for example, interferes with the parasite wasp *Encarsia formosa* (Gahan)'s biological control of the greenhouse whitefly. Cucumber hybrids now have half the number of hairs, resulting in 30% more *E. Formosa* movement and considerably higher parasitism levels on these experimental hybrids. Similarly, the establishment of pigeon pea lines lacking glandular trichomes may increase *H. armigera* egg parasitization on this crop (Van Emden, 1978; Isenhour and Wiseman, 1987, 1989).

Interaction between plant resistance and insect pathogens, depending on the kind of plant resistance, may help limit the efficacy of insect diseases (Schultz, 1983; Barbosa, 1988). *Bacillus thuringiensis* is more efficient against insect pests that have acclimated to high tannin levels (with a gut pH of 8.0 to 9.5). *B. thuringiensis* pathogenicity is influenced by secondary plant compounds *M. sexta* and *Trichoplusia ni* (Hub.). *Manduca sexta* can withstand greater doses of poisons like nicotine and is immune to entomogenous bacteria. The addition of dietary nicotine, on the other hand, increases insect mortality when *T. ni* is raised on diets containing greater concentrations of *B. thuringiensis.* Corn earworm larvae that eat tomatine from tomato plants lower the virulence of the fungus *Nomuraea ridley.* Maize cultivars resistant to the autumn armyworm *Spodoptera frugiperda* (J.E. Smith) and the nuclear polyhedrosis virus (NPV) have been found to have a synergistic relationship. The infectivity of NPV against *H. zea* is inhibited by rutin and chlorogenic acid. Thus, if antibiosis aspects of host resistance are compatible with insect pathogens, insect pathogens can be more successful in a pest control program (Hamm and Wiseman, 1986).

6.9 CULTURAL PRACTICES FAVOR FOR HPR

Insect-resistant cultivars, such as those that are resistant to pest damage, are extremely beneficial in pest control when used in conjunction with cultural techniques. The combined action of pesticide-resistant plants will have the same effect on population density of the species of insects mentioned. In the post-rainy season, for example, planting M 35–1 sorghum and Phule Yashoda can significantly reduce fly damage. In agro ecosystems, cultural management is a strong technique for controlling insect pests. There are two primary methods to this technique:

- To make the pest's habitat less hospitable.
- To provide a more favorable environment for the pest's natural enemies.

Although cultural management alone may not be enough to bring insect numbers below economic thresholds, it can aid in loss reduction via interaction with plant resistance. Crop resistance combined with cultural management can significantly reduce the need for pesticides. Plant development and pest numbers are no longer compatible: Pest-resistant plant varieties, as well as early planting, early ripening, leaf decay, stem damage, and deep cultivation, can effectively control boll weevil and cotton worms. This not only reduces pest damage but also reduces pests. The number of insects in the winter is extreme, leading to small crop losses in subsequent seasons. Negative cotton plants can reduce the incidence of pink worms by 50% and, when combined with traditional activities, can reduce the invasion by 16 times. The cropping pattern was carefully planned and can be modified so that the most vulnerable part of the plant is not exposed to the action of insects. Worm suppression, a combination of plant resistance and short-term cotton varieties, has proved to be very effective. This can be adjusted so that the most dangerous time of the plant is not exposed to insects. A mixture of plant resistance and pesticides is used to control worms and greatly reduces the need for pesticide application (Glass, 1975). See Table 6.2.

TABLE 6.2
Insect-Resistant Cultivars/Sources of Resistance Identified/Developed for Different Vegetables in India

Vegetables	Resistant to	Cultivars/Varieties
Brinjal	**Coccinellids**	Arka kusumkar, Arka Shirish, Hisar Sel 1, Vijay
	Jassids	Chaklasi Doli, Doli 5, Pusa Purple
	Shoot and fruit borer	Chaklasi Doli, Doli 5, Pusa Purple Long, Pusa Purple Round, SM 67, SM 68, Pant Samrat
	Whitefly	Pusa Purple
Chili	**Aphid**	HC 144, NP-76, UC-45
	Jassids	NP-76, UC-45
	Thrips	G5, K2, X 235, LIC-13, LIC-36, LIC-45, Bonapuri
	Whitefly	HC 144
Tomato	**Fruit borer**	BT 1, T 27, T 32, Punjab Kesari, Punjab Chuhashu, Pant Bahar, Azad Pusa Hybrid 4, DT-10, Parker, Bonus, VFN-8
Okra	**Shoot and fruit borer**	AE-30, AE 57, PMS 8, Parkins Long Green, PKX 9275, Karnal Special, Punjab-7, Reshmi, AC-3375, Sangli-2, Pusa Makhmali, IC-44526
	Whitefly	
	Jassids	Parbhani Kranti
Onion	**Thrips**	Punjab Selection, White Persian, Sweet Spanish, Crystal Wax
Potato	**Potato tuber moth**	QB/A 9-120
Pumpkin	**Fruit fly**	Arka Suryamukhi
Round melon	**Fruit fly**	Arka Tinda
Musk melon	**Fruit fly**	Arka Suryamukhi, Jaunpur Jewel
Bottle gourd	**Fruit fly**	Sultan's Long Green

6.9.1 Trap Crops

The physiological changes caused by cultural processes limit the suitability of plants that host phytophagous insects. Many of these methods are compatible with other pest control methods, such as HPR, and have long been linked to subsistence farming. Insects or other organisms are attracted to host plants, which reduces the incidence of pests on the target plant. Insect damage is minimized by preventing pests from targeting plants or by focusing on a specific area of the field where they can be easily eradicated. Pests of pests show special speculation about certain plant species, cultivars, or categories of plants, compared to resistant strains. Plant stands can be changed from time to time to provide troublesome pests with attractive plants found in an important development area. Insects congregate in the right place where the trap is planted, and as a result, the trap crop attracts insects. Each female moth is thought to lay 500 eggs, at a ratio of 1:1. In each generation, the larval parasite *Campoletis chloride* causes 20% larval mortality, as summarized by Romeis *et al.* (1997). In a cropping season, there are three generations. At the start of the season, the *Helicoverpa armigera* population is estimated to be 10 female moths ha'. In each generation, ICCV 37 has a 15% larval death rate, ICCV 2 has a 35% larval mortality rate, and ICC 506 has a 40% larval mortality rate. *Helicoverpa* population in regions with various cultivars plus natural enemies compared to cultivars alone is 0.51 (Hare, 1983). See Figure 6.2.

6.10 GENETIC DIVERSITY

Pest control can benefit from genetic variety. One approach to increase agricultural diversity is to use polyculture (growing many crops in the same place). Due to interspecific and intraspecific competition with insects and natural enemies, polycultures are ecologically complicated. Reduced predator and parasite populations may result from the removal of other habitats. A rise in the number of insect pest population densities is often lower in polycultures as a result of associational resistance or resource concentration, as well as natural adversaries' actions. Due to the low concentration of their food in the area and the increase in the activity of natural enemies, specialist insects are typically less numerous in a variety of environments. Natural enemies may benefit from plant diversity because it provides different prey, nectar, pollen, and breeding places. Because of the misleading or masking impact of chemical cues from non-host plants, physical obstacles to mobility, and changes in the micro-environment of the tarantula, a specialized insect is less likely to discover its host in varied plant communities. As a result, insect survival in polycultures may be lower than in monocultures. The population density of *Diaphania hyalinata* (L.) is lower in polyculture

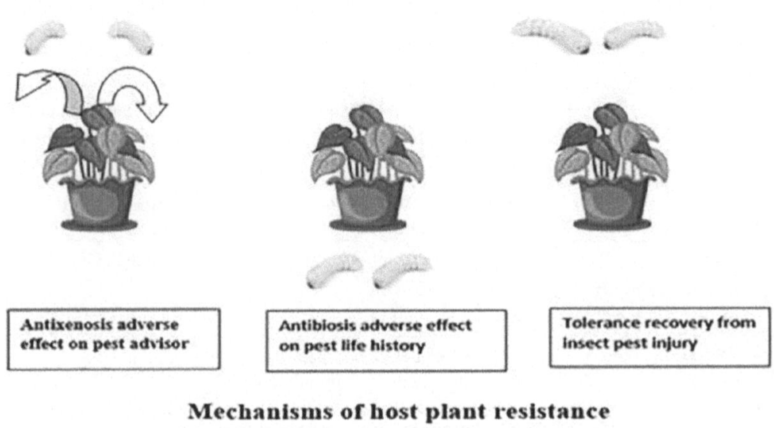

FIGURE 6.1 Mechanism of host plant resistance in insects.

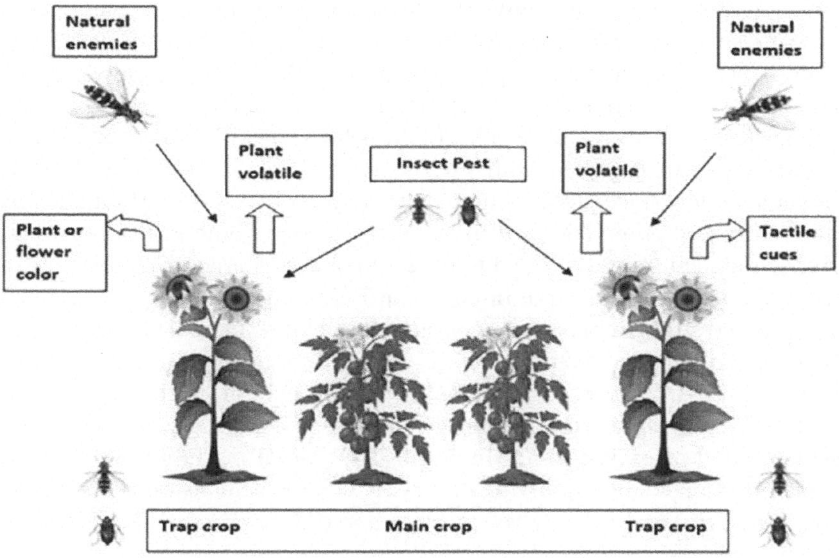

FIGURE 6.2 Role of trap crop to attract insect pests and natural enemies in a farming system.

(corn-cowpea-squash) than monoculture (squash alone). When calculated as an equal proportion of earth, the total plant production in polycultures is high. Also, depending on the variety of herbivores and the interactions between harmful and useful insects, the impact of plant variety on pest damage might alter across time and place (Van Emden, 1965). Since their introduction in 1996, genetically modified crops that produce insecticidal proteins from the soil bacterium *Bacillus thuringiensis* have been widely used in agriculture around the world (Abbas, 2018). This cry gene transformation technology involves the incorporation of specific DNA sequences or genes into crop plants through *Agrobacterium*-mediated transformation or particle bombardment (Juturu *et al.*, 2015). Other strategies for protecting plants from insect attack have also been investigated. Lectins, which are found in a variety of plants, bind to carbohydrates in the mid guts of phytophagous insects, causing the digestive system to malfunction (Vandenborre *et al.*, 2011). Transgenic techniques have also been used to deploy protease inhibitors, which prevent insects from digesting their food (Singh *et al.*, 2020). Similarly, transgenic plants expressing alpha-amylase inhibitors have been developed to be resistant to insects such as lepidopterans, coleopterans, dipterans, and hemipterans. Insect chitinase and chitinase-like proteins are also important in the degrading of chitin in the exoskeleton and gut linings of insects.

6.10.1 Cry Gene

Bacillus thuringiensis (Bt) is a Gram-positive soil bacterium that produces insecticidal crystalline proteins (ICPs) that are extremely toxic to certain pests (Panwar *et al.*, 2018). The genes encoding parasporal crystal protoxins are responsible for insecticidal activity in insect-resistant Bt crops (De Palma *et al.*, 2022). Transgenic plants' ICPs have had a significant impact on the successful evolution of insect resistance. The crystal contains a protoxin protein that is solubilized in the larval mid gut by alkaline pH and then cleaved enzymatically to form an active toxin. The toxin diffuses through the peritrophic membrane that covers the gut and binds to receptors present in the mid gut epithelium, causing pores to form in the mid gut epithelium (Paul and Das, 2021). The pest's gut becomes paralyzed, and he or she stops feeding and dies within two to three days. Bollgard I (BG I), a first-generation Bt cotton expressing cry1Ac, was commercialized and released in 2002 to control dominant bollworms such as *Pectinophora gossypiella*, *Earias vittella*, and *Helicoverpa armigera* in cotton-growing areas of India. Following that, in 2006, Bollgard II (BG II) was introduced as a second-generation Bt cotton with pyramided traits expressing cry1Ac and cry2Ab (MON15985 event) and is now grown in 95% of India's cotton sowing area. In comparison to BG I, which contained only cry1Ac, BG II, which contained multiple toxins, cry1Ac and cry2Ab, has greater pest management ability (Carrière *et al.*, 2015). Dow Agro Sciences approved another transgenic cotton, wide-strike cotton expressing cry1Ac and cry1F, in the United States in 2004. which increased crop yield as well as farmer income. Both BG II and wide-strike cotton expressing pyramided traits are found to have a greater potential to suppress a wide range of lepidopteran, coleopteran, and dipteran insects than BG I. Another transgenic cotton containing cry10Aa demonstrated strong resistance to cotton boll weevil (*Anthonomous grandis*), with 100% mortality observed in bioassay testing when T1 generation larvae consumed transgenic cotton leaves (Ribeiro *et al.*, 2017). Transgenic cotton expressing pyramided traits, cry1Ac and cry2Ab cloned in the T-DNA, was produced using *Agrobacterium*-mediated transformation, and the resulting plant demonstrated resistance to *S. litura* with 93% larval mortality (Siddiqui *et al.*, 2019). Transgenic rice lines created by expressing the cry2AX1 gene demonstrated resistance to the rice leaf folder (*C. medinalis*) and the rice yellow stem borer (*S. incertulas*) (Rajadurai *et al.*, 2018). See Table 6.3.

TABLE 6.3
Transgenic Crops Carrying Bt Genes for Insect Resistance

Sr. no.	Target insects	Transgene	Target crop	References
1	*Chilo suppressalis*, *Cnaphalocrocis medinalis*	cry1a(b)	Rice	Fujimoto *et al.* (1993)
2	*Scirpophaga incertulas* and *Chilo suppressalis*			Wünn *et al.* (1996)
3	*Scirpophaga incertulas*, *Cnaphalocrocis medinalis*			Ghareyazie *et al.* (1997)
4	*Scirpophaga incertulas*	cry1a(c)		Nayak *et al.* (1997)
		cry1a(b)/cry1a(c)		Tu *et al.* (2000)
5	*Cnaphalocrocis medinalis*, *Scirpophaga incertulas*	cry2a/cry1a(c)		Maqbool *et al.* (2001)
6	*Cnephalocrosis medinalis*	cry2A		Gunasekara *et al.* (2017)
7	*Scirpophaga incertulas*, *Cnaphalocrocis medinalis*	cry2AX1		Rajadurai *et al.* (2018)
8	*Helicoverpa armigera*	cry1Ab + NptII	Cotton	Khan *et al.* (2011)
		cry2AX		Sakthi *et al.* (2015)
		cry2AX1		Jadhav *et al.* (2020)
9	*Heliothis* sp.	cry1Ab		Khan *et al.* (2013)
10	*Anthamous grandis*	cry1Aa		Ribeiro *et al.* (2017)
11	*Spodoptera litura*	cry1AC + cry2Ab		Siddiqui *et al.* (2019)
12	*Holtrichia panallele*	cry 8 like	Soya bean	Qin *et al.* (2019)
13	*Achaea janata*, *Spodoptera litura*	cry1AC	Castor	Muddanuru *et al.* (2019
14	*Helicoverpa armigera*	cry1AC + cry2Aa	Pigeon pea	Ghosh *et al.* (2017)
		cry2Aa		Baburao and Sumangala (2018)
		cry2Aa		Singh *et al.* (2018)
15	*Helicoverpa armigera*	cry2Aa	Chickpea	Sawardekar *et al.* (2017)
16	*Tuta absoluta*	cry1Ac	Tomato	Selale *et al.* (2017)
		cry1Ab		Soliman *et al.* (2021)
17	*Spodoptera litura*	cry1Aa	Sweet potato	Zhong *et al.* (2019)

6.10.2 Multiline/Synthetics

Polyphagous insects' feeding preferences can be influenced by combining genetically diverse genotypes of comparable age and height. Mixed cultivation of beetle-resistant wheat varieties has been found to be more effective than pure varieties of any type in controlling the cereal leaf beetle, *Oulema melanoses* (L.). On a regional scale, simulated aphid predator growth on vulnerable plants in variety mixes also decreases the pace of virulent aphid biotype evolution. Varietal combinations and natural enemies working together are quite efficient at reducing pest populations (Wilhoit, 1991).

6.10.3 Vegetative Insecticidal Proteins Genes

The secretion of various proteins during different growth phases, the *Bacillus thuringiensis* bacteria, which can be found in a variety of ecological habitats, has natural entomo-pesticidal properties against a variety of economically important crop pests. Vegetative insecticidal proteins (Vip), which are produced during the vegetative growth phase of the plant and are recognized as an excellent toxic candidate due to sequence homology and receptor site differences from cry proteins, are one of the best-known families of Bt proteins. Vip proteins are classified into three subfamilies. Vip1 and Vip2 heterodimer toxins are effective against *Hemiptera* and *Coleoptera* pests, whereas Vip3, the most extensively studied family of Vip toxins, is toxic to lepidopteran and other insects. VIP proteins are second-generation insecticidal proteins that can be used alone or in combination with cry proteins to control a variety of pest insects (Gupta *et al.*, 2021). These Vip3 proteins are comparable to cry proteins in terms of toxicity potential against susceptible insects. They are reported to be toxic to pests, which cry proteins cannot control. They reduce the population of insect pests through osmotic lysis, which causes swelling and disruption of the mid gut epithelial cells. In transgenic crops such as maize and cotton, the Vip3 proteins have been successfully pyramided alongside cry proteins to overcome resistant pest populations and delay resistance evolution (Syed *et al.*, 2020). The presence of Vip genes that exhibit greater resistance to cotton bollworm (*H. armigera*) and tobacco budworm (*Heliothis virescens*) demonstrates that these genes are an excellent option for insect control. Transgenic cotton containing Vip3A alone, as well as another multitoxin line expressing Vip3A and cry1Ab (Vip Cot), had higher resistance to both *H. zea* and *H. virescens* (Bommireddy *et al.*, 2011).

6.10.4 Protein Fusion

Bt insecticides are widely used, accounting for up to 90% of all microbiological insect control products. As a result, there is a chance that insects will develop resistance to Bt toxins (Tabashnik, 1994; Ferré *et al.*, 1995), In response to repeated exposure to foliar sprays containing Bt proteins, the diamondback moth (*Plutella xylostella*) has evolved resistance in some open field populations. In laboratory selection experiments with other insect pests, recessive mutant alleles can confer resistance to multiple Bt toxins (Perez and Shelton, 1997; Tabashnik *et al.*, 1996). Two recent strategies to address potential limitations in conventional transgenic insect pest control are the stacking or pyramiding of multiple transgenes in the same transgenic plant and the use of hybrid toxins against insect pests. Thus, fusion proteins are created by combining different insecticidal proteins. After transcription and translation in the host plant system, these proteins combine to form single polypeptide units that are more effective against phytophagous insects. Dow Agro Sciences' transgenic cotton lines containing the hybrid fusion protein cry1Be + cry1Fa demonstrated increased resistance to the insects *S. litura* and *O. nubilalis* (Meade *et al.*, 2017). According to Chakraborty *et al.* (2016), transgenic rice containing the cry2AX1 (derivative of cry2Aa and cry2Ac) gene was resistant to lepidopteran insects Koerniati and colleagues (2020). Resistance to phytophagous aphids was achieved through *Agrobacterium*-mediated transformation of transgenic *B. juncea* expressing a fusion protein derived from lectin and a protease inhibitor (Rani *et al.*, 2017). Chang *et al.*

(2017) demonstrated in another case study that transgenic maize containing a cry1Ab/cry2Aj fusion protein in the kernel conferred higher mortality in *S. exigua*, a lepidopteran pest, and *Harmonia axyridis*, a coleopteran pest. Transgenic rice lines expressing cry1Ac + ASAL provided long-lasting and improved resistance to major insects such as yellow stem borer, leaf folder, and brown plant hopper (Boddupally *et al.*, 2018). Cry1Ab + vVp3A pyramiding demonstrated resistance to the rice leaf folder and Asiatic rice borer (Xu *et al.*, 2018). Transgenic rice lines containing the cry2Aa + cry1Ca protein, which is lethal to the Asiatic rice borer, *Chilo suppressalis*, showed a significant level of protein expression (Qiu *et al.*, 2019). Katta *et al.* (2020) investigated whether a transgenic plant created by transferring a triple gene construct containing cry2Ab + cry1F + cry1Ac genes into an elite cotton variety (Narasimha) conferred significant mortality to *H. armigera* and *S. litura* at T2 generation. Table 6.4 depicts an account of studies that used fusion proteins for insect resistance in crop plants for pest management. Pyramiding of two or more genes is a long-term strategy for controlling lepidopteran, coleopteran, and hemipteran insect pests. Furthermore, by delaying the development of insect resistance, these innovations may pave the way for the development of insect-resistant crops.

6.11 CONCLUSION

The development of strategies to test for pest resistance under natural and artificial infestation has made significant progress. The host plant resistance approach against insect pests and its sustainable management offer a promising solution for addressing the challenges of pest control in agriculture. By incorporating natural or induced resistance mechanisms into crop plants, HPR aims to reduce pest damage, decrease reliance on chemical pesticides, and promote sustainable agricultural practices. This approach has gained significant attention due to its potential environmental benefits and long-term viability. Through the utilization of physical, chemical, and physiological traits, crop plants can deter or tolerate insect pests, reducing the need for chemical insecticides. The enhancement of HPR traits through conventional breeding techniques, genetic engineering, and marker-assisted selection has provided valuable tools for developing resistant crop varieties. Host plant resistance represents a valuable approach for sustainable insect pest management in agriculture. By harnessing natural defense mechanisms, crop plants can reduce pest damage and decrease

TABLE 6.4
Fusion Proteins for Insect Resistance in Crop Plants

Sr. no.	Target pest	Transgene	Target pest	References
1	*Scripophaga incertulas*, *Cnaphalocrocis medalis*	cry2AX1 (cry2Aa + cry2Ac)	Rice	Chakraborty *et al.* (2016)
2	*Chilo suppressalis*	cry2Aa + cry1Ca	Rice	Qiu *et al.* (2019)
3	*Ostrinia furnacalis*, *Cnaphalocrocis medinalis*	cry1Ab + vip3A	Rice	Xu *et al.* (2018)
4	*Scirpophaga incertulas*, *Cnaphalocrocis medinalis*, *Nilaparvata lugens*	cry1AC + ASAL	Rice	Boddupally *et al.* (2018)
5	*Lygus* spp.	cry51Aa2	Cotton	Gowda *et al.* (2016)
6	*Spodoptera litura*, *Ostrinia nubialis*	cry1Be + cry1Fa	Cotton	Meade *et al.* (2017)
7	*Helicoverpa armigera*, *Spodoptera litura*	cry2Ab + cry1F + cry1AC	Cotton	Katta *et al.* (2020)
8	*Spodoptera exigua*, *Harmonia axyridis*	cry1Ab/cry2Aj	Maize	Chang *et al.* (2017)
9	*Scirphophaga excerptalis*	cry2Aa + cry1Ca, cry1Ab + cry1Ac	Sugarcane	Koerniati *et al.* (2020)
10	*Lipaphis erysimi*, lentil lectin (LL) and chickpea	protease inhibitor (CPPI) genes	*Brassica juncea*, mustard	Rani *et al.* (2017)

reliance on chemical pesticides. However, the successful implementation of HPR requires continuous research, development, and monitoring to address challenges such as the evolution of insect populations. By integrating HPR with sustainable management practices like IPM, we can achieve effective and environmentally friendly pest control, contributing to the overall sustainability and productivity of agricultural systems.

REFERENCES

Abbas MST (2018) Genetically Engineered (Modified) Crops (Bacillus thuringiensis Crops) and the World Controversy on Their Safety. *Egypt J. Biol. Pest Control*. 28 (1), 1–12. doi:10.1186/s41938-018-0051-2

Baburao TM, Sumangala B (2018) Development and Molecular Characterization of Transgenic Pigeon Pea Carrying cry2Aa for Pod Borer Resistance. *J. Pharm. Phytochem*. 75, 1581–1585.

Barbosa P (1988) Some Thoughts on "the Evolution of Host Range". *Ecology* 69 (4), 912–915.

Barbosa P, Saunders JA, Kemper J, Trumbule R, Olechno J, Martinat P (1986) Plant Allelochemicals and Insect Parasitoids Effects of Nicotine on *Cotesia congregata* (say) (Hymenoptera: Braconidae) and *Hyposoter annulipes* (Cresson) (Hymenoptera: Ichneumonidae). *Journal of Chemical Ecology*. 12, 1319–1328.

Bergman JM, Tingey WM (1979) Aspects of Interaction between Plant Genotypes and Biological Control. *Bulletin of the ESA*. 25 (4), 275–279.

Boddupally D, Tamirisa S, Gundra SR, Vudem DR, Khareedu VR (2018) Expression of Hybrid Fusion Protein (Cry1Ac: ASAL) in Transgenic Rice Plants Imparts Resistance against Multiple Insect Pests. *Sci. Rep*. 8 (1), 8458. doi:10.1038/s41598-018-26881-9

Bombosch S (1963) Untersuchungen zur Vermehrung von *Aphis fabae* Scop. in Samenrübenbeständen unter besonderer Berücksichtigung der Schwebfliegen (Diptera, Syrphidaé) 1. *Zeitschrift für Angewandte Entomologie*. 52 (104), 105–141.

Bommireddy PL, Leonard BR, Temple J, Price P, Emfinger K, Cook D, *et al*. (2011) Field Performance and Seasonal Efficacy Profiles of Transgenic Cotton Lines Expressing Vip3A and Vip Cot against *Helicoverpa Zea* (Boddie) and *Heliothis virescens* (F.). *J. Cotton Sci*. 15, 251–259.

Carrière Y, Crickmore N, Tabashnik BE (2015) Optimizing Pyramided Transgenic Bt Crops for Sustainable Pest Management. *Nat. Biotechnol*. 33 (2), 161–168. doi:10.1038/nbt.3099

Chakraborty M, Reddy PS, Mustafa G, Rajesh G, Narasu VML, Udayasuriyan V, *et al*. (2016) Transgenic Rice Expressing the cry2AX1 Gene Confers Resistance to Multiple Lepidopteran Pests. *Transgenic Res*. 25 (5), 665–678. doi:10.1007/s11248-016-9954-4

Chang X, Lu Z, Shen, Z, Peng, Y, Ye G (2017) Bitrophic and Tritrophic Effects of Transgenic cry1Ab/cry2Aj Maize on the Beneficial, Non Target *Harmonia axyridis* (Coleoptera: Coccinellidae). *Environ. Entomol*. 46 (5), 1171–1176. doi:10.1093/ee/nvx113

De Moraes CM, Lewis WJ, Pare PW, Alborn HT, Tumlinson JH (1998) Herbivore-Infested Plants Selectively Attract Parasitoids. *Nature*. 393 (6685), 570–573.

De Palma M, Scotti R, D'Agostino N, Zaccardelli M, Tucci M (2022) Phyto-Friendly Soil Bacteria and Fungi Provide Beneficial Outcomes in the Host Plant by Differently Modulating Its Responses through (In) Direct Mechanisms. *Plants* 11 (20), 2672.

Emden HV, Wearing CH (1965) The Role of the Aphid Host Plant in Delaying Economic Damage Levels in Crops. *Annals of Applied Biology*. 56 (2), 323–324.

Ferré J, Escriche B, Bel Y, Van Rie, J (1995) Biochemistry and Genetics of Insect Resistance to *Bacillus Thuringiensis* Insecticidal Crystal Proteins. *FEMS Microbiol. Lett*. 132 (1–2), 1–7.

Fujimoto H, Itoh K, Yamamoto M, Kyozuka J, Shimamoto K (1993) Insect Resistant Rice Generated by Introduction of a Modified δ-endotoxin Gene of *Bacillus Thuringiensis*. *Nat. Biotechnol*. 11 (10), 1151–1155. doi:10.1038/nbt1093-1151

Ghareyazie B, Alinia F, Menguito CA, Rubia LG, de Palma JM, Liwanag EA, *et al*. (1997) Enhanced Resistance to Two Stem Borers in an Aromatic Rice Containing a Synthetic cryIA (B) Gene. *Mol. Breed*. 3 (5), 401–414. doi:10.1023/a:1009695324100

Ghosh G, Ganguly S, Purohit A, Chaudhuri RK, Das S, Chakraborti, D (2017) Transgenic Pigeonpea Events Expressing Cry1Ac and Cry2Aa Exhibit Resistance to *Helicoverpa armigera*. *Plant Cell Rep*. 36 (7), 1037–1051. doi:10.1007/s00299-017-2133-0

Glass EH (1975) Integrated Pest Management. *Rationale, Potential, Needs, and Improvement. Annals of Entomological Society America*. 141.

Gould F (1991) The evolutionary potential of crop pests. *American Scientist,* 79 (6), 496–507.

Gould F, Kennedy GG, Johnson MT (1991) Effects of Natural Enemies on the Rate of Herbivore adaptation to Resistant Host Plants. *Entomologia Experimentalis et Applicata*. 58 (1), 1–14.

Gowda A, Rydel TJ, Wollacott AM, Brown RS, Akbar W, Clark TL, *et al.* (2016) A Transgenic Approach for Controlling Lygus in Cotton. *Nat. Commun.* 7 (1), 12213. doi:10.1038/ncomms12213
Gunasekara J, Jayasekera G, Perera K, Wickramasuriya, A (2017) Development of a Sri Lankan Rice Variety Bg 94–1 Harbouring cry2A Gene of *Bacillus thuringiensis* Resistant to Rice Leaf folder *Cnaphalocrocis medinalis* (Guenée). *J. Natn. Sci. Found. Sri Lanka.* 45 (2), 143. doi:10.4038/jnsfsr.v45i2.8180
Gupta M, Kumar H, Kaur S (2021) Vegetative Insecticidal Protein (Vip): Potential Contender from *Bacillus thuringiensis* for Efficient Management of Various Detrimental Agricultural Pests. *Front. Microbiol.* 12, 1139. doi:10.3389/mice2/9/2023.2021.659736
Hamm JJ, Wiseman BR (1986) Plant Resistance and Nuclear Polyhedrosis Virus for Suppression of the Fall Armyworm (Lepidoptera: Noctuidae). *Florida Entomol.* 541–549.
Hare JD, Andreadis, TG (1983) Variation in the susceptibility of Leptinotarsa decemlineata (Coleoptera: Chrysomelidae) when reared on different host plants to the fungal pathogen, Beauveria bassiana in the field and laboratory. *Environmental Entomology,* 12 (6), 1892–1897.
Isenhour DJ, Wiseman BR (1987) Foliage Consumption and Development of the Fall Armyworm (Lepidoptera: Noctuidae) as Affected by the Interactions of a Parasitoid, *Campoletis sonorensis* (Hymenoptera: Ichneumonidae), and Resistant Corn Genotypes. *Environ. Entomol.* 16 (5), 1181–1184.
Isenhour DJ, Wiseman BR (1989) Parasitism of the Fall Armyworm (Lepidoptera: Noctuidae) as Effected by Host Feeding on Silks of *Zea mays* v. Zaplote Chico. *Environ. Entomol.* 18, 394–397.
Jadhav MS, Rathnasamy SA, Natarajan B, Duraialagaraja S, Varatharajalu U (2020) Study of Expression of Indigenous Bt Cry2ax1 Gene in T3 Progeny of Cotton and Its Efficacy against *Helicoverpa armigera* (Hubner). *Braz. Archieves Biol. Technol.* 63 (2), 2020. doi:10.1590/1678-4324-2020180428
Juturu VN, Mekala GK, Kirti PB (2015) Current Status of Tissue Culture and Genetic Transformation Research in Cotton (*Gossypium spp.*). *Plant Cell Tiss. Organ Cult.* 120 (3), 813–839. doi:10.1007/s11240-014-0640-z
Katta S, Talakayala A, Reddy, MK, Addepally, U, Garladinne, M (2020) Development of Transgenic Cotton (Narasimha) Using Triple Gene Cry2Ab[1]Cry1F-Cry1Ac Construct Conferring Resistance to Lepidopteran. *Pest. J. Biosci.* 45 (1), 1–11. doi:10.1007/s12038-020-0006-0
Khan GA, Bakhsh A, Ghazanfar, M, Riazuddin, S, Husnain, T (2013) Development of Transgenic Cotton Lines Harboring a Pesticidal Gene (cry1Ab). *Emir. J. Food Agric.* 25 (6), 434. doi:10.9755/ejfa.v25i6.13133
Khan GA, Bakhsh A, Riazuddin S, Husnain T (2011) Introduction of cry1Ab Gene into Cotton (*Gossypium Hirsutum*) Enhances Resistance against Lepidopteran Pest (*Helicoverpa armigera*). *Span. J. Agric. Res.* 9, 296–302. doi:10.5424/sjar/20110901136-10
Koerniati S, Sukmadjaja D, Samudra IM (2020) C Synthetic Gene of CryIAb-CryIAc Fusion to Generate Resistant Sugarcane to Shoot or Stem Borer. *IOP Conf. Ser. Earth Environ. Sci.* 418 (1), 012069. doi:10.1088/1755-1315/418/1/012069
Kogan M (1986) Natural Chemicals in Plant Resistance to Insects. *Iowa State J. Res.* 60 (4), 501–527.
Kogan M, Kuhlman DE (1982) Soybean Insects: Identification and Management in Illinois. *Bulletin* (University of Illinois at Urbana-Champaign. Agricultural Experiment Station); no. 773.
Lal SS (1990) Insect Pests of Mung, Urid, Cowpea and Pea and Their Management. *Plant Protection in Field Crops* (pp. 185–201), M Veerabhadra Rao and S Sithanantham (eds.). Plant Protection Association of India, Hyderabad.
Maqbool SB, Riazuddin S, Loc N T, Gatehouse AMR, Gatehouse J A, Christou P (2001) Expression of Multiple Insecticidal Genes Confers Broad Resistance against a Range of Different Rice Pests. *Mol. Breed.* 7 (1), 85–93. doi:10.1023/a:1009644712157
Maxwell CA, Hartwig UA, Joseph CM, Phillips DA (1989) A Chalcone and Two Related Flavonoids Released from Alfalfa Roots Induce Nod Genes of *Rhizobium meliloti. Plant Physiology*. 91 (3), 842–847.
Maxwell JA, Davis MB (1972) Pollen Evidence of Pleistocene and Holocene Vegetation on the Allegheny Plateau, Maryland. *Quat. Res.* 2 (4), 506–530.
Meade T, Narva K, Storer, NP, Sheets JJ, Burton SL, Woosley AT (2017) US Patent No. 9,556,453. Washington, DC: US Patent and Trademark Office.
Muddanuru T, Polumetla, AK, Maddukuri L, Mulpuri, S (2019) Development and Evaluation of Transgenic castor (*Ricinus communis* L) Expressing the Insecticidal Protein Cry1Aa of *Bacillus thuringiensis* against Lepidopteran Insect Pests. *Crop Prot.* 119, 113–125. doi:10.1016/j.cropro.2019. 01.016
Nayak P, Basu D, Das S, Basu A, Ghosh, D, Ramakrishnan, NA, *et al.* (1997) Transgenic Elite Indica Rice Plants Expressing CryIAcz-Endotoxin of *Bacillus Thuringiensis* Are Resistant against Yellow Stem Borer (*Scirpophaga Incertulas*). *Proc. Natl. Acad. Sci. USA.* 94 (6), 2111–2116. doi:10.1073/pnas.94.6.2111
Panwar V, Jordan M, McCallum B, Bakkeren G (2018) Host-Induced Silencing of Essential Genes in *Puccinia triticina* through Transgenic Expression of RNAi Sequences Reduces Severity of Leaf Rust Infection in Wheat. *Plant Biotechnol. J.* 16 (5), 1013–1023.

Pathak RS (1991) Plant Genetics in Pest Management. *Int. J. Trop. Insect Sci.* 12 (5–6), 553–564.
Paul S, Das S (2021) Natural Insecticidal Proteins, the Promising Bio-Control Compounds for Future Crop Protection. *The Nucleus* 64, 7–20.
Perez CJ, Shelton AM (1997) Resistance of *Plutella Xylostella* (Lepidoptera: Plutellidae) to *Bacillus Thuringiensis* Berliner in Central America. *J. Econ. Entomology*. 90 (1), 87–93. doi:10.1093/jee/90.1.87
Qin D, Liu XY, Miceli, C, Zhang Q, Wang PW (2019) Soybean Plants Expressing the Bacillus Thuringiensis Cry8-like Gene Show Resistance to *Holotrichia Parallela. BMC Biotechnol.* 19 (1), 1–12. doi:10.1186/s12896-019-0563-1
Qiu L, Fan J, Zhang B, Liu L, Wang X, Lei C, *et al.* (2017) RNA Interference Knockdown of Aminopeptidase N Genes Decrease the Susceptibility of *Chilo suppressalis* Larvae to Cry1Ab/Cry1Ac and Cry1Ca-Expressing Transgenic Rice. *Journal of invertebrate pathology*. 145, 9–12.
Qiu L, Sun Y, Jiang Z, Yang P, Liu H, Zhou H, *et al.* (2019) The Midgut V[1]ATPase Subunit A Gene Is Associated with Toxicity to Crystal 2Aa and Crystal 1Ca-Expressing Transgenic Rice in *Chilo suppressalis. Insect Mol. Biol.* 28 (4), 520–527. doi:10.1111/imb.12570
Rabb RL, Bradley JR (1968) The Influence of Host Plants on Parasitism of Eggs of the Tobacco Hornworm. *Journal of Economic Entomology*. 61 (5), 1249–1252.
Rajadurai G, Kalaivani A, Varanavasiyappan S, Balakrishnan N, Udayasuriyan V, Sudhakar D, *et al.* (2018) Generation of Insect Resistant Marker-Free Transgenic Rice with a Novel cry2AX1 *Gene. Electron. Journ. Plan. Breed.* 9 (2), 723–732. doi:10.5958/0975-928x.2018.00086.8
Rani S, Sharma V, Hada A, Bhattacharya RC, Koundal KR (2017) Fusion Gene Construct Preparation with Lectin and Protease Inhibitor Genes against Aphids and Efficient Genetic Transformation of *Brassica juncea* Using Cotyledons Explants. *Acta Physiol. Plant* 39 (5), 1–13. doi:10.1007/s11738-017-2415-8
Rasmann S, Agrawal AA (2009) Plant Defense against Herbivory: Progress in Identifying Synergism, Redundancy, and Antagonism between Resistance Traits. *Current opinion in plant biology*. 12 (4), 473–478.
Rasmann S, Johnson MD, Agrawal AA (2009) Induced Responses to Herbivory and Jasmonate in Three Milkweed Species. *J. Chem. Ecol.* 35, 1326–1334.
Ribeiro TP, Arraes FBM, Lourenço-Tessutti IT, Silva MS, Lisei-de-Sá ME, Lucena WA, *et al.* (2017) Transgenic Cotton Expressing Cry10Aa Toxin Confers High Resistance to the Cotton Boll Weevil. *Plant Biotechnol. J.* 15 (8), 997–1009. doi:10.1111/pbi.12694
Romeis J, Shanower TG, Zebitz CP (1997) Volatile Plant Info Chemicals Mediate Plant Preference of *Trichogramma chilonis*. *Journal of Chemical Ecology*. 23 (11), 2455–2465.
Sakthi AR, Naveenkumar A, Deepikha PS, Balakrishnan N, Kumar KK, Devi EK, *et al.* (2015) Expression and Inheritance of Chimeric cry2AX1 Gene in Transgenic Cotton Plants Generated through Somatic Embryogenesis. *Vitro Cell. Dev. Biol.- Plant.* 51 (4), 379–389. doi:10.1007/s11627-015-9695-8
Sawardekar SV, Katageri IS, Salimath PM, Kumar PA, Kelkar VG (2017) Standardization of In-Vitro Genetic Transformation Technique in Chickpea (*Cicer Arietinum* L.) for Pod-Borer Resistance. *Adv. Agric. Res. Technol.* 1 (2), 198.
Schoonhoven LM, Jermy T, Van Loon JJA (1998) Host-Plant Selection: How to Find a Host Plant. *Insect-Plant Biol.* 121–153.
Schultz JC (1983) Impact of Variable Plant Defensive Chemistry on Susceptibility of Insects to Natural Enemies *ACS Symposium Series,* 37–54. doi:10.1021/bk-1983-0208.ch003
Selale H, Dağlı F, Mutlu N, Doğanlar S, Frary A (2017) Cry1Ac-mediated Resistance to Tomato Leaf Miner (*Tuta Absoluta*) in Tomato. *Plant Cell Tiss. Organ Cult.* 131 (1), 65–73. doi:10.1007/s11240-017-1262-z
Sharma HC (1993) Host-Plant Resistance to Insects in Sorghum and Its Role in Integrated Pest Management. *Crop Protection*, 12 (1), 11–34.
Sharma HC, Ortiz R (2002) Host Plant Resistance to Insects: An Eco-Friendly Approach for Pest Management and Environment Conservation. *Journal of Environmental Biology*. 23 (2), 111–135.
Sharma YK, Davis KR (1994) Ozone-Induced Expression of Stress-Related Genes in *Arabidopsis thaliana*. *Plant Physiol.* 105 (4), 1089–1096.
Siddiqui HA, Asif M, Asad S, Naqvi RZ, Ajaz S, Umer N, *et al.* (2019) Development and Evaluation of Double Gene Transgenic Cotton Lines Expressing Cry Toxins for Protection against Chewing Insect. *Pests. Sci. Rep.* 9 (1), 1–7. doi:10.1038/s41598-019-48188-z
Singh S, Kumar NR, Maniraj R, Lakshmikanth R, Rao KYS, Muralimohan N, *et al.* (2018) Expression of Cry2Aa, a *Bacillus thuringiensis* Insecticidal Protein in Transgenic Pigeon Pea Confers Resistance to Gram Pod Borer, *Helicoverpa armigera*. *Sci. Rep.* 8 (1), 1–12. doi:10.1038/s41598-018-26358-9
Singh S, Singh A, Kumar S, Mittal P, Singh IK (2020) Protease Inhibitors: Recent Advancement in Its Usage as a Potential Bio Control Agent for Insect Pest Management. *Insect Sci.* 27 (2), 186–201. doi:10.1111/1744-7917.12641

Smith SE, Smith FA (2012) Fresh Perspectives on the Roles of Arbuscular Mycorrhizal Fungi in Plant Nutrition and Growth. *Mycologia* 104 (1), 1–13.

Soliman HIA, Abo-El-Hasan FM, El-Seedy AS, Mabrouk YM (2021) Agrobacterium-Mediated Transformation of Tomato (*Lycopersicon esculentum* mill.) Using a Synthetic Cry1ab Gene for Enhanced Resistance against *Tuta absoluta* (Meyrick). *J. Microbiol. Biotechnol. Food Sci.* 7, 67–74. doi:10.15414/jmbfs.2017.7.1.67-74

Starks KJ, Burton RL (1977) Greenbugs: A Comparison of Mobility on Resistant and Susceptible Varieties of Four Small Grains. *Environmental Entomology,* 6 (2), 331–332.

Starks KJ, Muniappan R, Eikenbary RD (1972) Interaction between Plant Resistance and Parasitism against the Green Bug on Barley and Sorghum. *Annals of the Entomological Society of America.* 65 (3), 650–655.

Stipanovic RD (1983) Function and Chemistry of Plant Trichomes and Glands in Insect Resistance. *ACS Symposium Series,* 69–100. doi:10.1021/bk-1983-0208.ch005

Syed T, Askari M, Meng Z, Li Y, Abid MA, Wei Y, *et al.* (2020) Current Insights on Vegetative Insecticidal Proteins (Vip) as Next Generation Pest Killers. *Toxins (Basel).* 12 (8), 522. doi:10.3390/toxins12080522

Tabashnik BE (1994) Evolution of Resistance to *Bacillus thuringiensis. Annu. Rev. Entomol.* 39 (1), 47–79. doi:10.1146/annurev.en.39.010194.000403

Tabashnik BE, Malvar T, Liu YB, Finson N, Borthakur D, Shin BS, *et al.* (1996) Cross-Resistance of the Diamondback Moth Indicates Altered Interactions with Domain II of *Bacillus Thuringiensis* Toxins. *Appl. Environ. Microbiol.* 62 (8), 2839–2844. doi:10.1128/aem.62.8.2839-2844.1996

Tu J, Zhang G, Datta K, Xu C, He Y, Zhang Q, *et al.* (2000) Field Performance of Transgenic Elite Commercial Hybrid Rice Expressing *Bacillus thuringiensis* δ-endotoxin. *Nat. Biotechnol.* 18 (10), 1101–1104. doi:10.1038/80310

Vandenborre G, Smagghe G, Van Damme EJM (2011) Plant Lectins as Defense Proteins against Phytophagous Insects. *Phytochemistry.* 72 (13), 1538–1550. doi:10.1016/j.phytochem.2011.02.024

Van Emden HF (1965) The Role of Uncultivated Land in the Biology of Crop Pests and Beneficial Insects. *Scientific Horticulture.* 17, 121–136.

Van Emden HF (1978) Insects and Secondary Plant Substances – An Alternative Viewpoint with Special Reference to Aphids. *Biochemical Aspects of Plant and Animal Coevolution.* 309–323.

Van Emden HF (1990) The Interaction of Host Plant Resistance to Insects with Other Control Measures. In *Proceedings of the Brighton Crop Protection Conference, Pests and Diseases* (No. 3, pp. 939–948). British Crop Protection Council Publications.

Van Lenteren JC, De Ponti OMB (1990) Biological Control in Tritrophic System Approach. *Proc. Aphid-Plant Interactions: Populations to Molecules* (pp. 3–28), DC Peters *et al.* (eds.). Shelwater.

Vinson CC, Dal'Sasso TC, Sudré CP, Mangaravite E, de Oliveira LO (2015) Population Genetics of the Naturally Rare Tree *Dimorphandra wilsonii* (Caesalpinioideae) of the Brazilian Cerrado. *Tree Genetics & Genomes* 11, 1–10.

Vinson SB, Barbosa P (1987) Interrelationships of Nutritional Ecology of Parasitoids. *Nutritional Ecology of Insects, Mites, Spiders, and Related Invertebrates* (pp. 673–698). Wiley, New York.

Wilhoit LR (1991) Modelling the Population Dynamics of Different Aphid Genotypes in Plant Variety Mixtures. *Ecological Modelling.* 55 (3–4), 257–283.

Williams K, Field CB, Mooney HA (1989) Relationships among Leaf Construction Cost, Leaf Longevity, and Light Environment in Rain-Forest Plants of the Genus Piper. *The American Naturalist* 133 (2), 198–211.

Williams MH, Green PB (1988) Sequential Scanning Electron Microscopy of a Growing Plant Meristem. *Protoplasma* 147, 77–79.

Wünn J, Klöti A, Burkhardt PK, Biswas GCG, Launis K, Iglesias VA, *et al.* (1996) Transgenic Indica Rice Breeding Line IR58 Expressing a Synthetic cryl A (B) Gene from Bacillus Thuringiensis Provides Effective Insect Pest Control. *Bio Technology.* 14 (2), 171–176.

Xu C, Cheng J, Lin H, Lin C, Gao J, Shen Z (2018) Characterization of Transgenic Rice Expressing Fusion Protein Cry1Ab/Vip3A for Insect Resistance. *Sci. Rep.* 8 (1), 1–8. doi:10.1038/s41598-018-34104-4

Zhong Y, Ahmed S, Deng G, Fan W, Zhang P, Wang H (2019) Improved Insect Resistance against *Spodoptera litura* in Transgenic Sweet Potato by Overexpressing Cry1Aa Toxin. *Plant Cell Rep.* 38 (11), 1439–1448. doi:10.1007/s00299-019-02460-8

7 Sustainable Management Practices in Contrast to Pod Borer, *Helicoverpa armigera* (Hubner), in Chickpea

Sunilkumar Anilkumar Dwivedi and V.K. Sonawane

7.1 INTRODUCTION

Chickpea, *Cicer arietinum* L., is also called Bengal gram or gram, garbanzo, Egyptian pea, or chana. It is a leguminous crop that belongs to the family *Fabaceae*. It is consider the third-most important legume grain after beans (Merga and Haji, 2019). In the world it is enormously cultivated in 45 countries, including India (Sarwar *et.al.*, 2011). Cultivation of gram crop in sandy soil but recent in few years grown on irrigated land after harvesting of rice as rotation of cereal legume to maintain soil nutrients (Baloch and Zubair, 2010). Two types of chickpea viz Desi and Kabuli having differed on the parameter of colour, size, and shape of seeds. It constitutes 59.6 % carbohydrate, 24 % protein, and 3.2% minerals wherein nutritional vitality is about 1507kJ nutritive values for de-husked spilt chickpea are 21.8% protein, 5.8% fat, and 2.9% mineral. It contains a rich source of quality protein known as "Poor man's meat" (Zaman *et al.*, 2012). Cumming and Jenkins (2011) stated that its ability to nitrogen present in the atmosphere and also survive in high temperature during the crop period (Bell, 2014). In the production of chickpea, major restrictions are attack of pest and disease (Nadeem *et al.*, 2011). Pod borer losses caused in production this pest has become major problems to chickpea crop (Akhtar *et al.*, 2011) and losses yield up to 26.01%–40.08% (Sarwar, 2013) and 80 % (Ruttoh *et al.*, 2013). The economic threshold level of *Helicoverpa armigera*/m length of row with one larva (Zahid *et al.*, 2008). According to the ICRISAT (1992) reported in semi-arid tropics that the approximately crop losses about US$328 million. In the universe due to pod borer losses to the crop – i.e., namely vegetables, cotton, cereals, legumes (over US$2 billion), and insecticides required to the managing the insect's pest (approx. over US$1 billion per year) (Reed and Pawar, 1982).

On chickpea crop, two various species of pod borer are observed by the EPPO (2006), namely *H. armigera* and *H. punctigera*, which cause huge damage to the crop. Out of that, the *H. armigera* is the major pod boring pest in chickpea and it is spread in different regions like Oceania, Africa, Mediterranean regions, and Asia; the species *H. punctigera* is limited in Australia (southern part). Outbreak of pod borer was recorded in Italy, Switzerland, Slovakia, Romania, Spain, Hungary, UK, Switzerland, Sweden, etc. according to the CABI (2013) acclimatized global population of various species. Pod borer distribution is in Europe, Portugal, Bulgaria, Greece, Spain, and Romania, and restriction in France, Italy, Cyprus, and Hungary. Because of this pest causes considerable losses which are recorded in South Asia. Such as in India about 10% to 85% losses of yield in chickpea (Yadava and Lal, 1997; Reed, 1983; Ahmed, 1984; Qadeer and Singh, 1989; Das, 1987; Lal *et al.*, 1985a; Kumar and Sarada, 2015; Ahmad et al., 2019 and El Fakhouri *et al.*, 2022), Its damage is vulnerable in Bangladesh and Nepal on chickpea field ranges from 5% to 15% (Musa *et al.*, 1999; Pande and Rao, 2000; Ahmed *et al.*, 1986) and it was (Anonymous, 1987) reported that in the region of northern Pakistan *H. armigera* causes vulnerability to damage to the chickpea fields. Using

DOI: 10.1201/9781003269380-7

sustainable approaches like trap crop, intercropping, sowing date, deep ploughing, etc. has been noted to minimize the damage and survival of *H. armigera* (Romeis *et al.,* 2004). Apart from that if following practices such as land reclamation, introduction of new varieties, use of irrigation, crop rotation with a similar host crop, fertilizer, pest migration, etc. for enhancing the pest population such as pod borer, *H. armigera* (Rivnay, 1962; Talhouk, 1969; Elmosa, 1981; Hariri, 1981; White, 1987; Ahmed and Awan, 2013). Irrigation scheduling helps form new locations that encourage the migration of insects from one habitat to the other (Bhatnagar, 1987). Such areas where pigeon pea and cotton (host of *Helicoverpa armigera*) are cultivated in Central and South India have serious pest problems due to the migration from one host to another. In 2002 in the area of Pakistan the outbreak of pod borer which is reported by chickpea-grown farmers near the cotton field (Anonymous, 2002). According to the study by Anwar and Shafique (1993), the predators and parasites are effective as a bio-control agent against chickpea insect pests. Apart from that management practices used as a genetically modified crop as showing insecticidal gene which is obtained from soil i.e., *Bacillus thuringiensis* crop against larvae of the lepidopteran order (Shelton *et al.,* 2002). Use as botanicals like extracts from neem leaves, NSKE 5% which affects the feeding and survival of the *H. armigera* (Mesfin *et al.,* 2012).

7.2 INSECT PESTS ACT AS MAJOR CONSTRAINT IN CHICKPEA

On chickpea crop about 60 different nuisances feed on it: i.e., Gram pod borer – *H. armigera*, hairy caterpillar – *Spilarctia oblique*, black cutworm – *Agrotis ipsilon* (Hufnagel), *Spilarctia obliqua (Walker*), native budworm *H. punctigera* (Wallengren), leaf miners *Liriomyza cicerina* (Rondani), aphids, *Aphis craccivora* (Koch) chickpea stunt virus, bruchids (*Callosobruchus* spp), etc., are the main nuisance in the Indian region on chickpea crop, and in the Mediterranean region the leaf miner is a major pest on gram crop. In case of the overview of the world due to Bruchids, *Callosobruchus* spp. which causes more damage.

7.3 BIOLOGY OF POD BORER

At an average temperature of 28°C, *Helicoverpa armigera* completes its life cycle (from egg to adult) in four to five weeks (Patil *et al.,* 2017; Jai *et al.,* 2020). Its life cycle includes four main stages: egg, larva, pupa, and adult. Moths are adult insects with large thoraxes and robust bodies. Up to 3,000 eggs can be laid by a female moth. In general, leaves, pods, and flowers are where eggs are laid. Depending on the temperature and the host plant, the oviposition period may continue for 5–24 days, while the incubation time for the eggs is 3–5 days. Freshly placed eggs are yellowish-white when they are first laid, but before they hatch, they turn a shade of brown. Larvae with six different instars are released after hatching (one to six instars caterpillar). Larvae eat first on leaves, young twigs, and flowers, but later on they make holes from the pod base to enter the growing pods (Singh and Singh, 2007). Pre-pupa period occurs in 1 to 4 days. The pupal period typically lasts between 10 and 16 days, although it can take from 6 days at 35°C up to 30 days at 15°C depending on the temperature. It shows a facultative diapause under very low temperature (in winter) and very high temperature (in summer) to survive in difficult environmental circumstances. Pupae exposed to temperatures above 30°C develop into pale-colour adults (Patil *et al.,* 2017). Adult males and females differ in their colour patterns, which are brownish orange and greenish grey.

7.4 HOST RANGE OF POD BORER

The American bollworm is a significant pest that affects a variety of agricultural crops, including vegetables, horticultural plants, and fibre crops (Gajbe, 2020) According to the various researcher describes host ranges of *Heliothis/Helicoverpa* species (Fitt, 1989; King, 1994; Bai *et al.,* 1997) and host range of *H. armigera* is lentil, potato, okra, cotton, chickpea, sunflower, Lucerne, groundnut, sorghum, soybean, field bean, *Phaseolus* spp., tobacco, tomato, pearl millets, linseed, maize, forest trees and vegetable crops. For better larval development plant species, namely *Datura* spp., *Cleome amaranthus* (grain amaranthus), *Acanthospermum* spp., *Hibiscus* sp., *Gomphrena*, *Amaranthus* spp. *Gompherena celosioides*, *Acaylpha* (copper leaf), *Hyposcyamus*, etc. are recorded for host ranges (Table 7.1).

7.5 SUSTAINABLE APPROACHES TO CONTROL GRAM POD BORER, *H. ARMIGERA*

In the sustainable approaches content host plant resistance, varietal resistance, biological control measures, integrated approaches, agronomic practices, etc. are as follows.

7.5.1 Pod and Varietal Attributes Develop Resistance against *H. armigera*

Several structural features characteristics helpful in subsidize to antixenosis have applied to breed pod borer–resistant cultivar. External characters like number of pods per plants, pod length, thickness, trichome length and breadth and density, displayed effect on resistance of pod borer in gram (Ujagir and Khere, 1987; Altaf *et al*., 2008) for the host selection process by the insects is totally depends on the trichome exudates on plant surface. Jeffree (1986), David and Easwaramoorthy (1988), and Peter *et al*. (1995) reported that the kind of trichome, density, orientation and length have been relation to the minimize the insects damage to the various crops and apart from that the other feature are pod borer damage and pod wall thickness showing negative type of correlation, therefore recommended results proves the genotypes having the thick wall of pod express low incidence and loss cause by pod borer. For the resistance mechanism the pod breadth, pod length, and genotype area showing substantial effects with respect to the pod borer. In case of the pod area, pod length, pod borer damage, and pod breadth showing negative correlation and pod borer damage and pod per plant having positive correlation (Table 7.2)

7.5.2 Breeding Resistant Varieties

For the purpose of controlling insect pest developing resistant varieties as an integrated control system against arthropods due to the use of resistance plant which helps the farmers for attack of the pest population on the crop is reduce and no expenditure required to the farmer. Hence, breeding aim recognized attributed and applied genetic component that confers a long resistance to pod borer (i.e., multiple factor resistance). It is easy to grow improved pod borer resistant cultivars if a

TABLE 7.1
Host Range of Gram Pod Borer, *Helicoverpa armigera*

Sr. no.	Host range	Reference
1	*Hyptis sauvelolons* and *Jatropa gossypiifloia*	Wilson (1997)
2	*Canbis sativa*	Dippenaar *et al*. (1996)
3	*Asparagus officinalis*	Kay and Hardy (1999)
4	*Lagacea mollis*	Aherkar *et al*. (1999)
5	Sowthistle (*Sonchus oleracea*)	Gu and Walter (1999)
6	*Striga* spp.	Onu *et al*. (1997)
7	*Paulownia*	Kumar and Ahmad (1998)
8	Cape gooseberry (*Physalis peruviana*)	Mehta *et al*. (1996)
9	*Lathyrus sativus*	Pophaly and Gupta (1996)

TABLE 7.2
Mechanism of Resistance against Gram Pod Borer

Sr. no.	Crop	Type of mechanism	Characteristics
1	Chickpea	Non-preference	Pod shape, pod wall thickness, foliage colour, and glabrousness
2		Antibiosis	Malic acid, oxalic acid, crude fibre, non-reducing sugars, low starch, cellulose, hemicelluloses, lignin in the pod wall, trypsin inhibitors, and HG proteinase inhibitor
3		Escape	Earliness and cold tolerance

good resistance source is available and there is an accurate and realistic screening method which can deliver virtuous choice pressure. Standard selection methods can be adopted based on the genetic system of the crop. Approaches like pedigree, mass, and bulk selection have been auspiciously applied for grow susceptible or sensitive chickpea variety. In the breeding process, several characteristics of quantitative inheritance are pursued – i.e., recurrent selection methods are most suitable for terminate the unacceptable linkages blocks and for the gathering desirable alleles in a single genotypes. To create the new genotypes with the help of mutation breeding which cause the advantageous effects against pod borer resistance (Dua *et al.*, 2005). The resistance mechanisms like antibiosis, antixenosis, and avoidance which plays main role in pod borer resistance in chickpea crop (Clement *et al.*, 1992) (Table 7.3)

Resistance mechanism present in *H. armigera* is polygenic as well as loci may in contrasting segment on several sources of resistance, in the considering that the resistance to pod borer is polygenic and the loci may vary in different segments on different resistance sources. In spite of the fact that absolute resistance is unavailable, ICC 506 has always expressed less damage of pod over the years with increase yields under untreated plots (Gowda *et al.*, 1983). Pod borer–resistant lines are mostly highly susceptible for wilt disease caused by the pathogen *Fusarium oxysporum* f. sp. *citri*. The main issue is in the linkage connecting with pod borer resistance and sensitivity with *Fusarium* wilt has got the better of identification of the lines. Chickpea genotype with respect to the tolerance and resistance to pod borer (Borikar *et al.*, 1982; Dias *et al.*, 1983; Tripathi and Sharma, 1984; Whightman *et al.*, 1995; Hafeez and Kotwal 1996; Patnaik and Mohapatra 1997; Rashid *et al.*, 2003; Shafique *et al.*, 2009; Sarwar, 2013; Mansour and Mohamed, 2014; Choudhary *et al.*, 2015). As per results suggest that most of the genotypes are greatly susceptible to the *Fusarium* spp. resistance of chickpea pod borer doing experiment on the mechanism and heritage of resistance and using of biochemical, molecular, as well as morphological markers will be convenient for cumulative point and varying the basis of *H. armigera* (Sharma *et al.*, 2008) (Table 7.4)

7.6 AGRONOMIC MANIPULATIONS (CULTURAL PRACTICES) AGAINST GRAM POD BORER

7.6.1 Manipulation in Date of Sowing

After intercultural operation of soil, in chickpea crop early date of sowing acts as a significant factor for production and incidence of pests; abiotic parameters like maximum and minimum temperatures, sunshine hours, and wind velocity play a key role in the management of pod borer population. It was accounted for that the case *Helicoverpa armigera* caterpillar numbers are emphatically related with temperature, while humidity and precipitation restrain the larval population (Kumar

TABLE 7.3
RESISTANCE GENOTYPE AND PERCENT OF DAMAGE BY THE POD BORER

Sr. no.	Resistance variety	Percent of damage	Reference
1	ICC 506, ICC 10619, and ICCL 84205	Low resistance	Singh *et al.* (1991)
2	ICCV 7		Singh *et al.* (1997)
3	Vishal and Vijay	Higher resistance	Mishra *et al.* (2023)
4	ICCV 95992	1% damage	ICRISAT (1996)
5	ICC 506	7% damage	ICRISAT (1996)
6	ICC 506EB, ICC 10667, ICC 10619, ICC 4935, ICC 10243, ICCV 95992, and ICC 10817		Lateef (1985); Sharma *et al.* (2005)
7	ICCV 7, ICCV 10, and ICCL 86103		Sharma *et al.* (2005)
8	C-727		Rajput *et al.* (2003)
9	ICCL 86102, ICCL 86111, ICPX 730020-1-1-1H, IPC 94–93, IPC 94–94, and IPC 94–102	High resistance	Singh *et al.* (1990); Rheenen and Van Rheenen (1991); Chaturvedi *et al.* (1998)
10	CH 73/02	High resistance (8.2%)	Nadeem *et al.* (2011)
11	CH 72/02, CH 77/02, and CH 80/02	Moderate resistance	Nadeem *et al.* (2011)
12	CH 79/02, B 17/03, CH 65/02, and CH 60/02	Least resistance	Nadeem *et al.* (2011)
13	Atmore and Flip03–139c	Higher level	Mansour and Mohamed (2014)
14	Mattama, Hawata, Selwa, Wad Hamed, Jebel Marra, Flip03–127c, and Flip04–9c	Moderate resistance	Mansour and Mohamed (2014)
15	ICC 3137	High resistance (51.9 larvae/5 plants)	Pavani *et al.* (2019)

TABLE 7.4
Regional Tolerant Variety against Chickpea Pod Borer

Sr. no.	Tolerant genotypes	Site/Place	Reference
1	ICCV 92311, ICC 3362, and ICCV 95311	Kenya	Mulwa *et al.* (2010)
2	BCX-91040–3, ICCV-95138, and ICCV-96020	Bangladesh	Hossain (2009)
3	Atmore and Flip03–139c	Sudan	Mansour and Mohamed (2014)
4	ICCX 730008–8–1-IP-BP and ICC 506	Myanmar	Ahmed *et al.* (1990)
5	Vijay, ICCV 10, ICCL 86103, Vishal, and ICPL 88034	India	Sharma *et al.* (2014)
6	CM 188/01, CH 07/02, CH 20/02, and CH 84/02	Pakistan	Shafique *et al.* (2009)
7	ICCX 860043-BP and ICCX 900239-BP	Nepal	Thakur (1998)

and Bisht, 2013; Shinde *et al.*, 2013). The *Helicoverpa armigera* invasion on crop is more in cases where the crop is sown late contrasted with the earlier. According to scientists chickpea sowing time is dependent on the larval damage of *Helicoverpa armigera* – i.e., early sowing crop results in low damage (Akhtar *et al.*, 2014) in Indian environments (Garg, 1990; Choudhary *et al.*, 2015; Parmar *et al.*, 2015). If the crop is sowing is late cause the grain yield decreased, demonstrating relation with incidence of pests (Borah, 1998; Singh *et al.*, 2002). According to the Anonymous (2013) doing research results are population of *Helicoverpa armigera* is less in case of the crop is sowing in the month of October as compared to the sowing in November. The peak population of pod borer in northern India during the 10 to 16 standard week (SW) and therefore for the avoiding the peak population of pest, use of short duration variety or early sowing should be permit the maturity of crop. In case of sowing of chickpea is late month can also provide a host to subsequent generation of pod borer and permit a increasing resistance build-up (Cumming and Jenkins, 2011). When the starting of increasing the temperature at that time the moths of pod borer pupa are emerging from the diapausing stage in months of March or April due to that low population of pest with prolonged crop duration particular in hilly region. In April particular (second and third week) the oviposition and development of first instar caterpillar of pod borer takes place and these larvae feed on the leaves and after next generation (instar) of larvae feed on the other parts and its emergence is in the months of April (third and fourth weeks) and thus that time early sown crop pods are well developed and hence cause very small damage by the pod borer to the crop and hence the infestation of pods by the later instar larvae on early sown crops is escape while in late sowing crops attack of pest noticeable. In comparison, population of pod borer is high in the early sown crops (October 15 to November 1) in Bangladesh and in other case late sowing crop less infestation found (November 1 to 30) where the pest population decreased and then increased again. It was detected that both the early (October 15 to November 1) and late sown (December and onward) crops having increasing damage by the pod borer and cause low yield. November 8 to 30 is best sowing date for the less damage of the pod borer and high yield (Hossain *et al.*, 2008; Mapuranga *et al.*, 2015; Genç and Yücel, 2017)

7.6.2 Plant Density/Planting Geometry Effective against Pod Borer

The maximum loss occurred by the *H. armigera* is also affected by plant density. A denser population of plants typically favours increased damaged to pods (Reed et al. 1987; Naresh *et al.* 1989; Qadeer and Singh, 1989; Begum *et al.*, 1992). Crowded plant densities do not actually contribute to losses because the overall less pod formation (Sithanantham and Reed, 1979; Pimbert, 1990). In any event, growers are constrained in their ability to manipulate the population of plants because of inconsistent seed productivity, seed diseases, and unfavourable physical soil conditions for crop development. By developing a micro-climate *Helicoverpa* larvae mostly active in dark area where crowded plant canopy and planting structure are likely to favour pest population multiplication. On chickpea population dynamics of pod borer showed that higher populations of larvae and pupae and higher-density crops having more populations than low population of plants (Sithanantham *et al.*, 1981; Kant *et al*, 2007; Mahmood, 2021). Rising quantity of seed resulted in dense crop structure that promotes a greater population of pod borer. Percentage of pod damage is significantly improved by raising quantity of seed from 75 to 100 kg per ha. In comparison, a major increase in pod damage was not demonstrated by an increase in the seed rate from 100 to 125 kg ha^{-1}. However, more plant population density caused more damage of pod after applying high seed quantity didn't impact on production of chickpea, reimbursing for the enhanced damage of pods (Anilkumar *et al.*, 2011).

7.6.3 Balance Application of Nutrient against *H. armigera*

Fertilizers are mainly used to grow a crop with a high yield; fertilizer application can reflect an indirect impact on incidence of pests. Such influence can be advantageous or detrimental (Coaker, 1987). Significant stage of usage of fertilizers outcome in robust development of plants which making bushy in nature to the plants and making them extra vulnerable to *Helicoverpa armigera*. The bushy aspect of the plant allows the dark-loving pod borer better shelter, causing increased damage to the pod. Minimum to medium Nitrogen, Phosphorus, and Potassium (NPK) fertilizer doses resulted in lower pod borer infestation in chickpeas; however high NPK doses resulted in increasing the infestation of pod borer (Hossain 2009). Enhanced phosphorous quantity significantly decreased the occurrence of *H. armigera* as well as enhanced chickpea yield (Coaker, 1987). Due to uses of fertilizer directly effect on changing plant physiology and for the insects' pest its "active" as a host. Compared with the use of inorganic fertilizers, the use of organic manures such as Farm Yard Manure (FYM), neem base manure, and vermicompost effect less incidence pest population (Rao, 2003; Muddukumar, 2007; Singh and Singh, 2007) there was no major overall impact of rhizobium inoculation, but nitrogen application resulted in a remarkable rising in the damage of pod by target pest (Ramakrishnan *et al.*, 1983).

7.6.4 Effective Adaptation Inter/Mixed Cropping Systems in Pod Borer Management

In addition to its other benefits over single planting, intercropping in the conventional farming method offers protection against pests and abnormal weather conditions. It provides an outstanding incentive for sustainable manoeuvring of the faunal population by carrying in improvements in the crop canopy and the cropping system that can have a direct economic effect on the damage caused by pests. Chickpea intercropping with specific crops has been shown to minimize pod borer damage. That could be the cause of higher numbers of predators harbouring on the companion crop or non-preference for oviposition of *Helicoverpa* in a field containing the intercrop. It should be possible to minimize the efficacy of the host finding actions of the nuisance and to obstruct with its population growth as well as durability by tucking away a plant among other organisms that are unable to provide the same kind of stimulant (Pimbert, 1990).

Compared to chickpea sole crops, intercropping chickpeas with mustard, linseed, and wheat with other non-host crops has found significantly reduction pod loss (Ahmad, 2003; Patil *et al.*, 2018). Similarly, in chickpea plus wheat, a mixed cropping showing result on damage of *Helicoverpa* was reduced by 38.3 compare to the chickpea single cropping pattern (Ali *et al.*, 1998). Incidence of major chickpea pest got delay due to intercropping especially when linseed use as intercrop (Patil *et al.*, 2018, Mehta *et al.*, 1989; Prasad and Kumar, 2002; Lal *et al.*, 2002; Borah *et al.*, 2010). Less incidence caterpillars having maximum production of chickpea is observed in Bengal gram plus mustard, according to (Patil *et al.*, 2018; Reena *et al.*, 2009).

Parasitiod activity is promoted by *Coriandrum sativum* and rich source of nectar plants. Bengal gram intercropped with coriander contained less occurrences of *Helicoverpa* with minimum damage of pod (Sekar *et al.*, 1996; Nath and Chakravorty, 2005; Pandey and Ujagir, 2008; Singh and Pandey, 2014; Chandra *et al.*, 2014). It was found that *Campoletis illota* favoured the old larvae and *C. chlorideae* targeted the young larvae. Parasitization by *C. illota* to the older larvae in the range 7.0% to 12.5% (Nath and Rai, 1999) Oil seed crop such as *Carthamus tinctorius* and *Helianthus annuus* intercropped with chickpea caused decreased damage with the spread of larvae between chickpea and intercrop during the growth phases of chickpea and also due to the enhancement and conservation of bio agents (Sidde Gowda *et al.*, 2004; Pattar *et al.*, 2012).

7.6.5 Crop Rotation

This is another kind of cultural control that most farmers have employed. Rotating the primary crop with other crops, such as cotton may be rotated with other crops like maize, wheat, and soybean, helps interrupt the life cycle of pests and diseases. Crop rotation is beneficial for red and pink bollworms, which have constrained host ranges. Pink bollworms only consume cotton squares and bolls, and hence a widespread conversion to other crops has a huge impact on this pest (Mapuranga *et al.*, 2015).

7.6.6 Effectiveness of Trap Crops for Monitoring Pest Population

Trap crops are cultivated to attract pests from cash crops as a protection mechanism to prevent them against threats. Pests are either prohibited from consuming the crop or are intensive aside from the major crop in some areas of the region. For several types of plants, germplasm, or a specific level of development crop on the concept of trap cropping depend on attack of pests. Plants produce synthetic compounds, or volatiles, that draw in insect pests for repel and pollination of arthropods. Varying degrees of specific volatiles are produced by various species and cultivars, allowing some species or cultivars to beat back insect pests most intensely than others, create them ideal as traps crops. Mainly two methods used in trap crop are a) choice of the most favoured varieties of plants or cultivars cultivated simultaneously as the key crop and b) growing of the same species and cultivar as principal crop, timed prior to the main crop for the favoured level of development. It's imperative that the trap crop is more desirable than the main crop, whether it utilizes the same or a different cultivar. Trap cropping plays an important role in pest management programmes. If trap crops effectively attracted pest incidence, loss of main crops is reduced; thus, insecticide treatment is rarely needed for the main crops. When target pest population is prevalent in trap crop, they may be managed instead of treating the whole field in a concentrated region. The cost of managing crops does not produce economic profits can be greatly outweighed by benefits arising from decreased pest attacks and insecticide use. Their predicted marketable yield is also improved by reduced damage to main crops. The trap crops were grown in varying ratios and boundary rows using either marigold or sunflower (Table 7.5). When the trap crops raise in the ratio of 5:1 to 7:1 which cause the minimized the pod damage by the pod borer (Anonymous, 2009).

7.7 MECHANICAL PRACTICES

7.7.1 Bird Perches: Tools for Reduction Larval Population

A few species of insect-eating birds have been found to benefit from reducing arthropods, including *Helicoverpa armigera* larvae (Chakravarthy, 1988). The larval population in Punjab, India, has been known to decrease by 84%. Among the predatory birds, blue jays, rosy pastor, black drongo, cattle egret, mynah, house sparrows, etc. are common predators in a significant number against pod borer and other lepidoptera order insect pests of *Cajanus cajan*, chickpea, and *Arachis hypogaea* crops (Gokhale and Ameta, 1991). Although the planet is granted a more heretic of avian diversification (Ali and Dillon, 1983), beyond faunistic history, the profitable role of insectophagous birds to management

TABLE 7.5
Different Crops Are Used as Trap Crop for Controlling Insects Pests

Pest to be controlled	Main crop	Trap crop	Location	Reference
Brown plant hopper (*Nilaparvata lugens*)	Rice	Rice	Philippines	Saxena, RC (1982).
Onion thrips (*Thrips tabaci*)	Cotton seedlings	Onion, garlic	Egypt	Yang *et al.*, (2023)
Pod borer (*Helicoverpa armigera*)	Chickpea	Marigold, sunflower	–	Anonymous (2009)
Wireworms (*Agriotes* spp.)	Strawberry	Wheat	Switzerland	Doane *et al.*, (1975)
Serpentine leaf miner (*Liriomyza trifolii*)	Chrysanthemum	Bean	Canada	Herbert *et al.* (1984)
Cutworms (*Euxoa messoria*)	Rye	Tobacco	Ontario	Bucher and Cheng (1970)
Fall armyworm (*Spodoptera jrugiperda*)	Sorghum	Maize	Honduras	Castro, M., *et al.* (1988)
Legume pod borer (*Maruca testulalis*)	Cowpea	*Erotalaria* spp.	Nigeria	Jackai and Singh (1983)

TABLE 7.6
List of Botanical Pesticides with Their Application Rate

Sr. no.	Microbial pesticide	Dose	References
1	Neem oil 1% EC	100 mg azadirachtin/liter	Salman Ahmad *et al.* (2015)
2	NeemAzal	20 ppm	Mehta *et al.* (2010)
3	*Andira paniculata* extract	0.05%	Neto *et al.* (2018)
4	Bioneem	0.09% EC	Abedi *et al.* (2014)
5	Neemarin 0.15% EC	15 mg azadirachtin/liter	Salman Ahmad *et al.* (2015)
6	Jatropa extract	–	Gopalakrishnan *et al.* (2011)
7	Tobacco leaf extract	12.5 g/L	Rahman *et al.* (2014)
8	Mahogany oil	4 ml/L	Rahman *et al.* (2014)
9	Neem seed water	3% EC	Mamoon-ur-rashid *et al.* (2012)

of pest population has not got considerable attention. This mainly for the plant protection scenario, largely due to the extra influence of insecticide (Gopali, 1998; Gopali *et al.*, 2007) that's ensuring sustainable productivity, the establishment and application of environmentally acceptable management of the *Helicoverpa armigera* is of greatest importance. In the chickpea biome, the sunflower served as the most acceptable perch for live birds, as it is a very quickly growing crop and offered robust protection for disembark insect feeder bird's right from the germination stage up to harvesting stage. On the sunflower perch, major prey birds alighted, decreasing the number of larvae within the minimal time. On sunflower plants, predator wasps having a large amount of grubs were reported. The findings of a field experiment found that sunflower and sorghum (50 g seeds/ha) together with Bengal crop seeds had poorer larval population relative to the exclusion of live-bird landing place (Gopali *et al.*, 2009; Mehta *et al.*, 2010). Pattar *et al.* (2012) have confirmed the significance of intercrop with sunflower and chickpea in minimizing the occurrence of damage to the *Helicoverpa armigera.*

7.7.2 Monitoring and Mass Trapping of *Helicoverpa* through Pheromone Traps

A critical part of every integrated pest management (IPM) strategy is to help manage the pest through regular monitoring. A successful management technique often relies on monitoring of the insect damaging stage. In order to predict it is important to recognize the principal factors affecting the pest incidence through monitoring and management. Pheromone traps may be integrated to build protective methods intended to supply data on possible trends of eggs deposition with abundance of pod borer adult population is directly associated with larva numbers (Prabhakar *et al.*, 1998). Hossain (2008) recorded findings at the Pulses Research Centre in Ishurdi, Bangladesh, on the appearance and quickness with timely inconstancy of *H. armigera* moths for the implementation of successful pest management technologies. The findings showed that the number of moths caught steadily increased and peaked in April, then subsequently declined and decreased in the last week of July as well as eventually stuck at zero till the month of December. To efficiently control its population, integrated pest management against *H. armigera* should be implemented in the middle of January. In Nepal, the maximum adult moth of pod borer trapped was recorded in the end of February and the beginning week of April. From SW 45 to 50, the pheromone trap catches were zero, and again from SW 1 to 10 in March, pheromone trap attracts were null, but during the pre-winter months the larval population was noticed. Moth catches/traps began growing from SW 11 and during SW 15 to 16 in April it hit its peak. It declined further until the regular SW 19 to 20 in May, at the time of harvest. An increase population of pests between SW 9 and 15 during the post-winter months has been identified as a normal activity in the northern part of India (Ahmad, 2003). During April to May, higher moth catches and increased larval population in fields were observed (Reena *et al.*, 2009; Pal *et al.*, 2014).

7.8 BIOLOGICAL CONTROL

The effect of both abiotic and biotic variables on pod borer abundance correlates with poorly understood natural enemies. The egg parasitoids, *Trichogramma* spp., due to dense trichomes and their acidic exudates, are almost absent from the chickpea ecosystem in India. The ichneumonide *Campoletis chlorideae* (Uchida) is probably the larval parasitoid most important for pod borer in chickpea in India (Wubneh, 2016). It has also been reported that up to 54% of chickpea larvae are parasitized by *Carcelia illota* (Curran), *Goniophthalmus halli Mesnil*, and *Palexorista laxa* (Curran) (Yadava *et al.*, 1991; King, 1994), though only 3% of chickpea parasitism was recorded by Bhatnagar *et al.* (1983). Assassin bug (*Encarsia sublutea*, *Phonoctonus* spp., *Eretrocerus* spp., and *Aphidius* spp.) effect on egg stage of pod borer (Mapuranga *et al.*, 2015).

7.9 HOST PLANT RESISTANCE

The enhancement of insect-resistant crop varieties has great capability for integrated control of pests, especially in subsistence agriculture situations in underdeveloped nations (Sharma *et al.*, 1999). There is some genetic difference in several crop species, which could be manipulated to create cultivars less vulnerable to pod borer. Breeding for insect resistance, related with disease-resistant varieties, has not been accepted and developed as quickly. This may be partially due to the comparative convenience at which the use of insecticides is done to manage insects. The challenges involved in maintaining sufficient insect pressure for resistance screening have been another explanation for gradual progress in developing insect-resistant varieties. Nevertheless, with the creation of insecticide resistance by insects, the harmful impacts on natural enemies due to insecticides and awareness of community of environmental pollution. Initiation of International Agricultural Research Centres and the evaluation as well as collection of insect resistance seed germplasm have provided renewed impetus to the detection and using of host plant resistance as an essential part of pest control nationwide. Under subsistence farming conditions, Host Plant Resistance (HPR), along with biological and cultural management, is a necessary element of any pest control techniques. Several phonological and morphological characters such as leaf colour, pod shape, crop duration, pod wall thickness, and foliage colour appear to impact the pod borer invasion in chickpea.

7.10 BOTANICAL PRODUCTS AGAINST GRAM POD BORER

Farmers perceive plant products to be sustainable ecosystem, economically effective than conventional insecticides (Muzemu *et al.*, 2011), at the time of production or stored product by grower themselves, it is often difficult to adulterate. Azadirachtin, extracted from the leaves, bark, wood, seed, and fruits of the neem (*Azadirachta indica*), is the most known and widely used plant extract. Azadirachtin has both antifeedant and growth-retardant characteristics and can result in death at a certain point of the lifespan, presumably by dealing with the neuro endocrine regulation of insect metamorphosis (Mehta *et al.*, 2010; Vinutha *et al.*, 2013; Salman Ahmad *et al.*, 2015). The larvicidal, poisonous, repellent, ovicidal, antifeedant, and anti-oviposition effects of neem and garlic extract on insect pests are reported by Zhu *et al.* (2001). Applying NSKE 5% care decreased the density of chickpea pod borers (Gupta, 2007; Pachundkar *et al.*, 2013; Hussain *et al.*, 2016). Biopesticides and insect antifeedant properties of foliage bark along with extract of seed from *Annona squamosa* (McLaughlin *et al.*, 1997; Bisen and Bansal, 2014). Utilization of botanical pesticides with vermiwash is the greatest substitute to pesticides and synthetic chemicals. In India, after spraying of vermiwash plant product neem oil and leaf extract of custard apple gave resulted in a substantial reduction in the proportion of pod damage (Mishra *et al.*, 2013). Combination of animal dung and domestic waste with aqueous garlic extract, vermiwash found effective against infestation of the pod borer. For effective plant growth, crop production, and attack control of pod borer, the vermiwash achieve from buffalo dung and urban solid waste with neem oil and garlic extract was more successful. Integrated delivery of materials dependent on plants and animals such as cow urine (17%) + vitex leaf extract (20%) + clerodendrone extract (4%); GCK (0.5%) + cow urine (30%) + NSKE (10%) + pongamia leaf extract (10%) + aloe (0.5%) + GCA (2%); and vitex leaf extract (20%) + clerodendrone extract (4%) + cow urine (17%) reduced the determined number of larvae with a developed chickpea production (Mallapur and Ladji, 2010; Ladji *et al.*, 2011).

7.10.1 Microbial Insecticides

Entomopathogenic fungi have excellent potential for specificity control in gram pod borer without any adverse effect on activity of predators and useful parasites; utilization of appropriately selected fungi for pest control poses no risk to the health of ecosystems or mammals (Dwivedi and Tomer, 2021) Entomopathogens are effective bio agents against pod borer worldwide and are capable as alternate tools for pest control. Selected plant products can minimize adverse effect on microbes and become essential to became combination of botanical and biopesticide in comparison with their individual action (Reddy and Chowdary, 2021).

7.10.1.1 Bacterial Biopesticides

For the control of population of pod borer, microbial insecticides may be used, and their use can minimize dependency on harmful toxic substances spread into the crop ecosystem, e.g., phytotoxicity, degrading of soil fertility, contamination of air, and harmfulness to birds and mammals. The application of microbial insecticide preparations based on *Bacillus thuringiensis* (Bt) offers an safe solution to the typically harmful broad-spectrum chemical insecticides in the developing world (Ahmed *et al.*, 2013). By adding sufficient concentrations of salts, acids, oils, adjuvants, thuringenesis (exotoxin of Bt), and biopesticides, the potency of Bt may be improved (Ahmed *et al.*, 1989, 1990; Khalique and Ahmed, 2001, 2003). The application of Dipel 2X as well as Dipel ES at 1.6 and 1.5 l ha^{-1}, respectively, at the vegetative stages of crop infestation (first, second, and third instar caterpillar infected) with at least two treatments at weekly intervals resulted in improved production of chickpea (Ahmed *et al.*, 1994; Ahmed, 1999; Ahmed *et al.*, 2012). Bt-based insecticide preparations with namely Dipel, Biobit, and Delfin along with nuclear polyhedrosis virus (NPV) demonstrated minimum damage to the pods (Ahmed *et al.*, 2012). Bt-based insecticides tend to be able to serve as appropriate IPM implements if farmers are made aware of the crucial time and approach for their safe utilization.

7.10.1.2 Viral Biopesticides

Traditional methods of managing pod borer that utilize chemical pesticides develop resistance in pests, pollute the environment, and are dangerous to human health; therefore, new tactics are required to reduce the *H. armigera* population. Larvae of *H. armigera* occurred low infection with NPV it's unique by specific species. Larvae can be decimated without deleterious consequences on all other species if the degree of NPV infection can be increased (Thakur, 1998) applied a 1.5 ml^{l-1} NPV formulation utilization resulted higher crop production, not substantially different from that of a synthetic insecticide (Deltamethrin 2.8 EC added at 1.0 ml^{l-1}), but significantly more than an unsprayed application. In bio-agent and chemical insecticide treatments Geremew *et al.*, (2023) found maximum death of larvae. A 78.7% decrease in the larval population was caused by NPV at 300 LE/ha^{-1} resulting in 10% pod damage and high grain yield (1.86 ton/ha) while the chemical insecticide Endosulfan 35 EC at 1.2 l/ha caused a 70.9% decrease in the number of larvae, resulting in 11.2% pod damage and 1.86 ton/ha^{-1} grain yield. Compared to chemical insecticides and control plot, several other workers have recorded substantial declines in the pod borer larval population and thus less pod damage in chickpea from NPV utilization (Muddukumar, 2007; Magar *et al.*, 2023). An integrated spraying of Panchagavya with NSKE and HaNPV has caused lesser pod injury and formed developed yield of chickpea reported by Muddukumar (2007). *Heliothis virescens ascovirus* 3i (HvAV-3i) is a recently identified virus against lepidopteran larvae. The effects of HvAV-3i on *H. armigera* larvae following oral ingestion of HvAV-3i-containing hemolymph (about 1.0×10^{10} virus genome copies per larvae) and following injection of HvAV-3i-containing hemolymph by insertion of a needle. Following oral ingestion, first-instar to fifth-instar larvae grew and developed normally. Larvae injected with virus-containing hemolymph had extended survival times and were unable to complete the pre-pupal stage (Gong *et al.*, 2020).

7.11 BIOTECHNOLOGICAL METHOD

7.11.1 RNA Interference (RNAi) Technology

The *H. armigera* is a common lepidopteron pest that is particularly well known for its resistance to several common insect poisons. Thus, RNAi is a biotechnological method, begun by dsRNA, for managing this pest by silencing its specific fatal genes. By infusion, ingestion, or ingesting specially developed microbial forms expressing dsRNA, the dsRNA is delivered (Jing and Zhao-jun, 2014). Another biotechnological strategy for pest control is nanotechnology.

7.12 INTEGRATED MANAGEMENT PRACTICES FOR GRAM POD BORER

The excessive use of synthetic pesticides is a result of existing sensitivities related to the pest resurgence, environmental pollution, and human health. Agronomic practices, HPR, biopesticides, botanical, bio agents providing a potentially viable alternative for integrated control of pests. For non-target species and human beings, they are relatively safe. Included in the IPM module are pheromone traps, HaNPV, resistant varieties, NSKE and good agronomic practices, etc., and provides an eco-friendly option to efficient chickpea pod borer control. A most efficient IPM module, which is consisted of pheromone traps for monitoring with mass trapping and augmentation of bio agents (*Trichogramma chilonis* + *Bracon hebetor*) and spraying NSKE caused in the superlative performance for managing *Helicoverpa armigera*, followed by pheromone traps and sequential release of the natural enemies (*T. chilonis* + *B. hebetor*) (Anonymous, 2009). The IPM module applied from sowing of Bengal gram on November 15 and application of first dose HaNPV at 250 Larval Eqivalent/ha at gave result of 100% plant pod development and second dose applied after seven days with cypermethrin at 1 ml^{l-1} ensures good yield. Rahman *et al.* (2014) suggested that in addition to the concurrent release of bio-control agents and spraying HaNPV, the IPM package consisting of pheromone traps revealed the best value for minimizing pod destruction and giving the highest yield, by a greater income.

7.13 CONCLUSION

Production of chickpea sustainability can see success by integrated management of pod borer that utilize a proper integration of practices such as resistant varieties, effective agronomic practices, biological control, cultural methods such as early sowing of a resistant variety with optimum planting density and fertilizer levels and inter/trap crops and installing animated bird perches, along with chickpea seeds with T-perches at 2 m predatory zones, are optimum for high chickpea yields with sustainable control of *H. armigera*. Any single practice for pest control may not be highly effective; hence, the best alternative is the IPM approach, which is based on the principles of management of the pest rather than aiming at its complete eradication. In view of this, the present chapter concludes that the use of IPM options, along with other management practices such as growing resistant varieties, good agronomic practices, biological control, chemical control (if necessary), behavioural approaches, etc. reduce the negative impact of insecticides on the natural enemies that are present in the suitable ecological niche and will protect the ecosystem.

7.14 FUTURE SCOPE

This chapter revealed that monetary returns are directly affected the market value of chickpea seed as well as cost of production during previous year utilize IPM component. It also encouraging farmer to adopt this technology due to highly appreciable resulting effective high benefit cost ratio with eco-friendly management of pod borer with toxic-residue-free crop production.

7.15 CONFLICT OF INTEREST

Authors are declaring that they have no conflict of interest with respect to research, authorship, and publication.

7.16 ACKNOWLEDGEMENT

Authors wish to acknowledge the valuable efforts of reviewers to enhance the aspect, consistency, and content presentation.

REFERENCES

Abedi Z, Saber M, Vojoudi S, Mahdavi V, Parsaeyan E (2014) Acute, sublethal, and combination effects of *Azadirachtin* and *Bacillus thuringiensis* on the cotton bollworm, *Helicoverpa armigera. Journal of Insect Science*, 14(30): 1–9.

Ahmad A, Husain A, Mujeeb M, Khan SA, Najmi AK, Siddique NA, ... Anwar F (2013) A review on therapeutic potential of Nigella sativa: A miracle herb. *Asian Pacific journal of tropical biomedicine*, 3(5), 337–352.

Ahmad M, Rasool B, Ahmad M, Russell, DA (2019) Resistance and synergism of novel insecticides in field populations of cotton bollworm *Helicoverpa armigera* (Lepidoptera: Noctuidae) in Pakistan. *J. Econ. Entomol.*, 112: 859–871.

Ahmad R (2003) Insect-pests of chickpea and their management. *Chickpea Research in India*, 229–260.

Ahmad Salman, Sha M, Muslim M (2015) Toxic effects of neem based insecticides on the fi tness of *Helicoverpa armigera* (Hübner). *Crop Protection*, 68: 72–78.

Ahmed K (1984) Research on pulses entomology at national agricultural research center, Islamabad, Proceedings of the group discussion: Pulse pests management, 05–10 December 1983, ICRISAT Center, India. Patancheru A.P. 502 324, India, pp. 28–30.

Ahmed K, Awan MS (2013) Integrated management of insect pests of chickpea *Cicer arietinum* (L. Walp) in South Asian countries: Present status and future strategies-a review. *Pak. J. Zool.*, 45(4): 1125–1145.

Ahmed K, Khalique F, Afzal M, Malik BA (1986) Pulses entomology report. Food legumes improvement programme, Pakistan Agricultural Research Council, National Agricultural Research Center, pp. 4–10.

Ahmed K, Lal SS, Morris H, Khalique F, Malik BA, Waloby BJ, Hall SD (1989, December) Insect pest problems and recent approaches to solving them on chickpea in South Asia. In *Chickpea in the nineties: Proceedings of 2nd International Workshop on chickpea improvement* (pp. 4–8).

Ahmed K, Lal SS, Morris H, Khalique F, Malik BA (1990) Insect-pest problems and recent approaches to solving them on chickpea in South Asia. In: Proceeding of Second International Workshop on Chickpea Improvement, 4–8 December 1989, ICRISAT, Patancheru, India, pp. 165–168.

Ahmed K, Khalique F, Malik BA, Riley D (1994) Special Reference to Control of Chickpea pod Borer Helicoverpa (Heliothis) Armigera (Hübner). *Subtropical Plant Science,* 46, 38–44.

Ahmad M (1999) Initial frequencies of alleles for resistance to Bacillus thuringiensis toxins in field populations of Plutella xylostella and Helicoverpa armigera (Doctoral dissertation).

Akhtar LH, Pervez MA, Nasim M (2011) Genetic divergence and inter relationship studies in chickpea *Cicer arietinum* L. *Pak. J. Agri. Sci.*, 48: 35–39.

Ahmed K, Khalique F, Durrani SA, Pitafi KD (2012) Field evaluation of bio-pesticide for control of chickpea pod borer Helicoverpa armigera, a major pest of chickpea crop. *Pakistan Journal of Zoology,* 44(6), 1555–1560.

Akhtar MF, Ahmed I, Nadeem I, Abbas Q, Raza A, Yousaf MJ, Ahmed R, Niaz T (2014) Impact of different dates of sowing on gram pod borer (*Helicoverpa armigera*) infestation in chickpea crop. *World J. Zoo.*, 9(4): 270–275. doi:10.5829/idosi.wjz.2014.9.4.86153.

Ali A, Choudhury RA, Ahmad Z, Rahman F, Khan FR, Ahmad SK (2009) Some biological characteristics of Helicoverpa armigera on chickpea. *Tunisian Journal of Plant Protection,* 4(1), 99–106.

Ali M, Agarwal SC, Rathore YS, Mishra JP (1998) Agronomic options in the management of Biotic stresses in pulse crops. In: Proceedings of National Symposium on "Management of Biotic and Abiotic Stresses in Pulse Crops"- Kanpur, June 26–28, 1998.

Ali M, Kumar (2006) Pulse Production in India. Yojna. September, pp. 13–15.

Ali S, Dillon RS (1983) A pictorial guide to the birds of the Indian Subcontinent. Bombay natural history, Oxford, pp. 176–188.

Anilkumar R, Nandan B, Sharma JP, Kumar J (2011) Effect of phosphorus and seed rate on growth and productivity of bold seeded Kabuli chickpea in subtropical Kandi areas of Jammu and Kashmir. *Plant Archives,* 10(1): 125–129.

Anonymous (1987) Annual Technical Report, Food Legumes Improvement Project in Pakistan. Pakistan Agricultural Research Council (PARC), National Agricultural Research Center, pp. 121–133.

Anonymous (2002) Annual report, food legumes improvement programme. Pakistan Agricultural Research Council, National Agricultural Research Center, Islamabad, Pakistan.

Anonymous (2009) Annual report 2008–09, All India Coordinated Research Project on Chickpea (AICRPC), Directorate of Pulse Research, Kanpur, India.

Anonymous (2013) Annual report 2012–13. Indian Institute of Pulses Research (IIPR), Kanpur.

Bai LX, Sun HW, Sun YW, Su CE (1997) Study on host plant species of cotton bollworms and their fitness. *Acta Phytophylacica Sinica,* 24: 1–6.

Baloch M, Zubair M (2010) Effect of nipping on growth and yield of chickpea. *Journal of Animal and Plant Sciences,* 20(3): 208–210.

Begum N, Husain M, Chowdhury SI (1992) Effect of sowing date and planting density on pod borer incidence and grain yield of chickpea in Bangladesh. *Int. Chickpea Newsletter,* 27: 19–21.

Bell S (2014) The small but mighty chickpea. Phys.org., 8 October 2015.

Bhatnagar VS, Sithanantham S. Pawar CS, Jadhav DR, Rao VR, Reed W (1983) Conservation and augmentation of natural enemies with reference to integrated pest management in Chickpea (Cicer arietinum L.) andPigeonpea (Cajanus cajan (L.) millsp.); Proc. lnt. Workshop in Integrated Pest Control for Grain Legumes, *EMBRAPA Centro Nacional de Pesquisa-arroz,* Goiania, Brazil, pp 157–180.

Bhatnagar VS (1987) Conservation and encouragement of natural enemies of insect-pests in dry land subsistence farming: Problems, progress and prospects in the Sahelian Zone. *International Int. J. Trop. Insect Sci.,* 8: 791–795. doi:10.1017/S1742758400022931.

Bisen SK, Bansal SK (2014) Feeding deterrent activity of certain plant extracts against *Helicoverpa armigera* pest of *Cicer arietinum. JCBPS,* 4(4): 3296–3300.

Borah BK, Debnath MC, Sharma KK, Das B (2010) Effect of intercropping on incidence of gram pod borer, *Helicoverpa armigera* Hubner in chickpea. *Insect Environ.,* 15(4): 8–9.

Borah RK (1998) Influence of sowing dates on the infestation of *Helicoverpa armigera* and grain yield of chickpea (*Cicer arietinum*) in the hill zone of Assam. *Indian J. Entomol.,* 60(3): 416–417.

Borikar PS, Madansure AN, Jambhale ND, Gite ND, Missal MB (1982) Damage caused by *Heliothis armigera* Hubner (Lepidoptera: Noctuidae). *J. Entomol.,* 44: 290–292.

Bucher GE, Cheng HH (1970) Use of trap plants for attracting cutworm larvae. *Can. Entomol.,* 102: 797–798.

CABI (2013) Crop protection compendium, 2013 edition. CAB International Publishing, Wallingford.

Castro M, Pitre H, Meckenstock D (1988) Potential for using maize as trap crop for the fall armyworm, *Spodoptera Jrugiperda* (Lepidoptera: Noctuidae), where sorghum and maize are intercropped on subsistence farms. *Fla. Entomol.,* 71: 273–278.

Cavallito CJ, Balley JH (1944) Allicin, the antibacterial principle of Allium sativum. I. Isolation, physical properties and antibacterial action. *Journal of the American Chemical Society,* 66(11), 1950–1951.

Chakravarthy AK (1988) Bird predators of pod borers of field bean (*Lablab niger*). *Trop. Pest Manag.,* 34: 395–398. doi:10.1080/09670878809371285.

Chandra S, Rachappa V, Yelshetty S, Sreenivas AG, Biradar SA (2014) Performance of intercrops in reduction of gram pod borer, *Helicoverpa armigera* (Hubner) incidence on chickpea. *J. Exp. Zool.,* 17(2): 627–630.

Chaturvedi SK, Gurha SN, Sewak S, Ahmed R, Dikshit HK, Bahduoria P (1998) Possible combined resistance against *Fusarium* wilt and pod borer in chickpea (*Cicer arietinum* L.). *Indian Journal of Pulses Research,* 11: 117–119.

Choudhary OM, Anwala R, Sharma MM (2015) Studies on varietal screening and date of sowing of chickpea, *Cicer arietinum* (L.) against *Helicoverpa armigera* (Hub.). *Eco-Friendly Agric. J.,* 10(1): 58–61.

Clement SI. El-Din, Sharaf EIDN, Weigand S, Lateef SS (1992) Screening techniques and sources of resistance to insect-pests of cool season food legumes. (In En) (Summaries in En.) Program and abstracts.

2nd International Food Legume Research Conference, Cairo, Egypt, 12–16 April 1992. Cairo, Egypt. International Food Legume Research Conference, Organizing Committee, pp. 22.

Clement SL, Sharaf Eidihn N, Weigand S, Lateef SS, (1994) Research achievements in plant resistance to insect pest of cool season food and legumes. *Euphytica*, 73(1–2): 41–50.

Coaker TH (1987) Cultural methods. In: Burn AJ, Coaker TH, Jepson PC (eds) The crop in integrated pest management. Academic Press, London, pp. 69–88.

Cumming G, Jenkins L (2011) Chickpea: Effective crop establishment, sowing window, row spacing, seeding depth and rate. *Northern Pulse Bulletin*, 7: 6.

Das GP (1987) Efficacy of neem oil on the eggs and grub mortality of *Callosobruchus chinensis* Linn. (Bruchidae: Coleoptera). *Tropical Grain Legume Bulletin*, 34: 14–15.

David H, Easwaramoorthy (1988) Physical resistance mechanisms in insect plant interactions. In: Ananthakrishnan TN, Raman A (eds) Dynamics of insect plant in and future trends. Oxford and IBH Publication, New Delhi, pp. 45–70.

Dias CAR, Lal SS, Yadava CP (1983) Differences in susceptibility of certain chickpea cultivars and local collection to *Heliothis armigera* (Hubner). *Indian J. Agric. Sci.*, 53: 842–845.

Doane JF, Lee YW, Klingler J, Westcott ND (1975). The Orientation Response of Ctemcera Destructor and Other Wire Worms (Coleoptera: Elateridae) to Germinating Grain and to Carbon Dioxide1. *The Canadian Entomologist,* 107(12), 1233–1252.

Dua RP, CLL G, Shivkumar, Saxena KB, Govil JN, Singh BB, Singh AK, Singh RP, Singh VP, Kranthi S (2005) Breeding for resistance to *Heliothis/Helicoverpa*: Effectiveness and limitations. In: Sharma HC (ed) *Heliothis/Helicoverpa* management: Emerging trends and strategies for future research. Oxford and IBH Publishers, New Delhi, pp. 223–242.

Dua RR, Gowda CL, Kumar S, Saxena KB, Govil JN, Singh BB, Kranthi S (2005) Breeding for resistance to Heliothis/Helicoverpa: Effectiveness and limitations. In *Heliothis/Helicoverpa Management* (pp. 235–254). CRC Press.

Dwivedi SA, Tomer A (2021). Application of mycobiocontrol agent in biodergradation and pest management. *Mycoremediation and Environmental Sustainability: Volume 3*, 247–265.

El Fakhouri K, Boulamtat R, Sabraoui A, El Bouhssini M (2022) The chickpea pod borer, *Helicoverpa armigera* (Hübner): Yield loss estimation and biorational insecticide assessment in morocco. *Agronomy*, 12(12): 3017.

Elmosa H (1981) FAO/UNEP Near eastern inter-country program for the development and application of integrated pest control in cotton growing, 1980 report, p. 44.

EPPO (2006) Distribution maps of quarantine pests, *Helicoverpa armigera*. On-line available at www.eppo.org/QUARANTINE/insects/Helicoverpa_armigera/HELIAR_map.htm

FAOSTAT (2015) http://faostat.fao.org/site/567/default.aspx#ancor, Accessed 4 June 2015.

Fitt GP (1989) The ecology of *Heliothis* sps in relation to agro ecosystems. *Annu. Rev. Entomol.*, 34: 17–52.

Gajbe PU (2020) Integrated pest management strategies for the management of global pest, *Helicoverpa armigera*. Recent Trends in, 1, 45.

Garg DK (1990) Effect of sowing date on the incidence of *Helicoverpa armigera* (Hub) and yield of chickpea in Kumaon Hills, India. *Int. Chickpea Newsletter*, 23: 14–15.

Genç H, Yücel S (2017) Observation of *Helicoverpa armigera* Hübner (Lepidoptera: Noctuide) Infestation on *Gladiolus grandiflorus* (Iridaceae) in Çanakkale. *ÇOMÜ Zir. Fak. Derg.*, 5(1): 105–114.

Geremew D, Shiberu T, Leta A (2023) Isolation, morphological characterization, and screening a virulence of Beauveria bassiana and Metarhizium robertsii as bioagents. *F1000Research,* 12, 827.

Gokhale VG, Ameta OP (1991) Predatory behaviour of house sparrow, *Passer domesticus* L. in the population regulation of *Heliothis sp.* infesting chickpea, *Cicer arietinum. Indian J. Entomol.*, 53: 631–634.

Gong Chen, Hang Liu, Bo-Cheng Mo, Jue Hu, Shuang-Qing Liu, Carlos Bustos-Segura, Jing Xue, Xing Wang (2020) Growth and development of *Helicoverpa armigera* (Lepidoptera: Noctuidae) larvae infected by *Heliothis virescens* ascovirus 3i (HvAV-3i). *Front. Physiol.*, 1–7.

Gopalakrishnan S, Rao GVR, Humayun P, Rao VR, Alekhya G, Jacob S, Deepthi K, Vidya MS, Srinivas V (2011) Efficacy of botanical extracts and entomopathogens on control of *Helicoverpa armigera* and spodoptera litura. *African Journal of Biotechnology*, 10(73): 16667–16673.

Gopali JB (1998) Studies on integrated pest management of pigeon pea pod borer, *Helicoverpa armigera* (Hubner) with special reference to HaNPV and insectivorous birds. PhD Thesis, University of Agricultural Sciences, Dharwad, India.

Gopali JB, Teggelli R, Mannur DM, Yelshetty S (2009) Bird perches for sustainable management of pod borer, *Helicoverpa armigera* (Hubner) in chickpea ecosystem. *Karnataka J. Agric. Sci.*, 22(3-Spl. Issue): 541–543.

Gopali JB, Yelshetty S, Teggelli R, Mannur DM (2007) Eco-friendly management of pod borer, *Helicoverpa armigera* (Hubner) by encouraging predatory birds in chickpea ecosystem. In: National Symposium on Legumes for Ecological Sustainability: Emerging Challenges and Opportunities, 3–5 November 2007, Indian Institute of Pulses Research, Kanpur, p. 168.

Gowda CLL, Lateef SS, Smithson JB, Reed W (1983) *Breeding for resistance, to Heliothis armigera in chickpea.*

Gupta MP (2007) Management of gram pod borer, *Helicoverpa armigera* (Hubner) in chickpea with bio rational. *Nat. Prod. Radiance*, 16(5): 391–397.

Hafeez A, Kotwal DR (1996) Screening of chickpea varieties against gram pod borer, *Helicoverpa armigera* Hubner. *Ann. Plant Prot. Sci.*, 4: 171–172.

Hariri G (1981) The problems and prospects of *Heliothis* management in Southwest Asia. In: Proceedings of International Workshop on *Heliothis* Management. 15–20 November 1981, ICRISAT, India, pp. 369–373.

Herbert HJ, Smith RF, McRae KB (1984) Evaluation of Non-Insecticidal Methods to Reduce Damage to Chrysanthemums by the Leafminer Liriomyza Trifolii (Diptera: Agromyzidae) 1. *The Canadian Entomologist,* 116(9), 1259–1266.

Hossain MA (2003) Management of chickpea pod borer, Helicoverpa armigera (Hubner) through intercroppings and insecticide spraying. *Thai Journal of Agricultural Science* (Thailand), 36(1): 51–56.

Hossain MA (2008) Monitoring and evaluation of chickpea pod-borer *Helicoverpa armigera* (Hubner) (Lepidoptera: Noctuidae) by using pheromone traps. *BJSIB*, 43(3): 419–426.

Hossain MA (2009) Field screening of chickpea genotypes against pod borer. *BJSIB*, 34(3): 517–521.

Altaf H, Azizul H, Prodhan MZH (2008) Effect of pods characteristics on pod borer (Helicoverpa armigera, Hubner) infestation in chickpea. *SAARC Journal of Agriculture,* 6(1), 51–60.

Hossain MA, Haque MA, Prodhan MZH (2008) Incidence and damage severity of pod borer, *Helicoverpa armigera* (Hubner) in chickpea (*Cicer arietinum* L.). *J. Agric. Res*, 4: 221–224.

Howard R (1998) A winter record of *Helicoverpa armigera* (Hubner) (Lepidoptera: Noctuidae). *Entomologist's Gazette*, 49: 237–238. International Crops Research Institute for the Semi-Arid Tropics (ICRISAT).

Hussain M, Ahmad KS, Majeed M, Mehmood A, Hamid A, Yousaf MM, Chaudhry MS, Shah MJ, Hussain K, Ahmad B, Khan AQ (2016) Integrated management of *Helicoverpa armigera* on different genotypes of Kabuli chickpea in Punjab, Pakistan. *Int. J. Bio. Sci.*, 9(2): 110–119.

ICRISAT (1992) Mid-Term plan. International Crops Research Institute for Semi-Arid Tropics (ICRISAT), Patancheru.

ICRISAT (1996) Annual report, ICRISAT, Patancheru 502 304, Andhra Pradesh, India.

Jackai LE, Singh SR (1983) Varietal resistance in the integrated pest management of cowpea pests. *International Journal of Tropical Insect Science,* 4, 199–204.

Jai BK, Soni VK, Chandraker HK (2020). Surveillance of pod borer, *Helicoverpa armigera* (Hubner) and its natural enemies on chickpea at Sahaspur Lohara blocks. *J. Pharm. Phytochem.*, 9(3): 1995–2000.

CE J (1986) The cuticle, epicuticular waxes and trichomes of plants, with reference to their structure, functions and evolution. *Insects on the plant surface,* 23–64.

Jing Y, Zhao-jun HAN (2014). Efficiency of different methods for dsRNA delivery in Cotton Bollworm (*Helicoverpa armigera*). *Journal of Integrative Agriculture*, 13(1): 115–123. doi: 10.1016/S2095-3119(13)60511-0.

Kant K, Kanaujia KR, Kanaujia S (2007) Role of plant density and abiotic factors on population dynamics of *Helicoverpa armigera* (Hübner) in chickpea. *Ann. Plant Prot. Sci.*, 15(2): 303–306.

Khalique F, Ahmed K (2001) Synergistic interaction between Bacillus thuringiensis (Berliner) and Lambda-cyhalothrin (Pyrethroid) against, chickpea pod borer, Helicoverpa armigera (Huebner). *Pak. J. Biol. Sci,* 4(9), 1120–1123.

Khalique F, Ahmed K (2003) Impact of Bacillus thuringiensis subsp. kurstaki on biology of Helicoverpa zea armigera. *Pak. J. Biol. Sci,* 6, 615–621.

King ABS (1994) *Heliothis/Helicoverpa* (Lepidoptera: Noctuidae). In: Matthews GA, Tunstall JP (eds) Insect Pests of Cotton. CAB International, Wallingford, UK, pp. 39–106.

Kumar GVS, Sarada O (2015) Field efficacy and economics of some new insecticide molecules against lepidopteran caterpillars in chickpea. *Curr. Biot.*, 9: 153–158.

Kumar L, Bisht RS (2013) Population dynamics of *Helicoverpa armigera* (Hubner) on chickpea crop. *Pantnagar J. Res.*, 11: 35–38.

Kumar M, Ahmad M (1998) Noctuid infestation to the introduced species of Paulownia. *Ann. For. Sci.*, 6: 177–185.

Ladji R, Mallapur CP, Ambika DS, Amitha K, Rudraswamy SM, Thimmegowda PR (2011) Management of chickpea pod borer, *Helicoverpa armigera* (Hubner) using indigenous materials. *Int. J. Nat. Sci.*, 2(2): 263–265.

Narayanamma VL, Sriramulu M, Gowda CLL, Ghaffar MA, Sharma HC (2007) Tolerance to Helicoverpa armigera damage in chickpea genotypes under natural infestation. *Indian Journal of Plant Protection,* 35(2), 227–231.

Lal OP, Sinha SR, Srivastava YN (2002) Role of intercropping on population buildup of *Lipaphis erysimi* on cabbage under field conditions. *Ann. Plant Prot. Sci.*, 10: 377–378.

Lal SS, Yadava CP, Dias CAR (1985a) Assessment of crop losses in chickpea caused by *Heliothis armigera. FAO Plant Prot. Bull.*, 33: 27–35.

Lateef SS (1985) Gram pod borer (*Heliothis armigera*) (Hub.) resistance in chickpea, Agriculture. *Ecosystem Environment*, 14: 95–102.

Luckmann WH, Metcalf L (1975) Introduction to insect-pest management, Wiley, New York, pp. 3–35.

Magar PN, Shamim M, Ravisankar N, Chincholkar AS (2023) Field efficacy of HaNPV for the management of pod borer, *Helicoverpa armigera (Hubner) in chickpea.*

Mahmood MT (2021). An update on biology, extent of damage and management strategies of chickpea pod borer (*Helicoverpa armigera*). *Pakistan Journal of Agricultural Research*, 34(1): 91.

Mallapur CP, Ladji RN (2010) Management of chickpea pod borer, *Helicoverpa armigera* (Hubner) using indigenous materials. *Int. J. Plant Prot.*, 3(2): 194–198.

Mamoon-ur-Rashid M, Khattak MK, Abdullah K (2012) Evaluation of botanical and synthetic insecticides for the management of cotton pest insects. *Pakistan J. Zool.*, 44(5): 1317–1324.

Mansour AEA, Mohamed AA (2014) Evaluation of different chickpea genotypes for resistance against pod borer, *Helicoverpa armigera* (Hub.) (Lepdoptera: Noctuidae) under field conditions, Sudan. *Int. J. Agric. Innov. Res.*, 2(6): 1147–1149.

Mapuranga R, Chapepa B, Mudada N (2015) Strategies for integrated management of cotton bollworm complex in Zimbabwe : A review. *International Journal of Agronomy and Agricultural Research*, 7(1): 23–35.

Matthews M (1999) *Heliothine* moths of Australia, a guide to pest bollworms and related noctuid groups. Monograph on Australian Lepidopteran, Volume 7. CSIRO Publishing, P O Box 1139, 150 Oxford Street, Callingford, Victoria, 3066, Australia, pp. 320.

McLaughlin JL, Zeng L, Oberlies NH, Alfonso D, Johnson HA, Cummings B (1997) *Annonaceous acetogenins* as new natural pesticides: Recent progress. In: Hedin PA, Hollingworth RM, Masler EP, Miyamoto J, Thompson DG (eds) Phytochemicals for pest control. *ACS Symposium*, 658: 117–133.

Mehta DN, Singh KM, Singh RN (1989) Influence of intercropping on succession and population build-up of insect-pests in chickpea, *Cicer arietinum* L. *Indian J. Entomol.*, 50: 257–275.

Mehta KS, Patyal SK, Rana RS, Sharma KC (2010) Ecofriendly techniques for the management of *Helicoverpa armigera* (Hubner) in tomato. *Journal of Biopesticides*, 3(1): 296–303.

Mehta PK, Kashyap NP, Thakur SS (1996) The Cape gooseberry (*Physalis peruviana* Linn.) a new host of gram caterpillar in Himachal Pradesh. *Insect Environ.*, 2: 84.

Merga B, Haji J (2019) Economic importance of chickpea: Production, value, and world trade. *Cogent Food and Agriculture*, 5(1): 1615718.

Mesfin W, Emana G, Geremew T (2012) Neem *Azadirachta indica* (*A. Juss*) extracts negatively influenced growth and development of African bollworm, *H. armigera* (Hubner) (Lepidoptera: Noctuidae). *Acad. J. Entomol.*, 5(1): 22–27.

Mulwa RMS, Kimurto PK, Towett BK (2010, September). Evaluation and selection of drought and pod borer (Helicoverpa armigera) tolerant chickpea genotypes for introduction in semi-arid areas of Kenya. In *Second RUFORUM biennial meeting* (pp. 20–24).

Mishra K, Singh K, Tripathi CPM (2013) Management of pod borer (*Helicoverpa armigera*) infestation and productivity enhancement of gram crop (*Cicer arietinum*) through vermiwash with biopesticides. *World J. Agric. Res.*, 9(5): 401–408.

Mishra YK, Sharma AK, Sondhia S, Ramakrishnan RS, Panday AK, Gharde Y, Singh S (2023) Role of Oxalic Acid in Expression of Resistance against the Pod Borer Helicoverpa armigera (Lepidoptera: Noctuidae) in Chickpea Varieties. *Legume Research-An International Journal,* 46(3), 375–380.

Muddukumar (2007) Organic nutrient management and plant protection in green chickpea production system, M.Sc. (Agri.) Thesis, University of Agricultural Sciences, Dharwad, India.

Musa AM, Johansen C, Kumar J, Harris D (1999) Response of chickpea to seed priming in the High Barind Tract of Bangladesh. *International Chickpea and Pigeonpea Newsletter,* 6, 20–22.

Musa AM (2000). On-farm chickpea seed priming trials and demonstrations in the High Barind Tract of Bangladesh, 1999–2000. Research report.

Muzemu S, Mvumi BM, Nyirenda SPM, Sileshi GW, Sola P, Chikukura L, ... Stevenson P (2011) Pesticidal effects of indigenous plants extracts against rape aphids and tomato red spider mites. In *African Crop Science Conference Proceedings* (Vol. 10, pp. 169–171).

Nadeem S, Hamed M, Shafique M, Atta BM, Shah TM (2011) Evaluation for resistance in Kabuli chickpea genotypes against chickpea pod borer, *Helicoverpa armigera* (Hübner) (Lepidoptera: Noctuidae) under field conditions. *Sonklanakarin J. Sci. Tech.*, 33(3): 291.

Naresh JS, Malik VS, Kaushik SK (1989) Larval population and damage of *Heliothis armigera* (Hubn.) estimated on different plant densities of chickpea during five weeks. *Bull. Entomol. Res.*, 27: 70–73.

Nath P, Chakravorty S (2005) Effect of intercropping on the infestation of chickpea pod borer *Helicoverpa armigera* (Hubner). *Int. J. Plant Prot.*, 29(1): 86–91.

Nath P, Rai R (1999) Study of key mortality factors in the population dynamics of chickpea pod borer, *Helicoverpa armigera* (Hubner) (Noctuidae: Lepidoptera) infestation chickpea, *Cicer arietinum* L. *Trop. Ecology*, 40(2): 281–288.

Neto MDS, Cristina F, Cirilo A, Almeida DS (2018) Toxicity of andira paniculata (Fabaceae) extracts to *Helicoverpa armigera* (Lepidoptera : Noctuidae). *Journal of Agricultural Sciences*, 10(6): 264–271.

Pachundkar N, Kamble P, Patil P, Gagare P (2013) Management of gram pod borer *H. armigera* in chickpea with neem seed kernel extract as a natural pest management practice in Bhojdari village. *Int. J. Curr. Res.*, 5(10): 2934–2935.

Pal S, Chatterjee H, Senapati SK (2014) Monitoring of Helicoverpa armigera using pheromone traps and relationship of moth activity with larval infestation on carnation (Dianthus caryophyllus) in Darjeeling Hills. *Journal of Entomological Research,* 38(1), 23–26.

Pandey R, Ujagir R (2008) Effect of intercropping on *Helicoverpa armigera* (Hub.) infesting chickpea. *Ann. Plant Prot. Sci.*, 16(2): 320–324.

Parmar SK, Thakur AS, Marabi RS (2015) Effect of sowing dates and weather parameters on the incidence of *Helicoverpa armigera* (Hubner) in chickpea. *The Bioscan*, 10(1): 93–96.

Patil AA, Thakur S, Navale JS, Narode MK, Kolhe PS (2018) Influence of intercropping on incidence of gram pod borer (*Helicoverpa armigera*) in chickpea (*Cicer arietinum* L.). *Journal of Entomology and Zoology Studies*, 6(5): 12–15.

Patil SB, Goyal A, Chitgupekar SS, Kumar S, El-Bouhssini M (2017) Sustainable management of chickpea pod borer: A review. *Agron. Sustain. Dev.*, 37: 20.

Patnaik HP, Mohapatra R (1997) Performance of chickpea cultivars against *Helicoverpa armigera* (Hubner) under field conditions in central plateau region of Orissa. *Legume Research*, 18: 177–180.

Pattar PS, Mansur CP, Alagundagi SC, Karbantanal SS (2012) Effect of intercropping systems on gram pod borer *Helicoverpa armigera* hubner and its natural enemies in chickpea. *Indian J. Entomol.*, 74(2): 136–141.

Pavani T, Babu TR, Sridevi D, Radhika K, Sharma HC (2019) Effect of different sowing dates on pest incidence in chickpea. *International Journal of Current Microbiology and Applied Sciences,* 8(9): 627–637.

Peter AJ, Shanower TG, Romeis J (1995) The role of plant trichomes in insect resistance: A selective review. *Phytophaga*, 7: 41–64.

Pimbert MP (1990) Some future research directions for Integrated Pest Management in chickpea: A viewpoint. Chickpea in the nineties. In: Proceeding of Second International Workshop on Chickpea Improvement, 4–8 December 1989, ICRISAT, Patancheru, India, pp. 151–163.

Pinto M, Lo Agro A, Salerno G, Peri E (1997) Serious attacks of the tomato moth *Helicoverpa armigera* (Hubner). *Informatore Agrerio*, 53(9): 67–69.

Prabhakar M, Singh Y, Newpane SP (1998) Predicting *Helicoverpa armigera* of chickpea by using sex pheromone traps. *Indian J. Entomol.*, 60(4): 339–334.

Prasad D, Chand P (1989) Effect of intercropping on the incidence of Heliothis armigera (hub.) and grain yield of chickpea. *Journal of Research Birsa Agricultural University,* 1(1), 15–18.

Prasad D, Kumar B (2002) Impact of intercropping and Endosulfan on the incidence of gram pod borer infesting chickpea. *Indian J. Entomol.*, 64: 405–410.

Qadeer GA, Singh YP (1989) Some observations on outbreak of gram pod borer on grains during rabi 1987–88 in Haryana. *Plant Prot. Bull.*, 41(1–2): 24–25.

Rahman AKMZ, Haque MH, Alam SN, Mahmudunnabi M, Dutta NK (2014) Efficacy of Botanicals against *Helicoverpa armigera* (Hubner) in Tomato. *The Agriculturists*, 12(1): 131–139.

Rajput AA, Sarwar M, Ahmad N, Siddiqui QH, Toufiq M (2003) Evaluation for resistance in some local and exotic chickpea genotypes against *Helicoverpa armigera* (Hubner). *Pak. J. Biol. Sci.*, 6: 1612–1615.

Ramakrishnan C, Radhakrishnan T, Rama Doss G (1983) Effect of nitrogen, rhizobium inoculation and phosphorus level on damage of pigeon pea by *H. armigera*. *Int. Pigeon Pea Newsletter*, 2: 62.

Rao KR (2003) Influence of host plant nutrition on the incidence of *Spodoptera litura* and *H. armigera* on groundnut. *Indian J. Entomol.*, 65(3): 386–392.

Rashid A, Saeed HA, Akhtar LH, Siddiqi SZ, Arshad M (2003) Performance of advance chickpea strains against gram pod borer (*Helicoverpa armigera* Hubner). *Asian J. Plant Sci.*, 2: 418–419.

Reddy DS, Chowdary NM (2021) Botanical bio pesticide combination concept a viable option for pest management in organic farming. *Egypt J. Biol. Pest Control*, 31(23): 1–10.
Reed W (1983) Estimation of crop losses due to insect-pests in pulses, *Indian J. Entomol.*, 2: 263–267.
Reed W, Cardona C, Sithanantham S, Lateef SS (1987) Chickpea insect-pests and their control. In: Saxena M, Singh KB (eds) The chickpea. CAB International, Wallingford, pp. 283–318.
Reed W, Pawar CS (1982) Heliothis: a global problem. In: Reed W, Kumble V (eds) *Proceedings of International Workshop on Heliothis Management,* 15–20 November 1981. ICRISAT, Patancheru, pp. 9–14.
Reena, Singh SK, Sinha BK, Jamwal BS (2009) Management of gram pod borer, *Helicoverpa armigera* (Hubner) by intercropping and monitoring through pheromone traps in chickpea. *Karnataka J. Agri. Sci.*, 22(3): 524–526.
Rheenen HAV, Van Rheenen HA (1991) Chickpea breeding progress and prospects. *Plant Breeding Abstract*, 61: 997–1009.
Rivnay E (1962) Field crop pest in the near east. Netherlands Junk Publisher, The Hague, p. 450.
Romeis J, Sharma HC, Das KKS, Sharma BK (2004) The potential of transgenic chickpeas for pest control and possible effects on non-target arthropods. *J. Crop Prot.*, 23(10): 923–938.
Roy NK, Dureja P (1998) New eco-friendly pesticides for integrated pest management. Pesticides World, New Delhi, pp. 16–22.
Ruttoh EK, Mulwa RMS, Ngode L, Gohole L, Towett B, Njogu N, Silim S, Rao GVR, Kimurto PK (2013) Screening for host plant resistance to *Helicoverpa armigera* (Lepidoptera: Noctuidae) in selected chickpea (*Cicer arietinum* L.) genotypes in Kenya. *Egerton J. Sci. Techn.*, 13: 39–55.
Sachan JN, Katti G (1994) Integrated pest management, In: Proceedings of International Symposium on Pulses Research, 2–6 April 1984, IARI, New Delhi, India, pp. 23–30.
Saleem M, Younis A (1982) Host plants and nature and extent of damage of *Helicoverpa armigera. Pak. J. Agric. Sci.*, 3: 554–558.
Sanap MM, Jamadagni BM (2005) Screening of chickpea for resistance to pod borer, *Helicoverpa armigera* (Hubner) at Rahuri, Maharashtra, India. *International Chickpea Pigeon Pea Newsletter*, 12: 37–39.
Sarode SV (1999) Sustainable management of *Helicoverpa armigera* (Hubner). *Pestology*, 13(2): 279–284.
Sarwar M (2013) Exploration on resource of resistance in chickpea (*Cicer arietinum* L.) genotypes to gram pod borer *Helicoverpa armigera* (Hubner) (Lepidoptera). *Afr. J. Agric. Res.*, 8(26): 3431–3435.
Sarwar M, Nazir N, Toufiq M (2011) Identification of susceptible and tolerant gram (*Cicer arietinum*) genotypes against gram pod borer (*Helicoverpa armigera* Hubner). *Pak. J. Bot.*, 43: 1265–1270.
Saxena RC (1982) Colonization of rice fields by *Nilaparvata lugens* (Stal) and its control using a trap crop. *Crop Prot.*, 1: 191–198.
Sehgal VK, Ujagir R (1990) Effect of synthetic pyrethroids, neem ex-tracts and other insecticides for the control of pod damage by *Helicoverpa armigera* (Hübner) on chickpea and pod damage yield relationship at Pantnagar in Northern India. *Crop protection,* 9(1): 29–32.
Sekar PR, Vankataaiah M, Rao NV, Rao VR, Rao VSP (1996) Monitoring of insect resistance in *Helicoverpa armigera* (Hub.) from area receiving heavy insecticide application in Andhra Pradesh. *J. Entomol. Res.*, 20(2): 93–102.
Shabbir MZ, Arshad M, Hussain B, Nadeem I, Ali S, Abbasi A, Ali Q (2014) Genotypic response of chickpea (*Cicer arietinum* L.) for resistance against gram pod borer (*Helicoverpa armigera*). *Advancements in Life Sciences*, 2(1): 23–30.
Shafique M, Nadeem S, Hamed M, Atta BM, Shah TM (2009) Performance of some advance desi chickpea genotypes against pod borer, *Helicoverpa armigera* (Hubner) resistance. *Pak. J. Zool.*, 41: 277–280.
Shahzad K, Iqbal A, Khalil S, Khattak S (2005) Response of different chickpea (*Cicer arietinum*) genotypes to the infestation of pod borer (*Helicoverpa armigera*) with relation to trichomes. *Research Journal of Agri. and Biological Sciences*, 1(1): 120–124.
Sharma HC, Gaur PM, Srinivasan S, Gowda CLL (2014) Exploiting host plant resistance for pest management in chickpea. *Legume Perspectives*, 3: 25–28.
Sharma HC, Pampapathy G, Lanka SK, Ridsdill-Smith TJ (2005) Antibiosis mechanism of resistance to pod borer, *Helicoverpa armigera* in wild varieties of chickpea. *Euphytica*, 142: 107–115.
Sharma HC, Rajeev V, Gaur PM, Gowda CLL (2008) Potential for using morphological, biochemical, and molecular markers for resistance to insect-pests in grain legumes. *J. Food Legumes*, 21: 211–221.
Sharma HC, Saxena KB, Bhagwat VR, (1999) The legume pod borer, *Maruca vitrata:* Bionomics and management, International Crops Research Institute for the Semi-Arid Tropics (ICRISAT). Patancheru, India, *Info. Bull.*, No. 55.
Shelton AM, Zhao JZ, Roush TR (2002) Economic, ecological, food safety, and social consequences of the deployment of Bt transgenic plants. *Annu. Rev. Entomol.*, 47: 845–881.

Shinde YA, Patel BR, Mulekar VG (2013) Seasonal incidence of gram caterpillar, *Helicoverpa armigera* (Hub.) in chickpea. *Current Biotica*, 7: 79–82.

Sidde Gowda DK, Halle D, Sharanabasappa (2004) Evaluation of different IPM modules and intercropping systems for the management of pod borer in chickpea. *Karnataka Journal of Agricultural Sciences*, 17(3): 586–589.

Singh BK, Singh RP (2007) Effect of host plant nutrition and pesticidal management on larval population of *Helicoverpa armigera* (Hubner) in chickpea *Cicer arietinum*. *Indian J. Entomol.*, 69(4): 345–349.

Singh H, Singh I, Mahajan G (2002) Effect of different dates of sowing on the incidence of gram pod borer (*Helicoverpa armigera*) on different cultivars of chickpea (*Cicer arietinum*) *Agriculture Science Digest*, 22(4): 295–296.

Singh K, Pandey J (2014) Evaluation of intercropping for management of pod borer, *Helicoverpa armigera* Hub. in chickpea. *Bio Infolet*, 11(2c): 688–691.

Singh KB, Kumar J, Haware MP, Lateef SS (1990) Disease and pest resistance breeding: Which way to go in the nineties? In: Proceedings of the Second International Workshop on Chickpea Improvement, 4–8 December 1989. ICRISAT, Patancheru 502 324, Andhra Pradesh, India, pp. 233–238.

Singh O, Gowda CLL, Sethi SC, Lateef SS (1991) Inheritance of and breeding for resistance to *Helicoverpa armigera* pod borer in chick-pea. In: Golden Jubilee Symposium of Indian Society of Genetics and Plant Breeding, 4–8 February 1991, New Delhi, India.

Singh O, Sethi SC, Lateef SS, Gowda CLL (1997) Registration of ICCV 7 chickpea germplasm. *Crop Sci.*, 37: 295.

Sithanantham S, Reed W (1979) Plant density and pest damage in chickpea. *Int. Chickpea Newsletter*, 1: 9–10.

Sithanantham S, Tahhan O, Hariri G, Reed W (1981) The impact of winter sown chickpeas on insect-pest and their management. In: Saxena MC, Singh KB (eds) Proceedings of Workshop on Ascochyta Blight and Winter Sowing of Chickpeas. ICARDA, Aleppo, pp. 179–187.

Talhouk AA (1969) Insects and mites injurious to crops in Middle Eastern countries, Verlag Poul Parey, Hamburg, p. 239.

Thakur P (1998) *Helicoverpa* management through IPM approach on chickpea and pigeon pea, In: NARC/CLAN, Review and Planning meeting, Nawalpur, Sarlahi, Nepal, 25–27 May 1998, pp. 1–10.

Tripathi SR, Sharma SK (1984) Extent of damage and incidence of *Heliothis armigera* (Hubner) (*Noctuidae: Lepidoptera*) on different varieties of gram in Terai belt of eastern Uttar Pradesh, India. *Ann. Entomol.*, 2: 31–35.

Ujagir R, Khere BP (1987) Preliminary screening of chickpea genotypes for susceptibility to *Heliothis armigera* (Hub.) at Pantnagar, India. *Int. Chickpea Newsletter*, 17: 14.

Vinutha JS, Bhagat D, Bakthavatsalam N (2013) Nanotechnology in the management of polyphagous pest *Helicoverpa armigera*. *J. Acad. Indus. Res.*, 1(10): 606–608.

Whightman JA, Anders MM, Rao VR, Reddy L (1995) Management of *Helicoverpa armigera* (Lepidoptera: Noctuidae) on chickpea in southern India: Threshold and economics of host plant resistance and insecticide application. *Crop Prot.*, 14: 37–46.

White GF (1987) Environmental effects of arid land irrigation in developing countries. MAB Technical Series 8, UNEP/SCOPE/UNESCO, France, pp. 1–67.

Wilson CG (1997) phytophagous insect fauna of two weeds, *Hyptis suaveolens* (L) Poit, And *Jatropha gossypifolia* L., in Australia's Northern Territory. *Aust. Entomol.*, 24: 55–60.

Wubneh WY (2016) Biological control of chickpea pod borer, *Helicoverpa armigera* Hubner (Lepidoptera: Noctuidae): A global concern. *World Scientific News*, 45(2): 92–110.

Yadava CP, Lal SS (1997) Studies on host plant resistance against gram pod borer, *Helicoverpa armigera* in chickpea. In: Symposium on integrated pest management for sustainable crop production, 2–4 December 1997, Indian Agricultural Research Institute, New Delhi, India, p. 37.

Yadava CP, Sachan JN, Ahmad R, Lal SS (1991) Record of natural enemies of pests of chickpea and pigeon pea. *Biol. Control*, 5: 52–54.

Yang Y, Zhang Y, Zhang J, Wang A, Liu B, Zhao M, ... Lu Y (2023). Plant volatiles mediate Aphis gossypii settling but not predator foraging in intercropped cotton. *Pest Management Science*.

Zahid MA, Islam MM, Reza MH, Prodhan MHZ, Rumana Begum M (2008) Determination of economic injury levels of *Helicoverpa armigera* (Hubner) in chickpea. *Bangladesh J. Agri. Res.*, 33(4): 555–563.

Zaman K, Shah IA, Khan MM, Ahmad M (2012) Macroeconomic factors determining FDI impact on Pakistan's growth. *South Asian J. Bus. Stud.*, 1(1): 79–95.

Zhu BC, Henderson G, Chen F, Fei H, Laine RA (2001) Evaluation of vetiver oil and seven insect-active essential oils against Formosan subterranean termite. *J. Chem. Ecol.*, 27(8): 1617–1625.

8 Recent Trends in Antimicrobial Drug Resistance and Implications for the Needs of Microbial Toxicology Research

Rida Sagheer, Malik Nasibullah, and Nazia Iqbal

8.1 INTRODUCTION

Microorganisms, often known as microbes, are microscopic organisms that can survive as a single colony of cells. Van Leeuwenhoek was the first to identify microorganisms in the 1670s. Louis Pasteur discovered that bacteria were to blame for food rotting in the 19th century. Robert Koch was founded in the late years of the 19th century. There is a strong link between bacteria and diseases. Microorganisms have the potential to cause diseases and are known as pathogens. Diseases including tuberculosis, cholera, diphtheria, and anthrax are caused by microbes that have grown and invaded an individual or population. These microbes also cause harm to a person's vital systems.

Furthermore, researchers have been interested in answering the challenges of infectious diseases. Researchers are trying to find a chemical that would kill, prevent, or reduce the growth of these disease-causing microbes. Antibiotics are one of our most powerful weapons against potentially lethal infections.

Antibiotics treat infections, and the discovery, widespread use, and commercialization of antimicrobial drugs revolutionized modern medicine. These antimicrobial drugs also altered the therapeutic approach. Bacteria can protect themselves against an antibiotic's effects and are experts at surviving in changing environments. The competence of pathogens to resist antibiotics applied to eradicate them is a unique and significant challenge to the overall communal health of the modern era. However, bacterial resistance to antibiotics is accelerated through inappropriate use and abuse of currently available antibiotics in both community and vet care, besides in agriculture. However, dangerous antibiotic-resistant microorganism strains occurred only within the past 20 decades with alarming regularity (Medina and Pieper, 2016). People worldwide, including Americans, are living in a period when we are dying from untreatable contagions because of the occurrence and spread of antibiotic resistance (Fair and Tor, 2014). Tuberculosis, gonorrhoea, pneumonia, wound infections, septicaemia, and childhood ear infections are some ailments that are hard to treat with antibacterial drugs. Almost 30% of all deaths accounted for drug resistance; this is a frightening prospect. In the 20th century, antibiotic resistance was one of the three most significant threats to public well-being, as per the World Health Organization (Walsh, 2003). According to estimates, an additional $20 billion in healthcare and 8 million different clinic stays costs per year in the US are due to resistant infections (Fair and Tor, 2014; European Medicine Agency, 2009; Reynolds et al., 2005) and €1.6 billion more and 2.5 million extra hospital stays in the EU (Grundmann et al., 2006). Currently, hospital pharmacies in the US spend 30% more on antibiotics than on other drugs. It was obligatory to examine and assemble the present classification and functions of more or less all antibacterial agents in light of the vast growth in the number and classes of newly added antibacterial medicines. Therefore, these have been discovered as antibiotic resistance phenomena – for example, structurally different agents with a somewhat unlike form of activity. Inhibiting infections in the first place,

DOI: 10.1201/9781003269380-8

providing antibiotics more efficiently, and interrupting the spread of resistance when it does occur are the best ways to deal with this problem. The intention is for both academicians and medical researchers to gain from this approach.

8.2 CLASSIFICATION OF ANTIBIOTICS

Globally, microbial contagions are the foremost reason for disease and transience. "Naturally obtained chemicals" was used initially to describe drugs, but "antibiotic" is now used to label both organic and manufactured antimicrobial agents. The first antibiotic, penicillin, was discovered in the 1940s and derived from mould, replaced by sulpha medicines, which showed promise as potential treatments for medical conditions (Russell, 2004; Schlegel, 2003). However, antibiotic medications are grouped as antibacterial, antifungal, or antiviral based on the class of microbes they destroy (Mehramuz et al., 2020; Brooks et al., 2004; Russell, 2004). Antibacterials, a subclass of antibiotics, have previously been categorized in several ways. But, to make it more transparent, we can classify as per D. Kahne based on the origin of their method, the biochemical arrangement, the activity spectrum, and the function as well as the type of resistance (Kahne, 2005).

8.2.1 Classification of Antibacterials Based on Action

There are two categories of antibiotics depending on how they affect microbial cells; substances are either bacteriostatic or bactericidal. Bacteriostatic antibiotics suppress or prevent bacteria's growth by inhibiting protein synthesis (Walsh, 2003), as it might be challenging to draw a precise line between bacteriostatic and bactericidal medications. Bactericidal antibiotics can kill bacteria by focusing on the bacterial cell wall or cell membrane; these are aminoglycosides, cephalosporins, fluoroquinolones, metronidazole, penicillin, and vancomycin. Several bacteriostatic medications, including sulphonamides, erythromycin, and nitrofurantoin, may also turn cidal at higher concentrations (Kahne, 2005). Here are a few well-known instances of bacteriostatic and bactericidal antibacterial substances, along with a brief description of how they work.

8.2.1.1 Bacteriostatic Antibiotics

Bacteriostatic drugs primarily block pathogenic bacteria's reproduction; their inhibitory effect also interferes with some of the bacteria's metabolic pathways or protein translation. Bacteriostatic antibiotics such as sulphonamides, trimethoprim-chloramphenicol, clindamycin, erythromycin, sulfamethoxazole, and tetracycline may contain foliate inhibitors that suppress the cell growth and mechanism of action as well, as shown in Table 8.1.

8.2.1.2 Bactericidal Antibiotics

Most antibiotics that target an organism's cell wall in order to kill infections are considered bactericidal. Aminoglycosides, cephalosporins, fluoroquinolones, metronidazole, penicillin, and vancomycin prevent the formation of peptidoglycan in bacterial cell walls and membranes. Antibiotics from the bactericidal class and their mode of action are shown in Table 8.2.

8.2.1.3 Broad- and Narrow-Spectrum Antibiotics

Under a specific target, antibiotics can also be classified in different ways. Antibiotics come in two forms: broad-spectrum and narrow-spectrum antibiotics, both effective at treating infections (Figure 8.1). Those antibiotics that solely kill either gram-negative or gram-positive bacteria and are used to treat a specific subset of microbial infections are called narrow spectrum. Broad-spectrum antibiotics can treat a range of microbial illnesses, together with those affected by gram-positive and gram-negative bacteria (Chancey et al., 2012). However, taking wide-spectrum antibiotics when not needed can affect bacteria resistant to treatment and difficult to destroy. Antibiotics with a narrow spectrum are generally chosen over antibiotics through a broad spectrum. The purpose is that

TABLE 8.1
Different Bacteriostatic Antibiotics and Their Mechanism of Action

Bacteriostatic antibiotics	Mechanism of action
Sulfonamides Derivatives of sulfanilamide similar to PABA (***p* -amino benzoic acid (PABA)** are needed for folic acid synthesis in bacteria.	Sulfonamide reduces dihydrofolate reductase's substrate, dihydrofolate, by having a greater affinity for the microbial enzyme tetrahydropteroic acid synthetase. As a result, dihydropteroate can no longer contain PABA dihydrofolate reductase (DHFR) (Allison and Lambert, 2015; Bushby and Hitchings, 1968; Woods, 1940; Rickey and Brown, 1971).
Trimethoprim Trimethoxybenzyl-pyrimidine derivative inhibits folate synthesis.	Trimethoprim, a potent antibacterial DFR inhibitor, prevents the production of tetrahydro folic acid, needed for the creation of purine and deoxyribonucleic acid (Allison and Lambert, 2015; Bushby and Hitchings, 1968). When administered in a ratio of 1:20, trimethoprim and sulfonamides become bactericidal while inhibiting two consecutive critical processes in the production of folate (Capasso and Supuran, 2014; Grose et al., 1979).
Tetracycline is after catalytic dehalogenation of chlortetracycline obtained from *Streptomyces rimosus*.	Stops protein production by attaching to the 70S ribosomal protein in protozoa and the 30S ribosomal protein in bacteria to inhibit protein synthesis.
Doxycycline, minocycline is a member of tetracycline and synthesized from oxytetracycline.	The bacterial ribosome's 30S subunits reversibly bind to inhibit protein synthesis, which prevents amino acyl tRNA from attaching to the mRNA complex's acceptor site. For this, the peptide chain's capacity to incorporate amino acids is blocked, inhibiting the production of new proteins.
Tigecycline is the member of glycylcycline class and active against resistant bacteria.	
Erythromycin belongs to the class of macrolides, isolated from *Streptomyces erythreus*, while **a**zithromycin and **c**larithromycin are derivatives of erythromycin.	Protein synthesis constructed through RNA is stopped by the binding of macrolides towards rRNA-23S (ribosomal ribonucleic acid), a component of the 50S ribosomal subunit of bacteria
Lincomycin is the member of lincosamides class derived from *Streptomyces lincolnensis*; clindamycin is derived from lincomycin.	Suppresses protein synthesis through binding with 50S ribosomes and fights with macrolides for similar location.
Linezolid and tedizolid are the members of oxazolidinones obtained by chemical modification of 5-(halomethyl) 3-aryl-2 oxazolidinones.	Fixes 23S RNA via V-domain, a component of 50S unit of ribosomes, and prevents the production of proteins.

compared to broad-spectrum antibiotics, narrow-spectrum antibiotics do not eradicate numerous beneficial microbes. Additionally, microorganism resistance to narrow-spectrum antibiotics won't develop or will develop less.

Categorizing antibiotics according to their chemical structures is necessary because different skeleton-containing antibiotics exhibit diverse therapeutic behaviours. With similar patterns of toxicity, efficacy, and other associated qualities exhibited through similar structural units, this classification is particularly crucial.

Antibacterial is typically divided into two sets based on their structure: β-lactams (group A) and aminoglycosides (group B). However, antibacterials can be categorized in more detail, as in Figure 8.2.

8.2.2 Sulphonamides

Diseases like pneumonia, meningitis, dysentery, etc. could not be adequately treated before the discovery of sulpha medicines. The earliest synthetic antibacterial drugs were sulphonamides (Bushby and Hitchings, 1968). The *p*-amino benzene sulphonamides are broad-spectrum antibiotics that are bacteriostatic but may also exhibit bactericidal activity at higher concentrations. In 1935, a German company produced *p*'-sulphonyl chrysoidine, also known as 4-sulphonamide-2',

TABLE 8.2
Different Bactericidal Antibiotics and Their Mechanism of Action

Bactericidal antibiotics	Mechanism of action
Soil-dwelling bacteria produce aminoglycosides as secondary metabolites.	Through reversible binding to the 16S decoder acceptance site of the ribosome at the reverse transfer RNA site of messenger RNA (mRNA) and the prokaryotic 30S subunit of ribosomes, the cationic aminoglycosidic end fix with the anionic end of lipopolysaccharides alters the structure, leading to a depletion of cell wall that improves penetrability and restricts the formation of proteins. The displacement of the mRNA slows translation, stops the formation of proteins.
Tobramycin, gentamicin, amikacin, kanamycin, and neomycin are the mycins that are obtained from ***Streptomyces***. ***Micromonospora*** spp. is the source of netilmic in its ultimate form.	
β-lactam antibiotics contain the basic structure of β-lactam ring penicillins, such as pen V, procaine penicillin G, benzathine penicillin G, methicillin, oxacillin, cloxacillin, dicloxacillin, and flucloxacillinare member of β-lactams antibiotic class.	Kills sensitive bacteria by preventing the construction of peptidoglycan, the primary constituent of both gram-positive and gram-negative cell wall of bacteria. This hetero polymer comprises alternating N-acetylmuramic acid and 1, 4-linked N-acetyl glucosamine unit. They retarded the formation of bacterial cell walls.
Carbapenems are β-lactam/β-lactamase inhibitors such as imipenem, meropenem, aztreonam, ticaracillinclvulnate, and piperaciin- tazobactam; cephalosporin, e.g., cefotaxime, ceftriaxone, ceftazidime, and cefepimeare some other β-lactam inhibitors.	Attached to PBP3 of aerobic GNB and is dynamic contrary to all β-lactamases class as well as the major class of A and D β-lactamases, inhibiting the formation of cell walls and causing cell death.
Monobactams is semisynthetic derivative of a natural compound isolated from *Chromobacterium violaceum.*	These obstruct DNA gyrase and topoisomerase IV, preventing bacterial DNA replication
Quinolones derivative of chloroquine	These disrupt peptidoglycan crosslinkage of bacterial cell wall.
Flouroquinolones is the newer quinolones contains Fluorine, ciprofloxacin, oxifloxacin, and levofloxacin and is active against various infections.	These antibiotics disrupt gram-positive bacteria's cell membrane by binding with anionic lipopolysaccharide molecules that displaced calcium and magnesium while it is not effective in gram-negative bacteria on account of thin cell wall of lipid bilayer acting as "shield" that doesn't allow vancomycin to enter gram-negative cell (Kohsanki, 2010).
Vancomycine is a member of natural glycopeptide active against multi-resistant gram-positive infections.	It prevents final three-D hard structure of bacterial cell wall by inhibiting the penicillin-sensitive enzymes transpeptidases and carboxy-peptidases, which are necessary for peptidoglycan production of the bacterial cell wall.
Telavancin is Semisynthetic glycopeptides	
Polymyxin B and colistin are polymyxins	
Cephalosporins **β**-lactam antimicrobials isolated from ***Acremonium chrysogenum*** used against gram-positive and gram-negative bacteria	These are employed to combat gram-negative bacteria and microorganisms that have not reacted to first- or second-generation antibiotics.
First-generation: cefazolin, cephalexin, and cefadroxil active against gram positive.	They are used for severe infections or in those with impaired immunity and are active counter to gram-positive as well as gram-negative bacteria.
Second generation: cefprozil and cefuroxime, cephamycin: cefoxitin	
Third generation: cefdinir, cefditoren, cefixime, cefotaxime, cefpodoxime, ceftazidime, ceftibuten, ceftriaxone.	To treat MRSA infections that are resistant to other antibiotics, new generations used.
Fourth generation: Cefepime, Cefpirome	
Fifth-generation: ceftobiprole Ceftaroline, Siderophorecephalosporins: Cefiderocol	
Rifampin belongs to rifamycins isolated from Amycol atopsis mediterranei. Rifabutin, Rifapentine, and Rifaximin are semisynthetic reforms of rifamycin	Rifamycins connected strongly to DNA-dependent RNA polymerase and prevent the formation of new RNA.

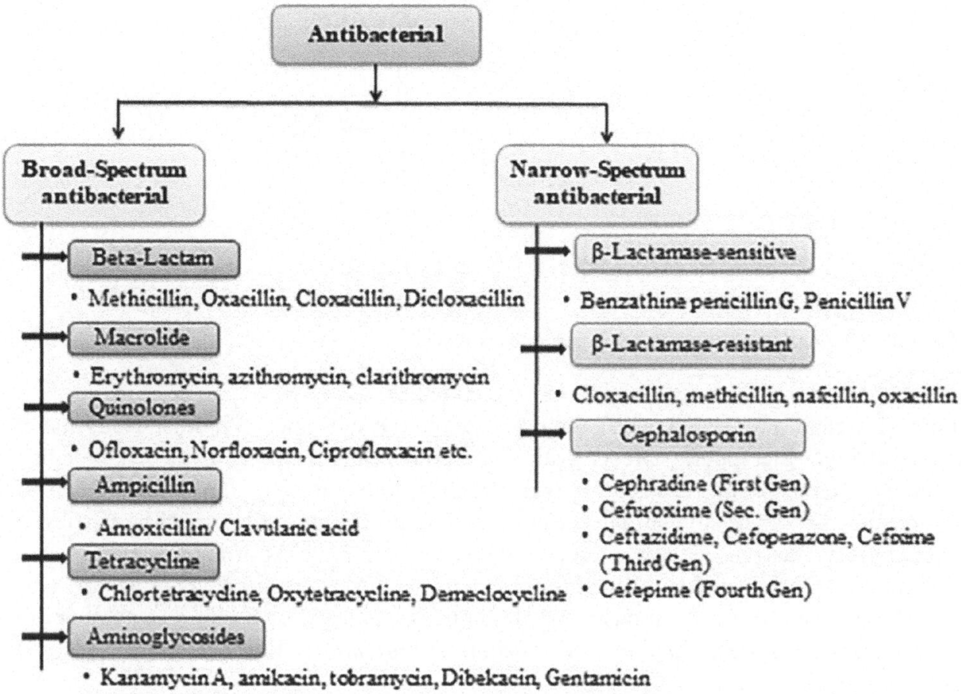

FIGURE 8.1 Antibacterial classification: broad-spectrum and narrow-spectrum antibacterials.

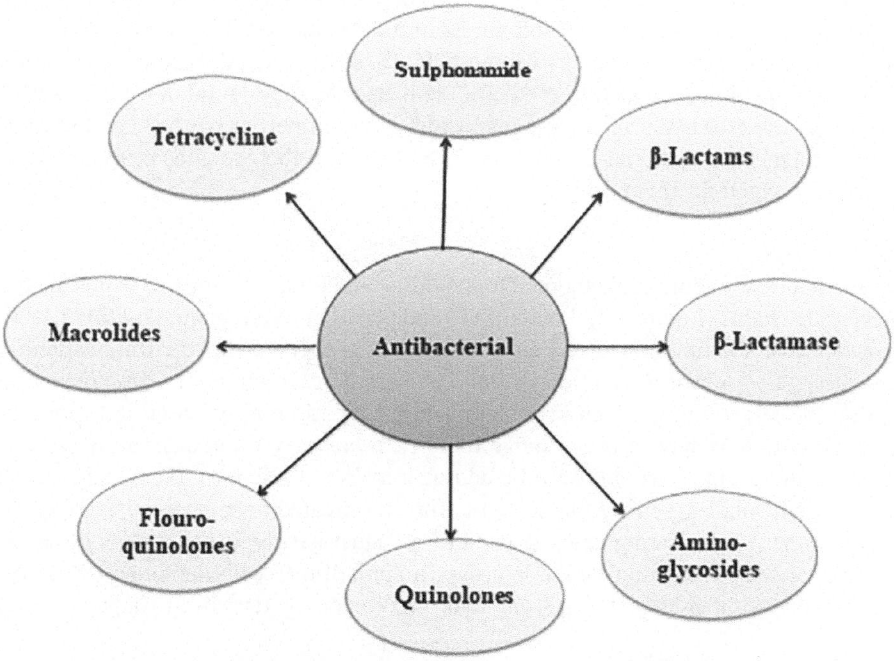

FIGURE 8.2 Different types of antibacterials.

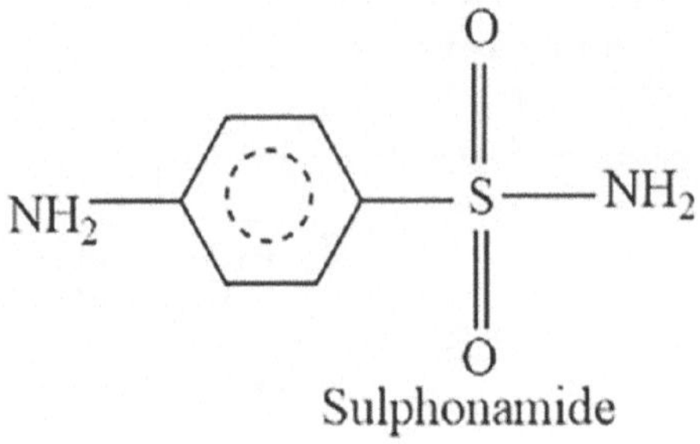

FIGURE 8.3 The basic structure of sulphonamide antibacterial.

4'-diamino-benzene, a red dye. Later, Domagk renamed the substance "Prontosil Red" and claimed it had considerable therapeutic effects against streptococcal infection in mice (Domagk, 1935). According to Tréfouël et al. (1935), Prontosil, whichdegrades in tissue into *p*-amino benzene sulphonamide, also known as sulphanilamide (Figure 8.3), it exhibits antibacterial properties.

Surprisingly, in the presence of microbes, the sensitivity patterns of antibacterials varied over time and space. Only a small number of *Staphylococcus aureus, meningococci, gonococci, pneumococci, Shigella*, and *E. coli* demonstrated antibiotic resistance. Furthermore, a number of strains of *Streptococcus pyogenes, Calymmatobacterium granulomatis, Vibrio cholerae, Haemophilus influenza*, and *H. ducreyi* are still sensitive.

8.2.3 β-Lactams

Penicillins and cephalosporins are part of this class of substances. Penicillin was the first antibiotic to be used in a therapeutic setting in 1941. *Penicillium notatum* and *Penicillium chrysogenum*, two different strains of fungus, were responsible for its initial production. The natural chemical mixture known as penicillin has the molecular formula $C_9H_{11}O_4N_2SR$. The advancement of a broad-range cepahlosporin semisynthetic β-lactam medicine with improved potential, a broader activity spectrum, longer half-lives for better medicating, and additional properties has been endorsed to combat resistance in the specific bacterial population. Even to retain their antibacterial activity, the ring must remain unchanged for these drugs.

8.2.3.1 Structure

The penicillin nucleus is composed of joined thiazolidine and β-lactam rings, to which an amide linkage connects side chains (Figure 8.4). Penicillin G and penicillin V are natural antibiotics. Penicillin G, a narrow-spectrum antibiotic with a benzyl side chain at R (benzyl penicillin), is predominantly dynamic contrary to gram-positive infections and a few others. However, the aminopenicillins such as ampicillin and amoxicillin, which are manufactured by addition of an amino group (-NH_2) in penicillin G (Figure 8.5), have a broad range of activity contrary to other gram-negative bacteria. Amoxicillin's acid stability was improved by adding a hydroxyl group (-OH), which signifies better oral absorption. Methicillin, semisynthetic penicillin, was created to combat the emergence of penicillinases, an enzyme inactivating existing penicillins. Surprisingly, the first penicillinase-resistant antibiotic was created by substituting the R group in penicillin G with the larger dimethoxyphenyl group, which protects the β-lactam ring from being enzymatically destroyed via penicillinases.

8.2.3.2 Mechanism of Action

In 1957, Park discovered that bacteria produce UDP-N-acetyl glucosamine and the "park nucleotide," a penta peptide of UDP-N-acetylmuramic acid. UDP is split off after the peptidoglycan

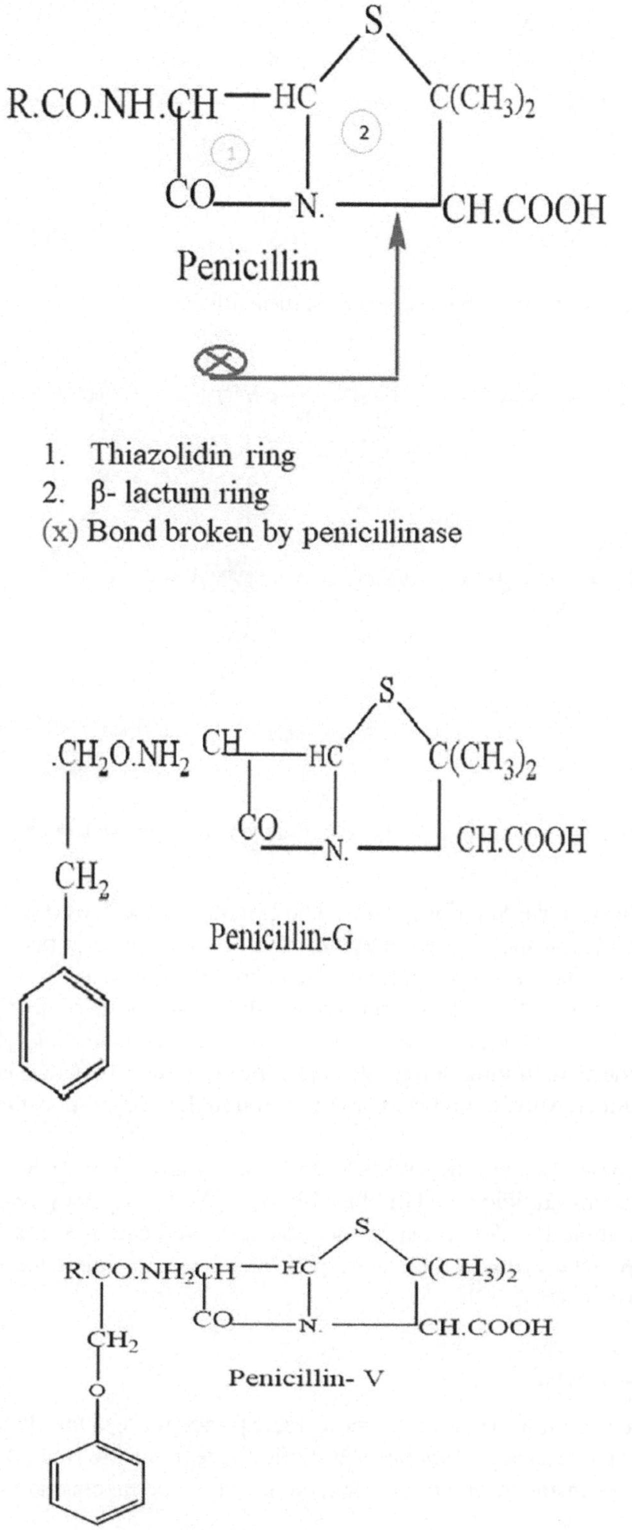

FIGURE 8.4 Basic structure of penicillin-containing β-lactam ring.

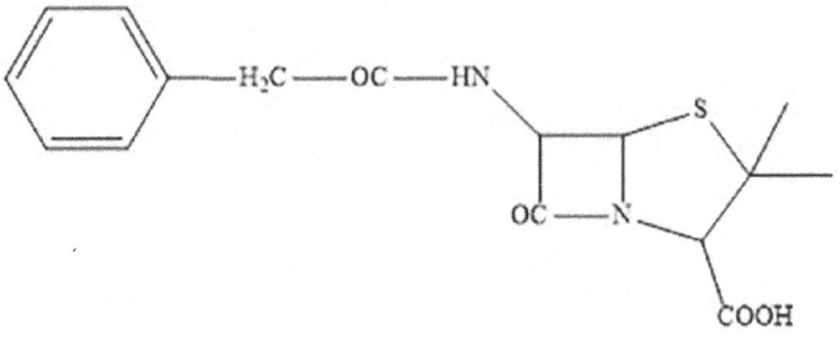

FIGURE 8.5 The basic structure of benzyl penicillin (penicillin G).

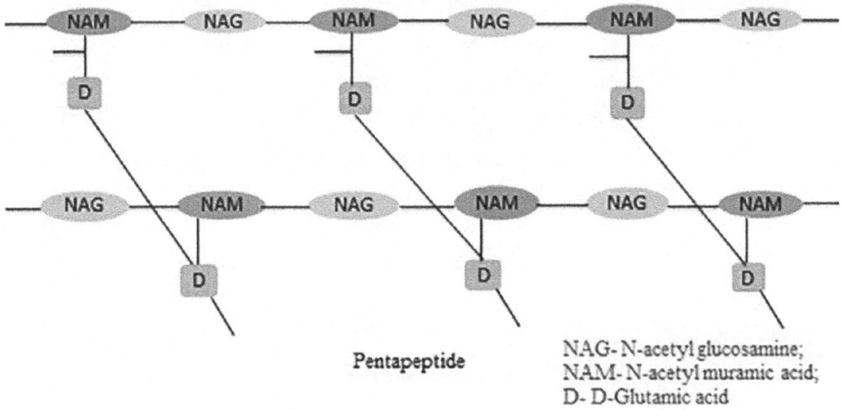

FIGURE 8.6 Bacterial cell wall through highly crosslinked peptidoglycan strands.

residues join together to form lengthy strands. The last step is the breaking of the peptide chains' terminal D-alanine by transpeptidases enzyme, also recognized as a penicillin-binding protein (PBP), present in the bacterial cell membrane. The energy thus released is used to create crosslinks between the peptide chains of nearby strands during the biosynthesis of new peptidoglycan (Bush, 2016). This crosslinking provides stability and stiffness to the bacterial cell wall. All β-lactam antibiotics prevent the crosslinking of peptide chains because their structure is similar to that of the peptidoglycan subunit constituent and is identified through the PBP responsible for the crosslinking (Figure 8.6).

Each organism possesses a number of PBPs, and PBPs from various species have varying affinities for various β-lactam antibiotics. This fact likely explains why they responded differently to different β-lactam antibiotics. Additionally, monobactams and carbapenems, two relatively recent antibiotics, have a β-lactam ring in their main pharmacological component to inhibit the peptidoglycan of the bacterial cell.

8.2.4 Anti-β-Lactamase Agents

Many gram-positive and gram-negative bacteria create β-lactamase, a family of enzymes dissolving the β-lactam ring that deactivates β-lactam antibiotics. Infections are treated with inhibitors of the lactamase enzyme, for instance, calvulanic acid, sulbactam, and tazobactam.

8.2.4.1 Cephalosporins

The β-lactam ring of cephalosporin restricts the trans-peptidase act of penicillin-binding proteins (Figure 8.7). Cephalosporins, unlike penicillins, have a six-membered dihydro thiazine

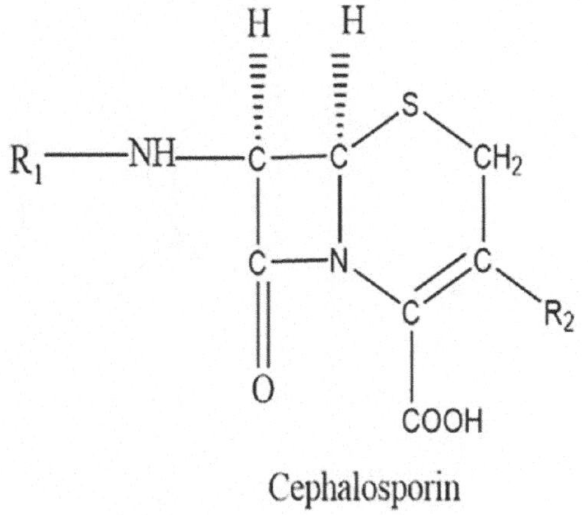

FIGURE 8.7 Basic structure of cephalosporin.

(7-aminocephalosporanic acid) ring structure. Cephalosporin-C is a semisynthetic antibiotic discovered in the fungus *Cephalosporium.*

Several semisynthetic products are formed by adding unlike side chains at position 7 of the β-lactam ring (activity spectrum modification) and the dihydro thiazine ring at position 3 (affecting pharmacokinetics). Because of this chemical change, cephalosporin is more resistant to enzymatic inactivation by β-lactamase (Hamilton-Miller, 1999). These were grouped into four generations based on the scope of their activities. A new fifth-generation cephalosporin has been constructed to fight methicillin-resistant *Staphylococcus aureus* (MRSA).

8.2.4.2 First-Generation Cephalosporin

Infections of the skin, bones, soft tissues, respiratory tract, and urinary tract caused by gram-positive bacteria are treated with first-generation cephalosporins, for instance, cefazolin (Figure 8.8), cephalexin, cephradine, and cephradoxil. Still, they are less effective against gram-negative bacteria.

8.2.4.3 Second-Generation Drug

Second-generation drugs produced after the first are better at combating gram-negative bacteria. All members can inhibit *P. aeruginosa*, even though some are also effective against anaerobes. The second generation of substances consists of cefuroxime, cefuroxime axetil, and cefaclor.

8.2.4.4 Third-Generation Cephalosporin

These substances can effectively eliminate gram-negative bacteria. However, they have less of an impact on anaerobes and gram-positive bacteria. Some of them include cefixime, cefpodoximeproxetil, cefdinir, ceftibuten, and ceftametpivoxil. Other examples include cefotaxime, ceftizoxime, ceftriaxone, and ceftazidime.

8.2.4.5 Fourth Generation

Cefepime, the sole fourth-generation cephalosporin that has been authorized for use, has an antibacterial spectrum similar to that of third-generation medications. However, due to its strong β-lactamase resistance potential, it is effective against many germs resistant to earlier medications. Both monobactams and carbapenems have a β-lactam ring in their core structures, preventing penicillin-binding proteins from acting as transpeptidase. Monobactam is a brand-new β-lactam with no extra rings and a restricted range of antibacterial activity. The monobactam class of substances includes aztreonam. It can kill *Pseudomonas* bacteria at low concentrations and inhibit

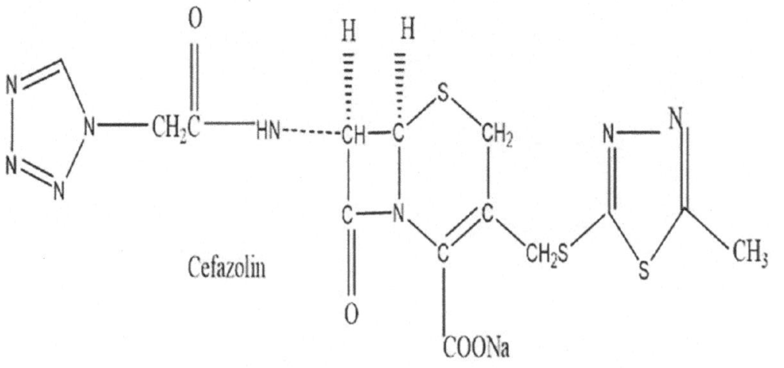

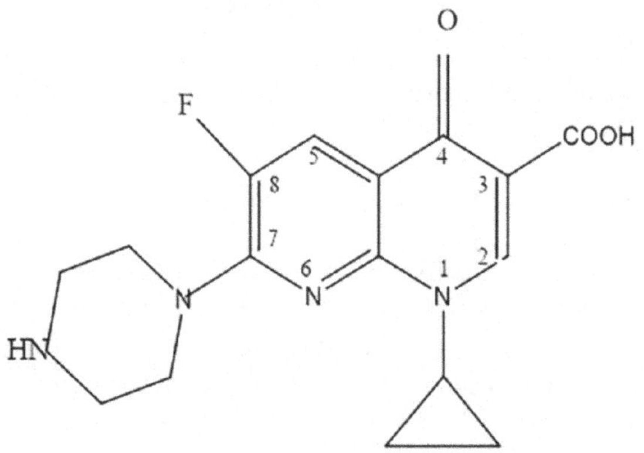

Ciprofloxacin: Fluoroquinolones

FIGURE 8.8 The basic structure of cefazolin and cephradine, a first-generation cephalosporin antibacterial.

gram-negative enteric bacilli and *H. influenza*. Ipenem, meropenem, and doripenem are medications that are a part of the carbapenem family (Walsh, 2007). Compared to both gram-positive and gram-negative bacterial infections, they are incredibly effective. Meropenem does not require cilastatin protection because, unlike imipenem, it is not hydrolyzed by renal peptidase. An oral β-lactam antibiotic called faropenem is effective against gram-positive and gram-negative bacteria and some anaerobic organisms.

8.3 AMINOGLYCOSIDES

These are a group of semisynthetic and natural antibiotics that contain essential carbohydrates linked glycosidically to two or more amino sugar residues; they include streptavidin, 2-deoxystreptamine, and glucosamine. These substances effectively combat gram-negative microorganisms. Streptomycin was the first aminoglycoside created by Waksman and his associates in 1944 and demonstrated to be highly effective against mycobacteria (Figure 8.9). All aminoglycosides produced through the biosynthesis of soil actinomycetes have a few traits in common. Bacteriocidal antibiotics such as aminoglycosides pass through the cytoplasmic membrane and bacterial cell wall. The suppression of protein synthesis results in binding to ribosomes. The researchers hypothesized that the number and size of amino groups on the hexoses and the binding site of the other rings to deoxy streptamine significantly stop the inhibition of protein formation. Furthermore, kanamycin B is a more effective antibiotic than kanamycin A or C. Its higher efficiency for inhibiting protein synthesis is due to the presence of a diamino hexose than the compound with only one amino group (Benveniste, 1973).

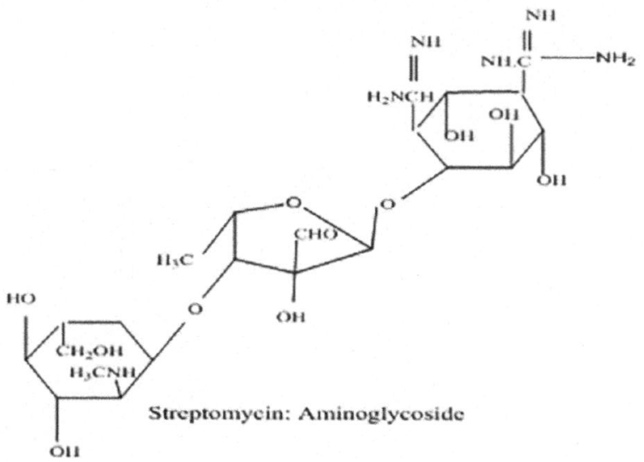

FIGURE 8.9 The basic structure of aminoglycoside antibacterials.

8.4 QUINOLONES

These are antimicrobials with quinolone structures, effective beside gram-negative bacteria. Nalidixic acid was first released in the middle of the 1960s (Figure 8.10), and due to its poor efficacy, it had limited application to urinary tract and gastrointestinal infections.

8.4.1 Fluoroquinolones

They are quinolone derivatives that have one or more fluorine substitutions. First-generation fluoroquinolones include norfloxacin, ciprofloxacin (Figure 8.11), ofloxacin, and pefloxacin, which all have one fluoro substitution. Second-generation fluoroquinolones include lomefloxacin, levofloxacin, sparfloxacin, gatifloxacin, and moxifloxacin and have additional fluoro and other substitutions.

8.5 MACROLIDES

These antibiotics contain sugars linked to a macrocyclic lactone ring. Erythromycin was the first macrolide revealed (Figure 8.12) in *Streptomyces erythreus* in 1952 and prevents translocation. Azithromycin and telithromycin are examples of semisynthetic macrolides. Azithromycin has a wider spectrum of activity, fewer side effects, and a long half-life than erythromycin

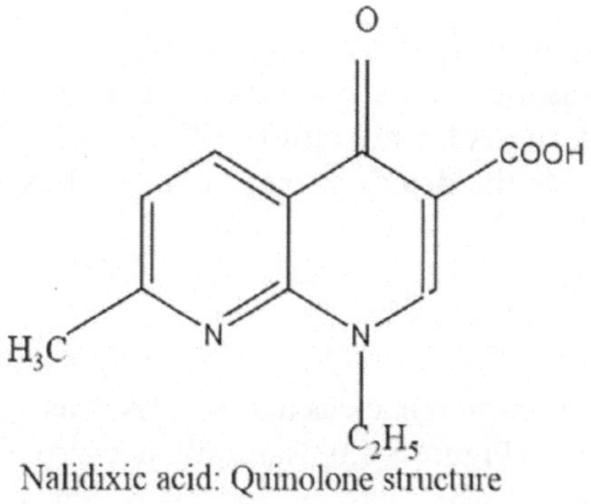

FIGURE 8.10 The basic structure of quinolone.

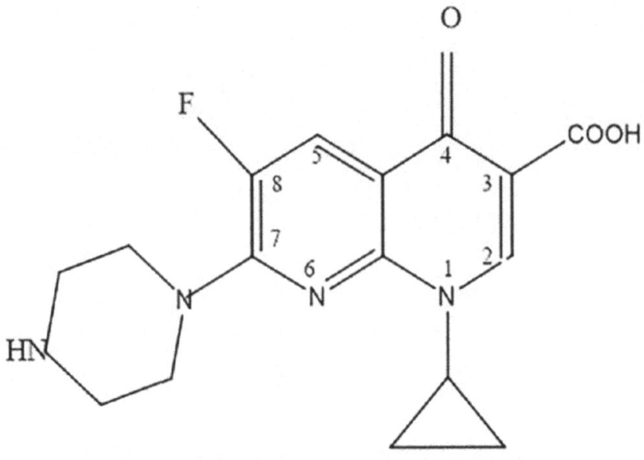

FIGURE 8.11 Structure of ciprofloxacin, a fluoroquinolone.

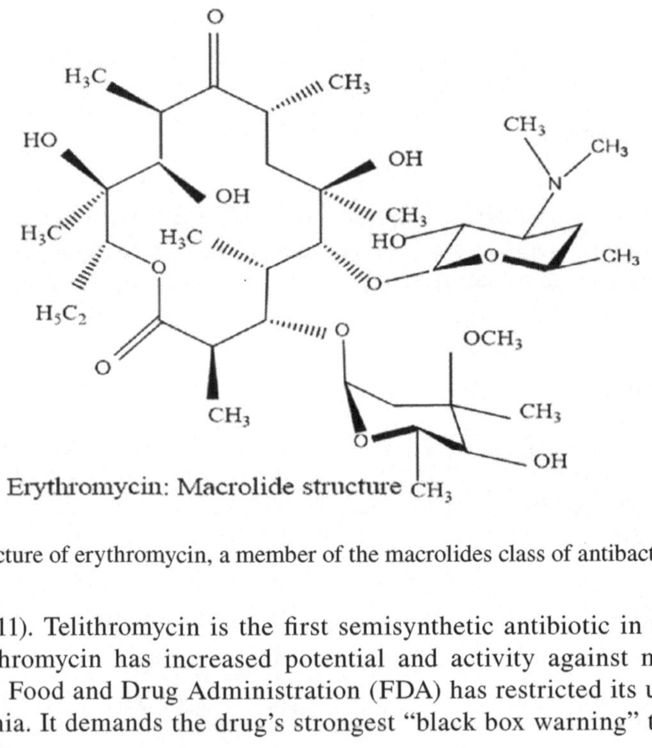

FIGURE 8.12 Structure of erythromycin, a member of the macrolides class of antibacterials.

(Mishra et al., 2011). Telithromycin is the first semisynthetic antibiotic in the ketolide class. Even though telithromycin has increased potential and activity against macrolide-resistant pathogens, the US Food and Drug Administration (FDA) has restricted its use to community-acquired pneumonia. It demands the drug's strongest "black box warning" title due to significant hepatotoxicity.

8.6 TETRACYCLINES

Tetracycline has a four-ring hydrocarbon nucleus and is classed as a polyketide with an octahydrotetracene-2-carboxamide skeleton (Figure 8.13). Naturally occurring tetracyclines were discovered in the 1940s by various strains of *Streptomyces*, and many semisynthetic tetracyclines, such as doxycycline and tigecycline, have also been formed. Large doses of tetracyclines can have a variety of negative effects, such as phototoxicity, permanent tooth discoloration during emergence, and

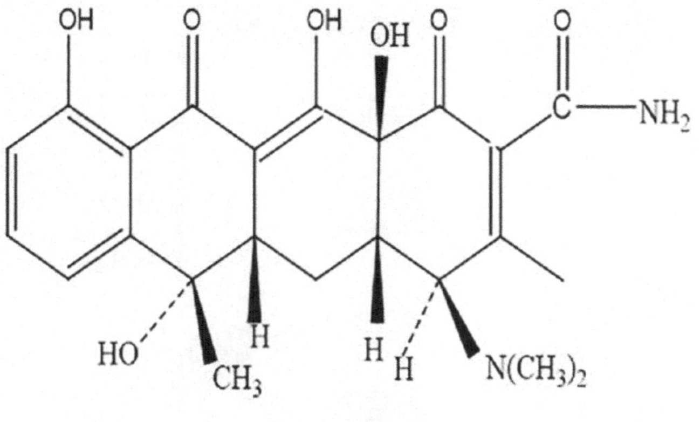

FIGURE 8.13 Structure of tetracycline antibiotic.

hepatotoxicity in patients with renal impairment. These side effects may limit the use of tetracyclines with high amounts in patients with kidney damage can limit their use.

8.7 ANTIBACTERIAL CLASSIFICATION BASED ON THE MODE OF ACTION

Each class of antibacterial medications used to treat microbial infections has a distinctive mode of action wherein a drug affects germs at the cellular level. They are typically categorized based on their main modes of action.

They are categorized into six important active methods of action:

1. Potent inhibitor of peptidoglycan formation,
2. Inhibiting membrane function,
3. Interference with protein synthesis,
4. Potent inhibitor of the synthesis of nucleic acid
5. Inhibiting a metabolic pathway, and
6. Inhibiting adenosine triphosphate (ATP) synthase.

Regarding their mechanism of action, antibacterial medications target protein synthesis, nucleic acid synthesis, and biological metabolic compounds.

8.7.1 Potent Inhibitor of Peptidoglycan Formation

The glycoprotein polymer constructs the skeleton of bacterial cells. These glycan chains are connected by glycosidic bonds and are composed of alternating N-acetyl glucosamine (NAG; G) and N-acetylmuramic acid (M) units (Figure 8.14). Transglycosidases crosslinked short peptide bridges between polysaccharide (glycan) chains to produce peptidoglycans.

Drugs containing β-lactam prevent the cleavage of the D-Ala terminal and transpeptidation. Further transpeptidation cannot occur because vancomycin firmly associates with the terminal D-Ala-D-Ala system and inhibits its discharge from the carrier. If peptidoglycan production is selectively inhibited by antibiotic activity, bacteria undergo several shape changes and significant damage to their cell walls, and they lyse and die. All eubacteria have peptidoglycan-based cell walls; human cells lack peptidoglycan cell walls exceptionally. The formation of peptidoglycan inhibition is an ideal target for uncovering antibacterial moiety since mammalian cells lack the peptidoglycan wall structure; therefore, it has no noticeable adverse effects on the mammalian host cells.

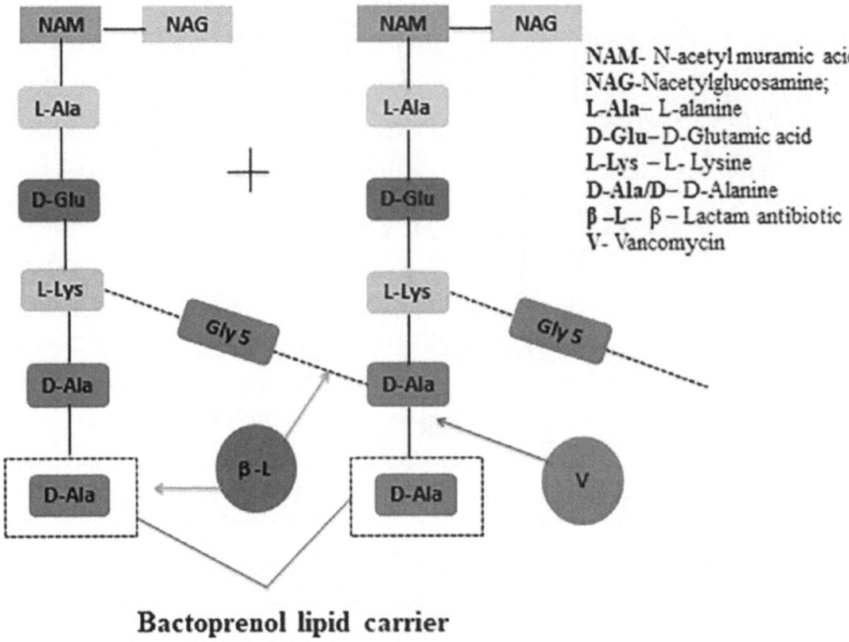

FIGURE 8.14 Bacterial cell wall synthesis and structure as the vancomycin and β-lactam antibiotic action site: Cleaving the D-alanine (D-Ala/D) terminal and transpeptiding it with a chain of five glycine (Gly5) residues, it is possible to link adjacent peptidoglycan residues.

The most effective antibiotics for preventing the synthesis of bacterial cell walls are named β-lactam drugs, which contain cephalosporins, monobactams, and carbapenems. Transpeptidase enzymes, also famous as PBPs, assist the final step in the production of peptidoglycan; it has been hypothesized that β-lactam drugs act as a false molecule D-alanyl D-alanine portion of the peptide chain that is usually crosslinked by glycine residues in the existence of PBP that obstructed transpeptidation reaction and peptidoglycan synthesis (Kahne, 2005; Reynolds, 1989). The lysis of bacterium occurs due to the disruption of the peptidoglycan layer (Ibrahim et al., 2001; Davis and Weiser, 2011; Džidić et al., 2008). PBPs are commonly divided by electrophoresis, allowing their size and numbered PBP1, PBP2, etc., starting from the class of maximum molecular weight (Hiramatsu, 2001). As for structural differences in their cell walls, gram-positive and gram-negative bacteria have varying degrees of susceptibility to β-lactam drugs. For example, gram-negative bacteria typically have lower susceptibility because these antibiotics are prevented from reaching the cell wall by their outer membrane. The availability of receptors, lipids, and peptidoglycan, as well as crosslinking characteristics and the activity of autolytic enzymes, all significantly impact the drug's activity, permeability, and incorporation. Penicillinases, cephalosporinases, cephamycinases, and carbapenemases are bacterially produced lactamases enzymes that can inactivate all β-lactam antibacterial due to the resistance practice. The D-alanyl D-alanine portion of the peptide side chain of the precursor peptidoglycan subunit is where the glycopeptides bind, and vancomycin, a potent drug, prevents this D-alanyl subunit from binding to the PBP; consequently, cell wall synthesis is inhibited (Hiramatsu, 2001; Grundmann, 2006).

8.7.2 Inhibiting Membrane Function

The bacterial membrane provides selective permeability for the transmission of metabolic energy and cellular homeostasis. Different antimicrobial medicines impact various targets and result in structural and functional membrane degradation by interacting with the bacterial membrane through a lipophilic moiety. Antibacterial drugs that focus on bacterial cytoplasmic membrane elements may harm both gram-positive and gram-negative bacteria.

The bacterial membrane provides selective permeability for cellular homeostasis and the transmission of metabolic energy. Several antimicrobial drugs affect various targets by interacting with the bacterial membrane through a lipophilic moiety and leading to structural and functional membrane breakdown (Kohanski et al., 2010). Polymyxin B is now used in over-the-counter topical antibiotic ointments (such as Neosporin) due to these severe side effects and poor absorption from the digestive tract. In contrast, oral colistin was previously only used for bowel decontamination to prevent infections caused by bowel microbes in immune-compromised patients or those undergoing certain abdominal surgeries. However, because of the rise and spread of multidrug-resistant microorganisms (Deak et al., 2016), hospitals now use intravenous colistin more often, sometimes even as a last option to treat dangerous infections (Trimble et al., 2016).

8.7.3 Interference with Protein Synthesis

Protein catabolism is a complicated, multistep process involving several enzymes, structural alignment, and cytoplasmic factors catalyzed by ribosomes. First, a process known as transcription uses the information in bacterial DNA to create a messenger RNA (mRNA) molecule. The translation is the process by which the macromolecular structure known as the ribosome creates the proteins found in mRNA. The 30S and 50S subunits of the ribonucleic protein make up the bacterial 70S ribosome (Yoneyama, 2006). However, most antibiotics restrict bacteria from generating proteins by interfering with the 30S or 50S subunits of the 70S bacterial ribosome (Vannuffel, 1996; Kapoor et al., 2017; Johnston et al., 2002).

Tetracycline is a member of a different group of antibiotics, along with doxycycline, that attaches to the 30S ribosome subunit and prevents amino acyl tRNAs from associating during translation to prevent protein synthesis. Both 70S and 80S (prokaryotes and eukaryotes) ribosomes can use them to suppress protein production.

Aminoglycosides are powerful, broad-spectrum antibiotics that bind to the 30S component of bacterial ribosomes and reduce the ribosomal complex's capacity for proofreading. Examples of aminoglycosides include the antibiotics streptomycin, gentamicin, neomycin, and kanamycin. Due to this defect, codons and anticodons don't match, leading to the synthesis of shorter proteins inserted into the cytoplasmic membrane and proteins with the wrong amino acids. These defective proteins destroy the cytoplasmic membrane, killing the bacterial cells. Aminoglycosides, however, are ototoxic, neurotoxic, and nephrotoxic. Additionally, several antibacterial medication families, including macrolides, interrupt translation by attaching to the 50S subunit of bacterial ribosomes.

Macrolides are broad-spectrum, bacteriostatic drugs synthesized in the body as secondary metabolites known as polyketides. They have a large, complex ring structure that is produced using a mechanism akin to that of fatty acid synthesis through the stepwise addition of two-carbon units repeatedly. By concentrating on the conserved regions in the peptidyl transferase core of the 23S rRNA of the 50S ribosomal subunit, these interfere with translocation, the first stage of protein synthesis. Macrolides stop proteins from extending by blocking the creation of peptide bonds between specific combinations of amino acids (Yoneyama, 2006; Wise, 1999).

8.7.4 Nucleic Acid Synthesis Inhibition

Some antibiotics function by preventing nucleic acid production. Quinolone medicines impede DNA replication by inhibiting the enzyme topoisomerase. Nalidixic acid is a synthetic antibacterial discovered as a by-product of the antimalarial medication chloroquine.

Second-generation quinolone medications, including levofloxacin, norfloxacin, and ciprofloxacin, are effective against gram-negative and gram-positive bacteria. However, the antimicrobial drugs doxorubicin and actinomycin D decrease RNA production by blocking RNA polymerases. Fluoroquinolones are the consequence of chemical changes to the original quinolone backbone.

When used to treat various diseases, such as urinary tract infections, lung infections, stomach infections, and skin infections, fluoroquinolones like ciprofloxacin and levofloxacin are effective against a broad spectrum of gram-positive or gram-negative bacteria. They prevent the bacterial DNA gyrase enzyme from nicking double-stranded DNA, adding negative supercoils and resealing the nicked ends. When the strands separate to allow for transcription or replication, this is required to prevent excessive positive supercoiling. The DNA gyrase is made up of two A and two B subunits. The A subunit nicks the DNA, the B subunit inserts negative supercoils, and the A subunit reseals the strands. Fluoroquinolones have a strong affinity for the A subunit and interfere with its strand-cutting and resealing activity. Topoisomerase IV, which nicks and separates the daughter DNA strand after DNA replication, is the main target of action in gram-positive bacteria. This enzyme's affinity may offer greater potency against gram-positive bacteria. Mammalian cells have topoisomerase II instead of DNA gyrase or topoisomerase IV, which has a low affinity for fluoroquinolones and is less hazardous to cells (Yoneyama, 2006; Wise, 1999; Spencer and Panda, 2023; Higgins et al., 2003).

Rifampin is a semisynthetic antibiotic of the rifamycin family. It inhibits bacterial RNA polymerase and is physically distinct from those found in eukaryotes; it has been used to treat various illnesses due to specific toxicity against bacterial cells. However, it is used against tuberculosis-causing *Mycobacterium* frequently in combination with other antibacterial medications. Despite the selectivity of its action, rifampin causes liver enzymes to accelerate the metabolism of other medications, resulting in hepatotoxicity and negatively influencing the bioavailability and therapeutic impact of other combination therapies. Antibiotics can then inhibit microbe multiplication, transcription, and folate production.

8.7.5 Inhibition of the Metabolic Pathway

Bacterial metabolic inhibitors are antibiotics that target the synthesis of nucleic acids and amino acids in bacteria. The coenzyme tetrahydro folic acid (TH4) synthesizes nucleic acids and some amino acids in all living things. Bacteria mainly produce their folic acid, which contains *p*-amino benzoic acid (PABA) and is taken from the medium. Bacterial metabolic inhibitors prevent the production of TH4 by bacteria via metabolic pathways. Some synthetic medications work as antimetabolites, or competitive inhibitors of bacterial metabolic enzymes, to treat bacterial infections.

Sulphonamides, which are structural counterparts of PABA, inhibit bacterial folate synthase, resulting in a lack of folic acid and several critical metabolic reactions. Sulphonamides block the creation of dihydropteroic acid from PABA and pteridine residue, which then combines with glutamic acid to generate dihydro folic acid (Figure 8.15). As a result, the creation of pyrimidine and purines essential for nucleic acid synthesis ceases. This action method provides bacteriostatic growth inhibition against a diverse range of gram-positive and gram-negative bacteria (Figure 8.16).

Human cells also require folic acid (FA), but they use prepared FA from food and are unaffected by sulphonamides. The following evidences support this mechanism of action of sulphonamides:

1. PABA, in tiny amounts, antagonizes the antibacterial effect of sulphonamides.
2. Sulphonamides are only toxic to microorganisms that generate their FA and cannot obtain it from the media.

8.7.6 ATP Synthase Inhibition

ATP synthase is the primary energy-generating enzyme in all species, from bacteria to vertebrates. Bacteria can create ATP by substrate-level phosphorylation of fermentable carbon sources or oxidative phosphorylation employing respiratory chains and ATP synthase. Several antibiotics have been discovered to hinder ATP synthase's oxidative phosphorylation, which kills bacteria.

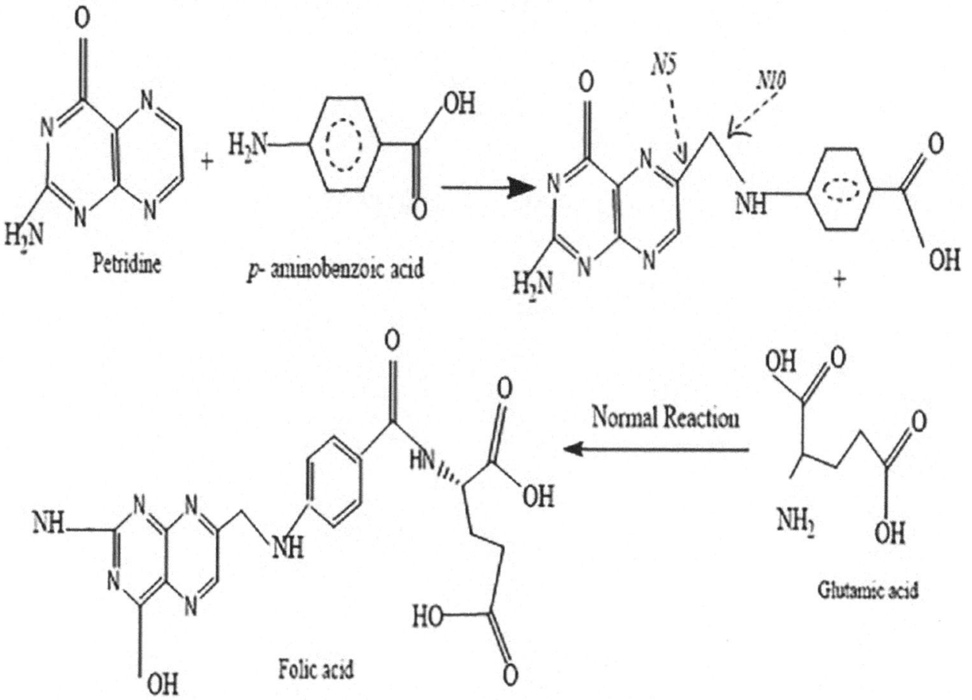

FIGURE 8.15 Formation of folic acid by dihydropteroic acid combining with glutamic acid.

8.7.7 Mechanism of Drug Resistance

Antimicrobial resistance is the typical practice of bacteria in nature against antibiotics synthesized by other germs; to survive, they constantly adapt to reduce the drug's effect. Humans' widespread clinical usage and advancement of antimicrobial medications have merely added pressure that drives future development (Mahon et al., 2014).

The progression of five critical factors can increase drug resistance: (1) drug interest prevention, (2) cellular inhibition, (3) chromosomal mutation, (4) overproduction of target or target mimicry, and (5) enzymatic drug adjustment (Chancey et al., 2012; Allison and Lambert, 2015).

8.7.8 Drug Interest Prevention

Bacteria naturally differ in their ability to resist antimicrobial interest. The shape and activities of the LPS (lipopolysaccharide) layer in gram-negative bacteria create a barrier to specific types of chemicals, providing intrinsic resistance of bacteria to particular categories of bulky antibiotics. Because mycobacteria have a high lipid outer layer, hydrophobic medications like rifampicin and fluoroquinolones can easily connect cells, but hydrophilic drugs have limited contact (Blair, 2014; Kumar, 2005; Lambert, 2002). Furthermore, because *Mycoplasma* and related species lack a cell wall, they are resistant to all cell wall–targeting medicines, including β-lactams and glycopeptides (Bébéar and Pereyre, 2005). Gram-positive bacteria are less common without an outer membrane, limiting medication interaction. Another gram-positive bacterium, *Staphylococcus aureus*, recently acquired vancomycin resistance, but enterococci naturally exhibited resistance to aminoglycosides due to polar molecules having difficulty penetrating their cell walls.

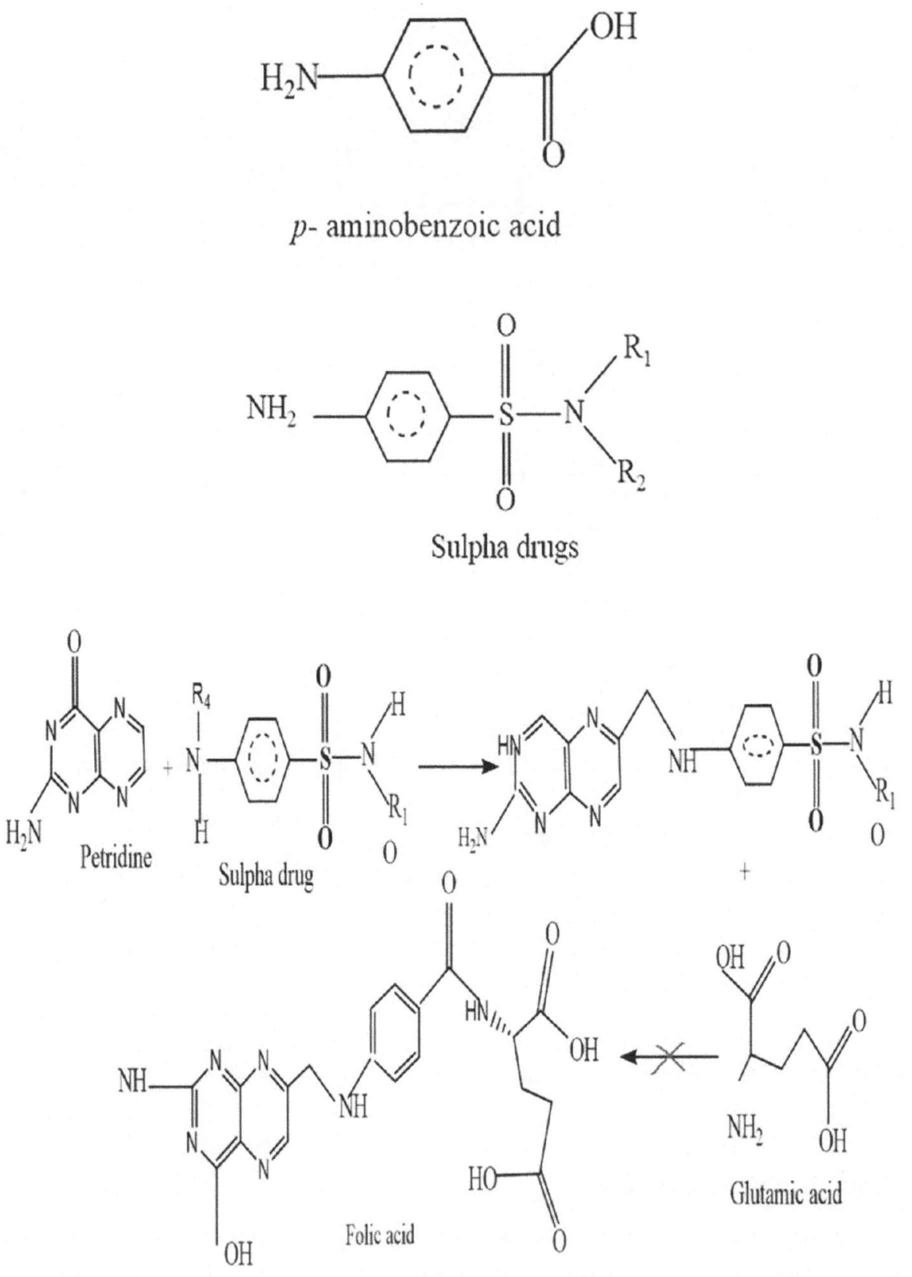

FIGURE 8.16 Folic acid inhibition of bacterial cell through sulphonamides, structural analogues of PABA.

8.7.9 Cellular Inhibition

Cellular inhibitions are the defense mechanisms that microbes typically develop prevent the antibiotic from reaching its intended cellular target. This tactic, which may involve modifications to the outer membrane lipid composition and porin channel selectivity, is widely used by gram-negative pathogens. Many gram-positive and gram-negative pathogens also possess efflux pumps, which actively transport antibiotics out of the cell and stop them from becoming antibacterial (Beceiro et al., 2013). Interestingly, one efflux pump is sufficient to transport several different antibiotic classes. Active efflux out of the cell is a common mechanism by which bacteria develop resistance to β-lactams, tetracycline, and fluoroquinolones (Higgins et al., 2003). The outer membrane porin D protein substrate is the essential entry portal for carbapenems in *Pseudomonas aeruginosa.* OprD-facilitated resistance is a common approach due to the reduction of transcriptional expression of OprD that interrupts protein activity, so carbapenem resistance occurs (Huseyin Agah Terzi et al., 2015).

8.7.10 Chromosomal Mutation

Antimicrobial medications have extremely specific targets; therefore, altering the structure of those targets can prevent the antibiotic from binding, which will cause the drug's effectiveness to fail. Chromosomal changes caused by pathogen exposure representing antimicrobial resistance can be vertically passed down to succeeding microbial generations and eventually take over in a microbial community exposed to the antibiotic frequently. Contrarily, bacteria have an evolutionary advantage that facilitates the acquisition of drug resistance, as demonstrated by the numerous drug resistance genes found on plasmids and by spontaneous changes in the genes encoding antibacterial drug targets. This technique for the development of drug resistance is fairly widespread (Džidić et al., 2008). The binding location of β-lactam medications is PBPs. Many of the drugs in this class can either inhibit or create resistance through gene changes. *Streptococcus pneumoniae* strains that alter their PBPs via genetic routes are often very likely to employ this strategy (O'Brien et al., 2009). However, certain strains of *Staphylococcus aureus* acquire a unique low-affinity PBP, leading to the emergence of methicillin resistance instead of structurally modifying their current PBPs (MRSA). Almost all β-lactam antibiotics, except the recent fifth-generation cephalosporin created to fight MRSA, were resistant to this novel low-affinity PBP (Reygaert, 2009). Tetracycline, aminoglycosides, and macrolides drug resistance brought on by ribosomal subunits are alterations to consider in this resistance approach (Sun et al., 2022). At the same time, resistance in polymyxins is produced by the LPS structure. The RNA polymerase provides resistance to the rifampin antibiotic; fluoroquinolones resistance occurs via DNA gyrase. However, metabolic enzymes provide resistance to sulpha drugs, sulfones, trimethoprim, and peptidoglycan subunits of peptide chains that provide resistance to glycopeptides.

8.7.11 Overproduction of Target or Target Mimicry

There are various ways that microbial resistance can emerge when an antimicrobial medicine acts as an antimetabolite, affecting a specific enzyme to decrease its activity. First, the microorganism may overproduce or create a bypass that eliminates the necessity for the functional target enzyme (Spratt, 1994). Both of these techniques have been identified as sulphonamide resistance mechanisms. Vancomycin resistance in *S. aureus* has been linked to reducing crosslinking of peptide chains in the bacterial cell wall, which increases the number of sites for vancomycin to bind to in the outer cell wall. Increased vancomycin binding in the outer cell wall creates a barrier that prevents free drug molecules from entering where they can inhibit new cell wall construction. Second, target mimicry is a recently discovered mechanism of resistance that involves the production of specific proteins that prevent drugs from binding to their bacterial cellular targets *Mycobacterium tuberculosis* is produced by a protein called *Mycobacterium* fluoroquinolone resistance protein A (MfpA) that resembles DNA and produces fluoroquinolone resistance (Huang et al., 2022). MfpA originates mimicry of DNA that results in DNA gyrase binding to MfpA, inhibiting the binding of fluoroquinolones to DNA gyrase shown in Figure 8.17.

8.7.12 Enzymatic Drug Target Adjustment

Drug resistance can develop when a drug's ability to adhere to its bacterial target is compromised, which can occur with various antimicrobials, including aminoglycosides. In the biosynthesis

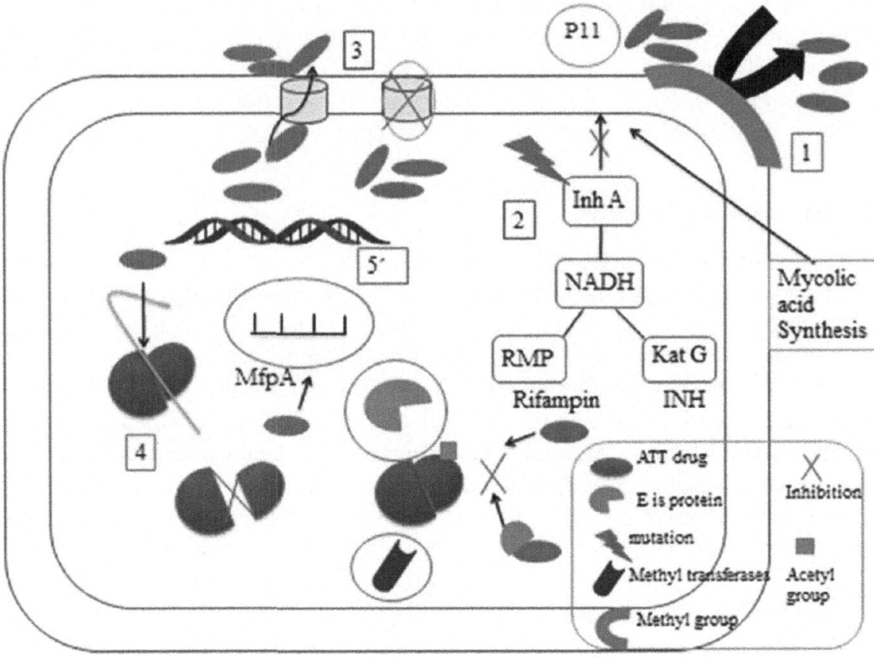

FIGURE 8.17 Drug resistance mechanism through target mimicry of protein production that inhibits cellular target.

pathway of folate, dihydropteroate synthase (DHPS) and dihydrofolate reductase (DHFR) enzymes are involved because sulphonamides and trimethoprim are structural analogues of their natural substrates, sulphonamides-p-amino-benzoic acid and trimethoprim-di hydro folate. Resistance and overproduction of resistance occur due to mutations in enzymes sulphonamides-DHPS and trimethoprim-DHFR. These medicines work by competitively inhibiting enzymes by binding at their active sites, causing structural changes in the enzyme that interfere with drug binding but still enabling the natural substrate to bind (Huovinen, 1995; Vedantam et al., 1998). However, bacterial resistance to β-lactams can involve the enzymatic hydrolysis of the β-lactam bond within the drug molecules of β-lactam ring. Resistance is mediated by β-lactamases, which can reduce or completely inhibit drug binding ability, and once the β-lactam bond is broken, the drug loses its antibacterial activity. However, fluoroquinolone antibiotics target the synthesis of nucleic acids (Hawkey, 2003; Redgrave et al., 2014), leading to changes in DNA gyrase or topoisomerase IV in gram-negative bacteria. Owing to alteration, the capacity of the medicine is reduced or eliminated to bind gyrase and topoisomerase that modify the structure of these enzymes. Nowadays, sharp upsurges in fluoroquinolones resistance have been observed due to modifications in target enzymes (DNA gyrase and topoisomerase IV) and the development of efflux mechanism in pathogens. Delafloxacin (DFX) is a novel broad-spectrum fluoroquinolone antibiotic that the FDA approved in 2017. DFX has demonstrated bactericidal solid activity against both aerobic gram-positive and gram-negative pathogens. Like other fluoroquinolone antibiotics, DFX disrupts the topology and replication process of bacterial DNA by inhibiting DNA gyrase and topoisomerase IV activity. It is a universal fact that pathogens could develop resistance to a newer generation of antibiotics with time, and there is a huge plausibility those pathogens will find a way to become DFX resistant in the near future (Abu Lila et al., 2022). However, the activity of rifampin is frequently reduced or sluggish by glycosylation, phosphorylation, or adenosine diphosphate ribosylation, and resistance to macrolides and lincosamides can also develop as a result of the drug's enzymatic inactivation or modification (El-Khoury et al., 2022; Davey et al., 2015)

8.8 WORLDWIDE SCENARIO OF DRUG RESISTANCE

A gram-positive bacterium called *S. aureus* is capable of causing a number of infections, most notably those of the skin, soft tissues, bones, and bloodstream. When penicillin was initially developed, it was a successful therapy for *S. aureus* infections. Toxic components of particular strains of *S. aureus* generate some specific symptoms, including toxic shock syndrome and food poisoning.

Since the 1940s, when the β-lactamase enzyme was developed, *S. aureus* has developed resistance to methicillin and other β-lactam antibacterial medications. This type of *S. aureus* is referred to as methicillin-resistant *S. aureus* (MRSA). These strains have a novel gene called mecA, which codes for an exclusive PBP (MRSA) (Reygaert, 2009). As a result, anti-resistance antibacterial medications might be produced by combining methicillin and cloxacillin, β-lactamase-stable, clavulanic acid, and sulbactam β-lactamase inhibitors.

S. pneumoniae (also known as pneumococci) is the most common cause of community-acquired pneumonia, which may include mild, self-limiting infections like acute otitis media and instances of invasive diseases with high death rates like meningitis. The oldest and youngest groups of the population have a disproportionate clinical burden of pneumococcal infection (O'Brien et al., 2009). *S. pneumoniae* acquire β-lactam antibacterial drug resistance via mutations in the genes encoding for PBPs, essential parts of the bacterial cell wall.

E. coli is a natural component of both human and animal intestinal flora. However, it causes urinary tract, kidney, and bloodstream infections at all ages, most frequently. Generally, auto-infection originates from the person affected with *E. coli* (World Health Organization, 2014). Still, disease-causing properties or resistance strains can also be transmitted through the food chain from animals or between individuals. Foodborne disease is the leading cause of *E. coli* infection worldwide; it is also associated with intra-abdominal infections like peritonitis.

Because it provides shelter and gives the cell its structure, the cell wall is necessary for the germ to survive. β-lactam antibiotics kill bacteria by attaching to proteins and preventing cell wall formation or stopping the bacterium from building one in the first place. Extended-spectrum β-lactamases (ESBLs) enzymes, which certain *Enterobacteriaceae* generates, break down and degrade β-lactam antibiotics and cause resistance to third-generation cephalosporins (Miller et al., 2014). Moreover, resistance to broad-spectrum drugs such as penicillin, e.g., ampicillin or amoxicillin, is caused by either acquisition of mobile genetic elements or fluoroquinolone resistance through mutations (Brown et al., 2012). A number of new β-lactams like ceftazidime-avibactam or ceftolozane-tazobactam have activity against these multidrug-resistant bacteria (Knothe et al., 1983).

Furthermore, carbapenem-containing combinations are good alternatives for the management of carbapenemase-producing bacteria. They can treat ESBL-producing bacteria, but resistance enzymes are on the rise, and destroying the antibiotics is the only treatment option for severe infections (Daikos et al., 2014). However, carbapenem resistance in *E. coli* mediated by metallo-β-lactamases is a newly emerging threat, which confers resistance to virtually all available β-lactam antibacterial drugs.

Genus *Klebsiella* is a frequent colonizer of the gut in humans and other vertebrates similar to *E. coli* bacteria. Gram-negative *K. pneumonia* is a common cause of urinary and respiratory tract infections in pre-term infants, and bloodstream infections are widespread in patients with impaired immune systems, diabetes, or alcohol-use disorders and those receiving advanced medical care (Munoz-Price et al., 2013). *K. pneumonia* produces resistance through ESBLs enzyme mainly through horizontal transfer of mobile genetic elements like transposons or plasmids to multiple antibacterial drugs like third-generation cephalosporin and carbapenems.

Like *E. coli*, it acquires resistance to penicillin with an extended spectrum, such as ampicillin and amoxicillin. Contrary to *E. coli*, *K. pneumonae* is also resistant to oral antibacterial drugs like cotrimoxazole and ciprofloxacin, members of the fluoroquinolones.

Salmonella bacteria, common in the intestines of many food-producing animals, including poultry and pigs, are a leading cause of foodborne illness worldwide. Consuming contaminated water or foods of animal origin, particularly undercooked meat, poultry, eggs, and milk, usually results in infection and can cause foodborne outbreaks. For instance, nontyphoidal *Salmonella* exhibits resistance to fluoroquinolones while *Salmonella enterica* serotype Typhimurium, the genomic element that carries resistance to five antimicrobials such as ampicillin, chloramphenicol, streptomycin, sulphonamides, and tetracycline, may spread horizontally among other serotypes and acquire additional resistance factors.

Fluoroquinolones-resistant *Shigella* strains, the leading cause of diarrhoea and dysentery, are spread globally through contact with infected people, contaminated water, or food (Agtini et al., 2005). Globally, it is estimated that there are 165 million *Shigella* incidents annually, 100 million more of which occur in developing countries and result in more than 1 million fatalities. The age group under five years old has the highest incidence, with 69% cases of *Shigella* infection and 61% death (Agtini et al., 2005; Kotloff et al., 2018, 1999; Lee et al., 2005). Cotrimaxazole used to be effective against some *Shigella* strains. However, as resistance has emerged to this antimicrobial, treatment recommendations have shifted to ciprofloxacin or azithromycin (Mishra et al., 2011).

8.9 RECENT ADVANCEMENTS IN NEW DRUG DISCOVERY

Nano medicines (NMs) have frequently been discovered to have novel physical and chemical properties compared to their larger-sized (bulk) counterparts, such as greater chemical reactivity, also when adjusted for the increase in the surface-to-volume ratio and valuable optical properties. New drugs for microbial infections are urgently needed due to the increased resistance of microorganisms to conventional antibiotics. When coated in polymer nanoparticles or enclosed in liposomes, dendrites, or hyper-branched polymers, antibiotic-functionalized nano molecules are efficient in killing multidrug resistance isolates of harmful bacterial species (Pornpattananangkul, 2011; Huang et al., 2011; Huh and Kwon, 2011). Additionally, recently developed metallic nanoparticles with antibacterial properties have become a viable platform for overcoming bacterial resistance (Mohammadi et al., 2011). DFX, a brand-new fluoroquinolone antibiotic, differs significantly from other fluoroquinolones by not having a proton table functional group. It gives the molecule a weakly acidic nature, enhancing its antibacterial activity in an acidic environment. According to a published study, minimum inhibitory concentration (MIC) values were reduced to less than 2 mg mL^{-1} when vancomycin-resistant *S. aureus* was treated in vitro with vancomycin conjugated to folic acid–tagged chitosan nanoparticles as contrasted to unconjugated vancomycin with MIC 50 $mgmL^{-1}$ (Chakraborty et al., 2010). Traditional antibiotics often have one specific target, such as blocking a particular enzyme, in contrast to functionalized NMs, which may have greater drug-delivery qualities than the molecular antibiotic alone. However, we think there is still a chance for resistance to emerge either by altering the target, degrading the antibiotic, or increasing regulation. Co-resistance may also be a factor if genes that enable resistance to natural antimicrobial products (NMs) are physically linked to the same genetic element and are subsequently "co-selected" when each gene offers a benefit against an antimicrobial. The offending organism would get a concentrated dosage of antibiotic rather than the sub-lethal amounts that may allow resistant populations to grow. Therefore, resistance may still be averted provided drug administration is very successful (Saroj, 2022; Davies et al., 2006). If the medicine's killing mechanism is complex or less focused than when it is laser focused, microorganisms may have a more challenging time developing drug resistance.

Furthermore, a recent finding by Mavridou's research group demonstrates that the bacterial protein DsbA helps antibiotic resistance proteins fold into the correct forms. By combining an antibiotic with a DsbA inhibitor, the researchers hope to restore the drugs' long-term ability to eliminate germs. The future development of therapeutically viable DsbA inhibitors may result in a brand-new technique for treating resistant infections by focusing on disulphide bond formation and protein folding across various pathogens and resistance mechanisms (Humphrey et al., 2021).

Researchers recently reported that when ciprofloxacin was combined with the widely used medicinal plant *Fagonia indica*'s (Zygophyllaceae) AgNPs, it had more of an impact directly on the bacteria *E. coli*, *Citrobacter amalonaticus*, *Shigella sonnei*, and *Salmonella typhi* than AgNPs alone did (Adil et al., 2019). Consequently, it is important to consider any potential interactions between antibiotics and AgNPs (Adil et al., 2019; Uttpal Anand et al., 2022).

8.10 CONCLUSION

People are relieved that bacteria will no longer exist in our world owing to the development of antibiotics. Antimicrobial medications have been increasingly resistant to bacteria over time due to their frequent use in clinical settings. The following biochemical resistance strategies have been embraced by bacteria: metabolic pathway bypass, altered permeability, target alteration, and antibiotic inactivation. Genetic analysis methods are helpful in identifying mutations that cause bacterial resistance to antibiotics in all classes. Additionally, to stop resistance, better drug delivery and knowledge of antibiotic development are required. Wide-spectrum microbial exposures and sublethal antibiotic doses become less likely, which may slow the emergence of resistance. Metallic-nanoparticle-mediated antibiotics could act as nano machines. NMs can disrupt multiple cellular processes in a less specific fashion. However, prior to introducing new techniques, it is essential to carefully consider both their effects and any potential threats. A targeted compound with pharmacokinetic and pharmacodynamic properties should be designed with reproducible activity, with a defined chemical structure(s) against one or more bacterial target(s). The target-to-lead optimization phase should focus on advancement while cytotoxicity one of the significant property, should be identified and performed. Also, the activity should be monitored through *in vitro* followed by *in vivo* models of infection.

8.11 FUTURE PROSPECTS

The synthesis of metallic NPs from plant extracts is one of the most notable trends, and it may be a good strategy for defining the mode of action of NPs. It must be a controlled formulation that is easily scaled up, guarantees ecological sustainability, and diminishes the process's impact on the environment. Also, Artificial intelligence technologies should lead developed and developing countries to improve upon the existing treatment mode. Communication and teamwork between academic researchers and pharmaceutical companies can accelerate NP- and AI-based target drug discovery.

8.12 ACKNOWLEDGEMENT

The authors are grateful to the Department of Chemistry, Integral University, and the Department of Applied Chemistry, University Polytechnic, Integral University, Lucknow, for facilitating research environment.

REFERENCES

Abu Lila, A. S., Huwaimel, B., Alobaida, A., Hussain, T., Rafi, Z., Mehmood, K., . . . & Ahmed, A. F. (2022). Delafloxacin-capped gold nanoparticles (DFX-AuNPs): an effective antibacterial nano-formulation of fluoroquinolone antibiotic. *Materials*, *15*(16), 5709.

Adil, M., Khan, T., Aasim, M., Khan, A. A., & Ashraf, M. (2019). Evaluation of the antibacterial potential of silver nanoparticles synthesized through the interaction of antibiotic and aqueous callus extract of Fagonia indica. *AMB Express*, *9*(1), 1–12. https://doi.org/10.1186/s13568-019-0797-2

Agtini, M. D., Soeharno, R., Lesmana, M., Punjabi, N. H., Simanjuntak, C., Wangsasaputra, F., . . . & Clemens, J. D. (2005). The burden of diarrhoea, shigellosis, and cholera in North Jakarta, Indonesia: Findings from 24 months surveillance. *BMC Infect. Dis.*, *5*(1), 1–11. DOI: 10.1186/1471-2334-5-89.

Anand, U., Carpena, M., Kowalska-Góralska, M., Garcia-Perez, P., Sunita, K., Bontempi, E., . . . & Simal-Gandara, J. (2022). Safer plant-based nanoparticles for combating antibiotic resistance in bacteria: A comprehensive review on its potential applications, recent advances, and future perspective. *Sci. Tot. Environ.*, 153472. https://doi.org/10.1016/j.scitotenv.2022.153472.

Bébéar, C. M., & Pereyre, S. (2005). Mechanisms of drug resistance in Mycoplasma pneumoniae. *Cur. Drug Tar. Infec. Dis*, *5*(3), 263–271.

Beceiro, A., Tomás, M., & Bou, G. (2013). Antimicrobial resistance and virulence: A successful or deleterious association in the bacterial world? *Clin. Microbio. Rev.*, *26*(2), 185–230.

Benveniste, R., & Davies, J. (1973). Structure-activity relationships among the aminoglycoside antibiotics: Role of hydroxyl and amino groups. *Antimicro. Agent. chemo.*, *4*(4), 402–409. DOI: 10.1128/AAC.4.4.402

Blair, J. M., Richmond, G. E., & Piddock, L. J. (2014). Multidrug efflux pumps in Gram-negative bacteria and their role in antibiotic resistance. *Fut. Microbio.*, *9*(10), 1165–1177.

Brooks, G. F., Butel, J. S., & Morse, S. A. (2004). *Jawetz, Melnick & Adelberg's Medical Microbiology* (23rd ed.). McGraw Hill Companies.

Brown, A. N., Smith, K., Samuels, T. A., Lu, J., Obare, S. O., & Scott, M. E. (2012). Nanoparticles functionalized with ampicillin destroy multiple-antibiotic-resistant isolates of Pseudomonas aeruginosa and Enterobacter aerogenes and methicillin-resistant *Staphylococcus aureus. App. Environ. Microbio.*, *78*(8), 2768–2774.

Bush, K., & Bradford, P. A. (2016). β-Lactams and β-lactamase inhibitors: An overview. *Cold Spr. Har. Pers. mede.*, *6*(8), a025247.

Bushby, S. R., & Hitchings, G. H. (1968). Trimethoprim, a sulphonamide potentiator. *Br. J. Pharmacol. Chemother.*, *33*(1), 72.

Capasso, C., & Supuran, C. T. (2014). Sulfa and trimethoprim-like drugs–antimetabolites acting as carbonic anhydrase, dihydropteroate synthase and dihydrofolate reductase inhibitors. *J. Enzyme Inhibi. Med. Chem.*, *29*(3), 379–387.

Chakraborty, S. P., Sahu, S. K., Mahapatra, S. K., Santra, S., Bal, M., Roy, S., & Pramanik, P. (2010). Nanoconjugated vancomycin: New opportunities for the development of anti-VRSA agents. *Nanotechnology*, *21*(10), 105103.

Chancey, S. T., Zähner, D., & Stephens, D. S. (2012). Acquired inducible antimicrobial resistance in Gram-positive bacteria. *Fut. Microbio.*, *7*(8), 959–978.

Daikos, G. L., Tsaousi, S., Tzouvelekis, L. S., Anyfantis, I., Psichogiou, M., Argyropoulou, A., . . . & Skoutelis, A. (2014). Carbapenemase-producing Klebsiella pneumoniae bloodstream infections: Lowering mortality by antibiotic combination schemes and the role of carbapenems. *Antimicro Agents and Chemothe*, *58*(4), 2322–2328.

Davey, P., Wilcox, M. H., Irving, W., & Thwaites, G. (2015). *Antimicrobial Chemotherapy*. Oxford University Press.

Davies, J., Spiegelman, G. B., & Yim, G. (2006). The world of sub inhibitory antibiotic concentrations. *Curr. Opi. Microbio.*, *9*(5), 445–453.

Davis, K. M., & Weiser, J. N. (2011). Modifications to the peptidoglycan backbone help bacteria to establish infection. *Infect. Immun.*, *79*(2), 562–570.

Deak, D., Outterson, K., Powers, J. H., & Kesselheim, A. S. (2016). Progress in the fight against multidrug-resistant bacteria? A review of US food and drug administration – approved antibiotics, 2010–2015. *Ann. Inter. Med.*, *165*(5), 363–372.

Domagk, G. (1935). A contribution to the chemotherapy of bacterial infections. *DMW-Ger. Med. Weekly*, *61*(7), 250–253.

Džidić, S., Šušković, J., & Kos, B. (2008). Antibiotic resistance mechanisms in bacteria: Biochemical and genetic aspects. *Food Tech. Biotech.*, *46*(1).

El-Khoury, C., Mansour, E., Yuliandra, Y., Lai, F., Hawkins, B. A., Du, J. J., . . . & Groundwater, P. W. (2022). The role of adjuvants in overcoming antibacterial resistance due to enzymatic drug modification. *RSC Med. Chem.*, *13*(11), 1276–1299.

European Medicine Agency. (2009). The bacterial challenge: time to react. A call to narrow the gap between multidrug-resistant bacteria in the EU and the development of new antibacterial agents. *Technical Report.* Available at www.ema.europa.eu/docs/en_GB/document_library/Report/2009/11/WC500008770.pdf. Accessed: December 20, 2013.

Fair, R. J., & Tor, Y. (2014). Antibiotics and bacterial resistance in the 21st century. *Persp. Med. Chem.*, *6*, PMC-S14459. PMC-S14459. DOI: 10.4137/PMC.S14459.

Grose, W. E., Bodey, G. P., & Loo, T. L. (1979). Clinical pharmacology of intravenously administered trimethoprim-sulfamethoxazole. *Antimicro. Agen. Chem.*, *15*(3), 447–451.

Grundmann, H., Aires-de-Sousa, M., Boyce, J., & Tiemersma, E. (2006). Emergence and resurgence of meticillin-resistant *Staphylococcus aureus* as a public-health threat. *The Lancet, 368*(9538), 874–885.

Hamilton-Miller, J. M. T. (1999). β-Lactams: Variations on a chemical theme, with some surprising biological results. *J. Antimicrob. Chemther., 44*(6), 729–734. DOI: 10.1093/jac/44.6.729Hawkey, P. M. (2003). Mechanisms of quinolone action and microbial response. *J. Antimicrob. Chemother., 51*(suppl_1), 29–35.

Higgins, P. G., Fluit, A. C., & Schmitz, F. J. (2003). Fluoroquinolones: Structure and target sites. *Curr. Drug Tar., 4*(2), 181–190.

Hiramatsu, K., Cui, L., Kuroda, M., & Ito, T. (2001). The emergence and evolution of methicillin-resistant Staphylococcus aureus. *Tren. Microbio., 9*(10), 486–493.

Huang, C. M., Chen, C. H., Pornpattananangkul, D., Zhang, L., Chan, M., Hsieh, M. F., & Zhang, L. (2011). Eradication of drug resistant Staphylococcus aureus by liposomal oleic acids. *Biomaterials, 32*(1), 214–221.

Huang, Y., Yan, S., Li, Y., Ai, X., Yu, X., Ge, Y., . . . & Xie, J. (2022). Mycobacterium fluoroquinolone resistance protein D (MfpD), a GTPase-activating protein of GTPase MfpB, is involved in fluoroquinolones potency. *Microbiol. Spectr., 10*(6), e02098-22.

Huh, A. J., & Kwon, Y. J. (2011). "Nanoantibiotics": A new paradigm for treating infectious diseases using nanomaterials in the antibiotics resistant era. *J. Contr. Rel., 156*(2), 128–145.

Humphrey, M., Larrouy-Maumus, G. J., Furniss, R. C. D., Mavridou, D. A., Sabnis, A., & Edwards, A. M. (2021). Colistin resistance in Escherichia coli confers protection of the cytoplasmic but not outer membrane from the polymyxin antibiotic. *Microbio., 167*(11), 001104. DOI: 10.1099/mic.0.001104

Huovinen, P., Sundström, L., Swedberg, G., & Sköld, O. J. A. A. (1995). Trimethoprim and sulfonamide resistance. *Antimicro. Agen. chemothe., 39*(2), 279–289.

Ibrahim, H. R., Thomas, U., & Pellegrini, A. (2001). A helix-loop-helix peptide at the upper lip of the active site cleft of lysozyme confers potent antimicrobial activity with membrane permeabilization action. *J. Biol. Chem., 276*(47), 43767–43774.

Johnston, N. J., Mukhtar, T. A., & Wright, G. D. (2002). Streptogramin antibiotics: Mode of action and resistance. *Curr. Drug Targ., 3*(4), 335–344.

Kahne, D., Leimkuhler, C., Lu, W., & Walsh, C. (2005). Glycopeptide and lipoglycopeptide antibiotics. *Chem Rev., 105*(2), 425–448.

Kapoor, G., Saigal, S., & Elongavan, A. (2017). Action and resistance mechanisms of antibiotics: A guide for clinicians. *J. Anaesthesiol. Clin. Pharmacol., 33*(3), 300.

Knothe, H., Shah, P., Krcmery, V., Antal, M., & Mitsuhashi, S. (1983). Transferable resistance to cefotaxime, cefoxitin, cefamandole and cefuroxime in clinical isolates of Klebsiella pneumoniae and Serratia marcescens. *Infection, 11*(6), 315–317.

Kohanski, M. A., Dwyer, D. J., & Collins, J. J. (2010). How antibiotics kill bacteria: From targets to networks. *Nat. Rev. Microbio., 8*(6), 423–435.

Kotloff, K. L., Riddle, M. S., Platts-Mills, J. A., Pavlinac, P., & Zaidi, A. K. (2018). Shigellosis. *The Lancet, 391*(10122), 801–812.

Kotloff, K. L., Winickoff, J. P., Ivanoff, B., Clemens, J. D., Swerdlow, D. L., Sansonetti, P. J., . . . & Levine, M. M. (1999). Global burden of Shigella infections: Implications for vaccine development and implementation of control strategies. *Bull. Wor. Health Org., 77*(8), 651.

Kumar, A., & Schweizer, H. P. (2005). Bacterial resistance to antibiotics: Active efflux and reduced uptake. *Adv. Drug Del. Rev., 57*(10), 1486–1513.

Lambert, P. A. (2002). Cellular impermeability and uptake of biocides and antibiotics in Gram-positive bacteria and mycobacteria. *Jour. Appl. Microbio., 92*(s1), 46S–54S.

Lee, H., Kotloff, K., Chukaserm, P., Samosornsuk, S., Chompook, P., Deen, J. L., . . . & Wanpen, C. (2005). Shigellosis remains an important problem in children less than 5 years of age in Thailand. *Epidemiol. Infect., 133*(3), 469–474.

Mehramuz, B., Taghizadeh, S., Kafil, H. S., Zonouzaq, G. Y., Khiabani, M. S., Sheikhsaran, E., . . . & Dehghani, L. (2020). High rate of contamination with Staphylococcus aureus in traditional Koozeh cheeses. A molecular typing approach. *An. Ig.: Med. Prev. Comunita., 32*(2).

Miller, W. R., Munita, J. M., & Arias, C. A. (2014). Mechanisms of antibiotic resistance in enterococci. *Exp. Rev. Anti-Inf. Ther., 12*(10), 1221–1236.

Mishra, M. K., Kotta, K., Hali, M., Wykes, S., Gerard, H. C., Hudson, A. P., . . . &Kannan, R. M. (2011). PAMAM dendrimer-azithromycin conjugatenanodevices for the treatment of Chlamydia trachomatis infections. *Nanomedicine: Nanotech, Biol. Med, 7*(6), 935–944.

Mohammadi, G., Nokhodchi, A., Barzegar-Jalali, M., Lotfipour, F., Adibkia, K., Ehyaei, N., & Valizadeh, H. (2011). Physicochemical and anti-bacterial performance characterization of clarithromycin nanoparticles as colloidal drug delivery system. *Colloids Surf. B: Biointerfaces, 88*(1), 39–44.

Munoz-Price, L. S., Poirel, L., Bonomo, R. A., Schwaber, M. J., Daikos, G. L., Cormican, M., . . . & Quinn, J. P. (2013). Clinical epidemiology of the global expansion of *Klebsiella pneumoniae carbapenemases. The Lanc. Infect. Diseas.*, *13*(9), 785–796. DOI: 10.1016/S1473-3099(13)70190-7.

O'Brien, K. L., Wolfson, L. J., Watt, J. P., Henkle, E., Deloria-Knoll, M., McCall, N., . . . & Cherian, T. (2009). Burden of disease caused by Streptococcus pneumoniae in children younger than 5 years: Global estimates. *The Lancet*, *374*(9693), 893–902. DOI: 10.1016/S0140-6736(09)61204-6.

Pornpattananangkul, D., Zhang, L., Olson, S., Aryal, S., Obonyo, M., Vecchio, K., . . . & Zhang, L. (2011). Bacterial toxin-triggered drug release from gold nanoparticle-stabilized liposomes for the treatment of bacterial infection. *Jour. Ame. Chem. Soci.*, *133*(11), 4132–4139.

Redgrave, L. S., Sutton, S. B., Webber, M. A., & Piddock, L. J. (2014). Fluoroquinolone resistance: Mechanisms, impact on bacteria, and role in evolutionary success. *Tren. Microbio.*, *22*(8), 438–445.

Reygaert, W. (2009). Methicillin-Resistant Staphylococcus Aureus (MRSA): Prevalence and epidemiology issues. *Clin. Labor. Sci*, *22*(2), 111.

Reynolds, P., Bosma, N., Autio, E., Hunt, S., De Bono, N., Servais, I., & Chin, N. (2005). Global entrepreneurship monitor: Data collection design and implementation 1998–2003. *Small Bus. Econ.*, *24*, 205–231.

Reynolds, P. E. (1989). Structure, biochemistry and mechanism of action of glycopeptide antibiotics. *Euro. Jour. Clin. Microbio. Infec. Dis.*, *8*, 943–950.

Rickey, D. P., & Brown, G. M. (1971). Hydroxymethyldihydropteridine pyrophosphokinase and dihydropteroate synthetase from Escherichia coli. *Methods Enzymol.*, *18B*, 765–771.

Saroj, S. D. (Ed.). (2022). *Antimicrobial Resistance: Collaborative Measures of Control.* CRC Press.

Schlegel, H. G. (2003). *General Microbiology* (7th ed.). Cambridge University Press.

Spencer, A. C., & Panda, S. S. (2023). DNA gyrase as a target for quinolones. *Biomedicines*, *11*(2), 371.

Spratt, B. G. (1994). Resistance to antibiotics mediated by target alterations. *Science*, *264*(5157), 388–393.

Sun, Q., Liao, X., Wang, C., Jiang, G., Yang, J., Zhao, J., . . . & Li, H. (2022). In vitro activity of fidaxomicin against nontuberculosis mycobacteria. *J. Med. Microbiol.*, *71*(6), 001549.

Terzi, H. A., Kulah, C., Atasoy, A. R., & Ciftci, I. H. (2015). Investigation of OprD porin protein levels in carbapenem-resistant Pseudomonas aeruginosa isolates. *Jund. Jour. Microbio.*, *8*(12). PMID: 26865937. DOI: 10.5812/jjm.25952

Tréfouël, J., Tréfouël, T. J., Nitti, F., & Bovet, D. (1935). Activité du p-aminophénylsulfonamide sur les infections streptococciques expérimentales de la souris et du lapin. *CR Soc. Biol. (Paris)*, *120*, 756–758.

Trimble, M. J., Mlynárčik, P., Kolář, M., & Hancock, R. E. (2016). Polymyxin: Alternative mechanisms of action and resistance. *Cold Spr. Harb. Perspec. Med.*, *6*(10), a025288.

Vannuffel, P., & Cocito, C. (1996). Mechanism of action of *streptogramins and macrolides. Drugs*, *51*(Suppl 1), 20–30.

Vedantam, G., Guay, G. G., Austria, N. E., Doktor, S. Z., & Nichols, B. P. (1998). Characterization of mutations contributing to sulfathiazole resistance in Escherichia coli. *Antimicro. Agen. chemo.*, *42*(1), 88–93.

Walsh, C. (2003). *Antibiotics: Actions, Origins, Resistance*. American Society for Microbiology (ASM).

Walsh, F. (2007). Doripenem: A new carbapenem antibiotic a review of comparative antimicrobial and bactericidal activities. *Therapeu. Clin. Risk Manag.*, *3*(5), 789–794.

Wise, R. (1999). Pharmacokinetics and pharmacodynamics of antimicrobials. In *First International Moxifloxacin Symposium: Berlin, 1999* (pp. 89–93). Springer.

Woods, D. D. (1940). The relation of p-aminobenzoic acid to the mechanism of the action of sulphanilamide. *Brit. Jour. Exper. Path.*, *21*(2), 74.

World Health Organization. (2014). *Antimicrobial Resistance: Global Report on Surveillance*. World Health Org.

Yoneyama, H., & Katsumata, R. (2006). Antibiotic resistance in bacteria and its future for novel antibiotic development. *Biosci., Biotech., Biochem.*, *70*(5), 1060–1075.

9 Applications of Artificial Intelligence in Antimicrobial Research

An Emerging Tool for Translational Research

Ravindra Kumar, Rajrani Ruhel, and Sohani Singh Jain

9.1 INTRODUCTION

Antimicrobial resistance (AMR) is a continuous process in which microbes develop the ability to survive against antimicrobial agents designed to kill them. Antibiotics are chemical compounds used to treat bacterial infections, but constant exposure to these agents has led to the emergence of drug-resistant bacteria and the threat of super-resistant strains. AMR does not mean that our bodies are resistant to antimicrobial agents; instead, it implies that the microbes causing an infection are resistant to these agents. AMR can be defined as the genetic ability of bacteria to encode resistance genes that counteract the inhibitory effects of potential antibiotics, enabling their survival. This can occur through natural recombination or horizontal gene transfer (Khan et al., 2019). Different bacteria use different mechanisms of action to evade antibiotics and develop resistance against them (Reygaert, 2018), such as (i) reducing intracellular antibiotics accumulation by decreasing permeability and/or increasing active efflux of the antibiotics, (ii) alteration of drug target site, and (iii) modification of the antibiotics resulting in its inactivation. Antibiotic resistance is a major threat to global health, with recent research indicating that 4.95 million deaths were associated with bacterial antimicrobial resistance in 2019 (Murray C.J.L. et al., 2022). Antibiotic resistance is a growing problem with significant implications for public health and the economy. According to estimates, it is expected to cause approximately 10 million deaths annually by 2050, and the economic cost of antibiotic resistance is predicted to exceed $100 billion (de Kraker et al., 2016). These projections highlight the urgency of addressing the issue and finding solutions to mitigate the impact of antibiotic resistance. However, some researchers have dismissed this forecast, but they agree that there is a significant clinical and public health burden associated with AMR. Nevertheless, it is challenging to quantify the excess morbidity and mortality associated with it (de Kraker et al., 2016).

Due to the extremely high mortality rate of antibiotic-resistant infections, appropriate antibiotic therapy is crucial for reducing the mortality associated with these infections. According to the State of the World's Antibiotics Report in 2001, global antibiotics consumption in humans increased by 65% between 2000 and 2015, and consumption in animals is expected to increase by 11.5% from 2017 to 2030. If these trends continue, worldwide antibiotic consumption is likely to increase by 200% from 2015 to 2030. Hence, there is a pressing need to implement new policies and make research efforts to address the antimicrobial resistance crisis (Lv et al., 2021).

Because traditional high-throughput screenings of antibiotics are very expensive and time consuming, the aforementioned challenges could largely be overcome by using computational approaches. Machine learning (ML), a subset of artificial intelligence (AI), is a rapidly evolving

DOI: 10.1201/9781003269380-9

field and is increasingly being harnessed by many pharmaceutical companies into the drug development process to automate repetitive data processing and also for analysis tasks.

ML tasks are broadly categorized into three types: supervised learning, unsupervised learning, and sequential learning. The data used in ML can be divided into two categories: labelled data and unlabelled data. Supervised learning relies on a labelled data that acts as a trainer, teaching the model. After training, the model can make predictions and decisions as new data is received. Deep learning, random forest, and support vector machines fall under the category of supervised learning. Deep learning uses artificial neural networks (ANNs) to identify highly complex patterns in large datasets (Webb, 2018). Unsupervised learning identifies the relationships or patterns in unlabelled data. The model learns through observation and creates clusters of the observed patterns and relationships in the dataset.

Machine learning techniques can accelerate the identification of novel antibiotics/antimicrobial peptide (Durrant and Amaro, 2015). It is well suited to assist in drug discovery for antimicrobial research due to the following reasons: (a) It has a remarkable ability to pre-process and analyze data quickly. (b) It can uncover important biological features in complex biological datasets, which leads to the development of an accurate prediction model and saves a great deal of effort, time, and money. (c) It is easy to train a machine learning model to uncover hidden information from data. A crucial aspect of working with machine learning models is choosing the right architecture and selecting an appropriate machine learning model, as well as the data on which the model is trained.

However, most of the previous reviews about antibiotic resistance have been made from a structural and molecular mechanism perspective. This chapter is focused on (1) the basics of machine learning algorithm, (2) commonly used machine learning tools in AMR, and (3) the challenges and limitations of machine learning in AMR, and finally we conclude with (4) the trends and future directions in this field along with the machine learning techniques.

9.2 BASIC CONCEPT AND STEPS OF MACHINE LEARNING

Every machine learning has five major steps: (1) data collection, (2) data pre-processing, (3) feature extraction, (4) training and model development, and (5) model evaluation (Figure 9.1). A crucial step of every machine learning model is the collection of high-quality datasets. The quality and quantity of the data dictates how accurate the model is. After data collection the dataset pre-processing is important. The real-world data has unorganized, missing, and noisy elements; thus, before feeding the data into machine learning, pre-processing is done to clean the dataset by removing duplicates, missing values, etc. This is a very critical step, and people need to be very careful when working with data pre-processing. The models developed using noisy data will fail or will be unreliable on real-life data.

After cleaning/pre-processing, we need to convert the raw data into a machine readable format and identify the most useful features for a given task. These selected features are described as a vector $x = [x_1, x_2, x_3, x_4 \ldots, x_n]$, where n is the dimension of the feature vector (He et al., 2021). The transformation of these raw data into a machine readable format is called feature extraction. Finally, the dataset is split into training and test sets. The training set is used to train the model while the test set is used to validate the model.

The goal of the training is to answer a biological question for which we feed the appropriate dataset into the machine learning algorithm and train the model. In terms of uses, the machine learning model can be divided into two types: (1) classification model and (2) regression model. The classification model is generally used to output the label or category of the given input variables, for example, prediction and classification of beta-lactamase enzyme from non-beta-lactamase enzyme (Kumar et al., 2015; Srivastava et al., 2018), predicting the susceptibility of a microorganism towards an anti-infective agent. A regression model is commonly used to predict continuous values such as prediction of antibacterial capacity of nanoparticles (Mirzaei et al., 2021). Regression analysis is an important tool for modelling and analyzing the data. Here, we fit a curve to the data points in such a way that the difference between the distance of data points from the curve

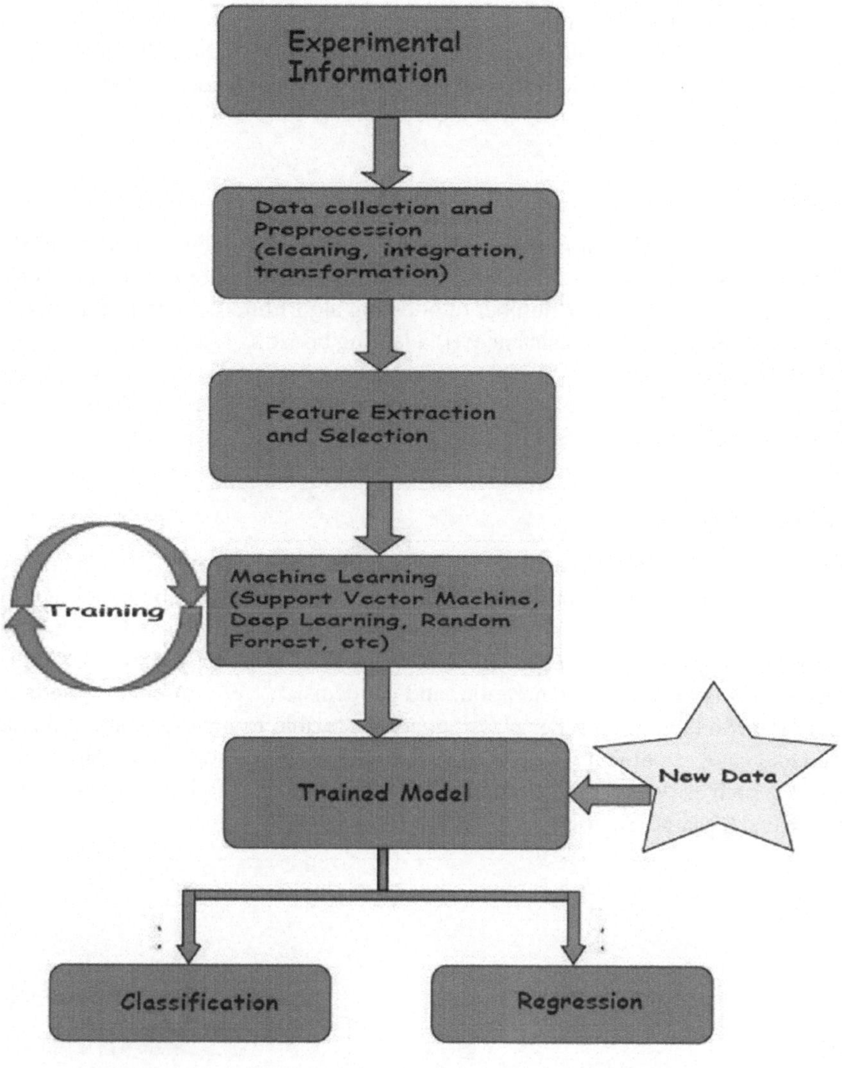

FIGURE 9.1 Different steps of machine learning.

is minimized. The performance of the classification model is evaluated on the basis of different measures such as accuracy, sensitivity, specificity, Matthews correlation coefficient (MCC), area under ROC curve (AUC), precision-recall plot, recall, and F-measure while a regression model is evaluated on the basis of mean squared error (MSE), mean absolute error (MAE), correlation, and R^2 for performance evaluation.

9.3 COMMONLY USED MACHINE LEARNING ALGORITHMS FOR AMR

Over the past few decades, very few antibiotics have been developed, and most of those newly approved antibiotics are slightly different variants of existing drugs. Traditional culture-based methods are costly, time consuming, and are usually limited to a narrow spectrum of chemical diversity (Burnham et al., 2017). Machine learning has proven to be a powerful method to predict antimicrobial resistance without using prior knowledge for selected bacterial species–antimicrobial combinations. There are several antimicrobial databases available publicly for data collection and model development (Table 9.1). A machine learning model can explore large chemical spaces that can be prohibitively expensive by traditional experimental approaches. Several machine learning algorithms are used in antimicrobial research, as shown in Figure 9.2 and Table 9.2. The choice of algorithm will depend on the specific research questions, the quality and availability of data, and the computational resources available.

9.3.1 Boost

Boost is a type of ensemble learning which combines multiple models, called weak learners, and generates the final model into strong learners. The objective of these predictive models is to improve the overall accuracy rate. There are a number of boosting algorithms available (Rahman et al., 2020) used in prediction of antimicrobial resistance: (i) adapting boosting (AdaBost), (ii) extreme gradient boosting (XGBoost) (Van Camp et al., 2020), (iii) gradient boosting, which is highly customizable to the particular needs of the applications (Natekin and Knoll, 2013), (iv) CatBoost, and (v) light gradient boosting machine (LightGBM).

9.3.2 Decision Tree

Decision trees are rule-based machine learning algorithms (He et al., 2021) which are used for classification problems. This is a non-parametric algorithm, and it classifies a population into bench-like segments that construct an inverted tree with a root node, internal node, and leaf nodes (Song and Lu, 2015). The learning process of a decision tree consists of three basic steps (Lv et al., 2021): (1) feature selection, (2) decision tree generation, and (3) decision tree pruning. Sometimes, decision trees are more interpretable than other classifiers; it is flexible to handle a large complex mixture of real valued and categorical features as well as the data with missing values. However, it also has some drawbacks, such as that a small change in the input data can lead to large changes in the constructed tree (Kingsford and Salzberg, 2008).

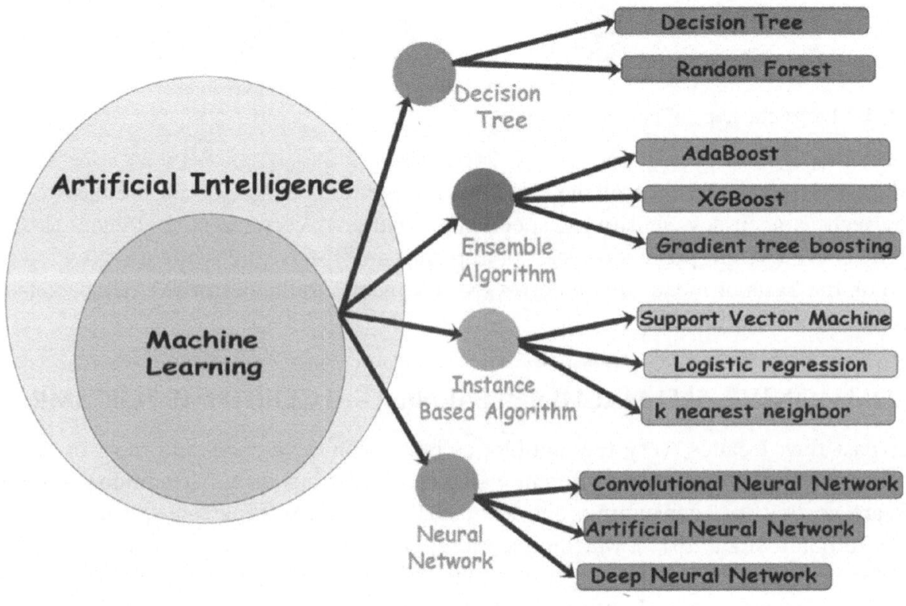

FIGURE 9.2 Schematic representation of popular machine learning used in antibiotics research.

TABLE 9.1
PUBLICLY AVAILABLE ANTIMICROBIAL DATABASE

Database	Description	Website
Antimicrobial Peptide Database (APD3) (Wang et al., 2016)	The Antimicrobial Peptide Database (APD) contains 3,569 antimicrobial peptides from six life kingdoms including bacteria, archaea, protists, fungi, plants, and animals.	https://aps.unmc.edu/
The Comprehensive Antibiotics Resistance Database (CARD) (Alcock et al., 2023)	A database of resistance genes, their products, and associated phenotypes.	https://card.mcmaster.ca
MEGARes (Bonin et al., 2023)	A comprehensive database of published AMR gene sequences for antimicrobial drugs, biocides, and metals. AMR++ v3.0, an update to customized bioinformatics pipeline for high-throughput analysis of metagenomics data.	https://megares.meglab.org
Antimicrobial Chemotherapeutics Database (ACD) (Azam et al., 2020)	Database of synthetic antimicrobial compounds and active antimicrobial peptides.	http://amdr.amu.ac.in/acd
Data Repository of Antimicrobial Peptides (DRAMP) (Shi et al., 2022)	Manually curated database of AMPs including sequences, structures, activities, physicochemical, patent, clinical, and reference information.	http://dramp.cpu-bioinfor.org
Community for Open Antimicrobial Drug Discovery (COADD) (Desselle et al., 2017)	Contains chemical structures and antimicrobial activity data to help screen compounds for antimicrobial activity for academic research and generate a public knowledge database for the development of novel agents for the treatment of microbial infections.	https://db.co-add.org
Collection of Anti-Microbial Peptides ($CAMP_{R3}$) (Waghu and Idicula-Thomas, 2020)	It is a database of AMP from 45 different families based on conserved sequence signatures captured as patterns and hidden Markov models (HMMs).	www.camp3.bicnirrh.res.in
dbAMP (Jhong et al., 2019)	It is an integrated database that collect AMPs from public database and literature.	http://csb.cse.yzu.edu.tw/dbAMP/
Shared Platform for Antibiotic Research and Knowledge (SPARK) (Thomas et al., 2018)	It is a cloud-based virtual laboratory that enables scientists to share data. It integrates chemical and biological data from published studies as well as previously unpublished data and provides an opportunity for real-time collaboration among scientists.	www.pewtrusts.org/en/research-and-analysis/articles/2018/09/21/the-shared-platform-for-antibiotic-research-and-knowledge
AntibioticDB (Farrell et al., 2018)	It is an open access database of antibacterial compounds that indicates the stage of development and those in clinical trials.	https://antibioticdb.com
FEAMR (Mishra et al., 2022)	It provides manually collected and curated information on the prevalence of AMR in food and the environment.	https://feamrudbt-amrlab.mu.ac.in
Database of Antimicrobial Activity and Structure of Peptides (DBAASP) (Pirtskhalava et al., 2021)	It is a manually curated database which contains >15,700 AMPs.	https://dbaasp.org
Linking Anti-Microbial Peptide (LAMP) database (Ye et al., 2020)	LAMP is an integrated database which contains 23,253 unique AMP sequences from 16 public AMP databases.	http://biotechlab.fudan.edu.cn/database/lamp/index.php

TABLE 9.2
Publicly Available Tools for Predicting Antimicrobial Peptides/Proteins

Prediction tool	Used algorithm	Description	Website
AntiBP2 (Lata et al., 2010)	SVM	Predict antibacterial peptides in protein sequences	www.imtech.res.in/raghava/antibp/
AVPpred (Thakur et al., 2012)	SVM	Predict antiviral peptides	http://crdd.osdd.net/servers/avppred
ClassAMP (Joseph et al., 2012)	RF, SVM	Predict AMP domains in protein sequences	www.bicnirrh.res.in/classamp/
$CAMP_{R3}$ (Waghu et al., 2016a)	SVM, RF, ANN, DA	Predict AMPs	www.camp3.bicnirrh.res.in/index.php
CAMPSign (Waghu et al., 2016b)	HMM	Identification of antimicrobial peptides.	www.campsign.bicnirrh.res.in
iAMPpred (Meher et al., 2017)	SVM	Predict AMPs	http://cabgrid.res.in:8080/amppred/
AMPscanner (Veltri et al., 2018)	DNN	Predict AMPs	www.dveltri.com/ascan/v2/about.html
dbAMP (Jhong et al., 2019)	RF/BLASTP	Predict AMPs	http://140.138.77.240/~dbamp/index.php
AMAP (Gull et al., 2019)	SVM, XGBoost	Predict biologically active peptides and AMPs	http://amap.pythonanywhere.com/
AntiVPP1.0 (Beltran Lissabet et al., 2019)	RF	Predict antiviral peptides	https://github.com/bio-coding/AntiVPP
Meta-iAVP (Schaduangrat et al., 2019)	k-NN, glm, RF, XGBoost, SVM	Predict antiviral peptides	http://codes.bio/meta-iavp/
ACEP (Fu et al., 2020)	DNN	Predict AMPs	https://github.com/Fuhaoyi/ACEP
AniAMPpred (Sharma et al., 2021)	SVM, DNN	Identify the antimicrobial function of proteins	https://aniamppred.anvil.app/

SVM: support vector machine; RF: random forest; HMM: hidden Markov model; ANN: artificial neural network; DA: discriminant analysis; DNN: deep neural network; KNN: K-nearest neighbors.

9.3.3 Random Forest

Random forest is a supervised machine learning technique which is used for both regression as well as classification problems. Random forests consist of multiple decision trees that operate as an ensemble to improve accuracy and reliability. Each individual tree in the random forest gives a class prediction, and the class with the most votes become the prediction model. Random forest models have been commonly used to predict antibiotic combination (Chandrasekaran et al., 2016).

Random forest consists of an ensemble of decision trees created by using bootstrap samples of the training data and random feature selection in tree induction. Prediction is made by aggregating the predictions of the ensemble (Svetnik et al., 2003).

9.3.4 Support Vector Machine

Support vector machine (SVM) is the most widely used supervised machine learning algorithm, and it efficiently classifies linear as well as nonlinear high-dimensional complex biological data by using different kernel functions (Kumar et al., 2015). It is a kernel-based nonlinear classifier that finds a separating hyperplane between the classes to minimize the risk of classification error (Said Agrebi, 2020). Kernels are the functions that transform low-dimensional input space to a high-dimensional space where the classes can be separated. The following parameters of SVM can be used to train the model: penalty parameter ($1.0 \leq C \leq 10.0$) and kernel function such as linear, polynomial, radial basis function, and sigmoid (Ivanenkov et al., 2019). SVM models have been extensively used in antimicrobial research such as beta-lactamase protein prediction and classification by Kumar et al. (2015) to predict the beta-lactamase enzyme and to further classify into its different classes. Liu et al. (2020) developed an SVM-based model for identifying strains resistant or susceptible to certain antibiotics, with more than 90% accuracy, to predict the resistance of five drugs, tetracycline, ampicillin, sulfisoxazole, trimethoprim, and enrofloxacin. SVM models have been considered a most potential tool in antimicrobial research due to their ability to handle high-dimensional data and their robust performance in the presence of noisy data.

9.3.5 Deep Neural Network

Deep learning is a class of machine learning algorithms that uses ANNs with many layers of nonlinear processing units for learning data representations. ANN is a computational model that attempts to mimic the structure and function of neural networks in the human brain (Jamal et al., 2019). There are three basic layers in an ANN: input layer, hidden layer, and output layer (Chen et al., 2018).

A large amount of research has been done using deep neural networks for antimicrobial research such as prediction of antibiotic resistance in *E. coli* from large scale pan-genome data (Moradigaravand et al., 2018). In 2020, Aytan-Aktug et al. (2020) developed an ANN-based multi-output and species independent model for prediction of antimicrobial resistance. They combined multiple bacterial species to develop species-independent machine learning models that can predict resistance profiles for multiple antimicrobials and species with high performance. They found that species-independent models can predict multi-AMR profiles for multiple species without losing any robustness.

In 2020, researchers at MIT, Stokes and co-author (Stokes et al., 2020), discovered an antibiotic, halicin, by using a deep neural network. It is the first discovered antibiotic identified by artificial intelligence. Researchers trained a deep neural network model to spot molecules that inhibit the growth of *E. coli*, using a collection of 2,335 molecules for which the antibacterial activity was known. This model learned the relationship between chemical structure and antibacterial activity in a manner that allowed it to identify the sets of chemicals it had never seen before, and it could then make predictions about whether these new molecules possessed antibacterial activity against *E. coli* or not.

In 2021, Pascucci et al. (Pascucci et al., 2021) presented a convolutional neural network–based fully offline smartphone application for antibiogram analysis, a broadly used test for antibiotic resistance in bacterial infections. The application captures images with the phone's camera, and the users are guided throughout the analysis on the same device by a user-friendly graphical interface.

In 2021, Das et al. (Das et al., 2021) developed a computational tool based on AI and molecular dynamics simulations, deep generative autoencoder, for generation of antimicrobials with desired attributes and its validation on discovering broad-spectrum, low toxicity, and low resistance antimicrobial peptides. Using this model, they have identified, synthesized, and experimentally tested 20 antimicrobial peptide candidates. Out of these 20, 2 peptides displayed high potency against diverse gram-positive and gram-negative bacteria. This study demonstrated that the combination of deep learning and molecular dynamics may accelerate the discovery of potent and selective broad-spectrum antimicrobials.

9.4 MACHINE LEARNING IN ANTIBIOTICS RESISTANCE PREDICTION

To minimize the burden of AMR on society, it is important to identify the etiological pathogen, differentiate the bacterial and viral infections, and find out which antimicrobial agent should be used for the treatment (Vasala et al., 2020). There is a need of rapid and accurate AMR diagnostic methods since antimicrobial susceptibility testing is time consuming. ML has been used to predict antibiotic resistance in bacteria, proving fast and cost effective compared to traditional methods. ML algorithms, such as support vector machine, random forests, decision trees, artificial neural network, and deep neural network, can be used to train on large datasets. Recently many studies highlighted the potential of machine learning in predicting AMR (Boolchandani et al., 2019; Feretzakis et al., 2021; Lv et al., 2021; Liu et al., 2020). For instance, Yang et al. (2018) and Ren et al. (2021) developed machine learning–based methods to predict AMR with high accuracy using whole-genome sequencing data. They used different machine learning methods such as support vector machine, random forest, logistic regression, and convolutional neural network to train the data. Weis et al. (2022) developed a machine learning approach to predict antimicrobial resistance directly from MALDI-TOF mass spectra profiles from clinical isolates. They used more than 300,000 mass spectra and more than 750,000 antimicrobial resistance phenotypes from four different medical institutes. The deep learning algorithm also showed significant potential for prediction of AMR gene, AMR peptides, and antibiotics drugs (Arango-Argoty et al., 2018; Stokes et al., 2020; Veltri et al., 2018).

One of the key advantages of using ML in antimicrobial resistance prediction is its ability to handle large and complex data, including genomic and phenotypic information, as well as clinical and epidemiological data. This information can be used to build predictive models that can accurately identify the likelihood of microbial resistance to a particular antimicrobial.

9.5 LIMITATIONS OF MACHINE LEARNING

Although machine learning works well, it still has some limitations that make it difficult to translate into healthcare: (1) A major limitation with machine learning is that machine learning needs lot of good quality data to train a model. Depending on the complexity of the machine learning architecture used, the demand of the input data might increase. So there is a need to develop a large and comprehensive databases with combined data standards and frequent update rates that can improve the prediction performance (Lv et al., 2021). (2) Another limitation is methodological bias that these systems may present as they are often based on studies, databases, and guideline from different countries that may not be representative of all patients across the globe (Fanelli et al., 2020). (3) Data need to be pre-processed and of high quality. A low-quality data cannot build a good prediction model. Sometimes during data pre-processing, we lose a lot of important and relevant information. So, we must be very careful when processing and filtering the datasets. (4) Data imbalanced is also one of challenges in machine learning. In the imbalanced data, the distribution of examples across the known class is biased. This imbalanced data results in poor performance, specifically for the minority class. (5) Overfitting and underfitting are a big issue when working with machine learning. Both cause poor performance of the model. Overfitting happens when the model learns the details from the training data which is not actually useful, and this training negatively impacts the performance of the model on unknown data. This overfitting problem can be overcome by removing unimportant features. Underfitting refers to the model which is not performing well either on training data or validation data. This can be removed either by trying alternate machine learning or by using different performance metrics. (6) ML may not generalize well to new and unseen data, leading to poor performance when applied to real-world problem. This can be problematic in antimicrobial research, where the goal is to make predictions about the new and unseen data.

9.6 TRENDS AND FUTURE DEVELOPMENT OF MACHINE LEARNING IN ANTIMICROBIAL RESEARCH

The application of machine learning in AMR is continuously expanding. To assess the state of publications on machine learning for antimicrobial research, we measured trends among papers in PubMed on four human diseases: cancer, cardiovascular diseases, antimicrobial research, and diabetes. To explore the extent to which research interest has changed over time, we queried PubMed on machine learning from the last two decades using the terms "machine learning in human diseases", "machine learning in antimicrobial research", "machine learning in cancer", "machine learning in cardiovascular disease", and "machine learning in diabetes" (Figure 9.3). We found that cancer research is in the top position in which machine learning has been applied while antimicrobial research is in the bottom. We also found that the application of machine learning in antimicrobial rese arch has grown very fast over the last few years; however, it still needs more attention from worldwide researchers.

The pharmaceutical industry is working on drug discovery in antibiotics research using machine learning and deep learning. Machine learning methods generally need large datasets for training; however, the human brain has the capability of learning through only a few examples. How to learn with only a small amount of available data is therefore one of the hottest topics in machine learning (Chen et al., 2018). People also need to work in close collaboration with different skill sets such as clinical pharmacology, bioinformatics, biostatistics, and computer science and integrate such tools that maximize the use of the data. It is also important to develop an interpretable and reproducible model for further uses. Use of multi-omics data such as genomics, transcriptomics, proteomics, and metagenomics will also provide more comprehensive understanding of the underlying mechanism of resistance and will lead to more accurate predictions.

AI is helping scientists discover novel antibiotics more accurately and quickly than the traditional methods. In February 2020 (Stokes et al., 2020), researchers from the Massachusetts Institute of Technology (MIT) discovered a new antibiotic, halicin, using deep learning. The researchers developed a prediction model using 2,500 molecules and tested on 6,000 compounds from Broad Institute's Drug Repurposing Hub and also identified one molecule which has a strong antibacterial activity. They further developed a machine learning model to demonstrate that the discovered molecule has low toxicity for humans. This halicine molecule was found to be effective against a number of drug-resistant bacterial strains such as *Clostridium difficile*, *Acinetobacter baumannii*, and *Mycobacterium tuberculosis*. They also found that *E. coli* did not develop any resistance to halicin over 30 days. Researchers also plan to use their model to design new antibiotics and to optimize existing molecules.

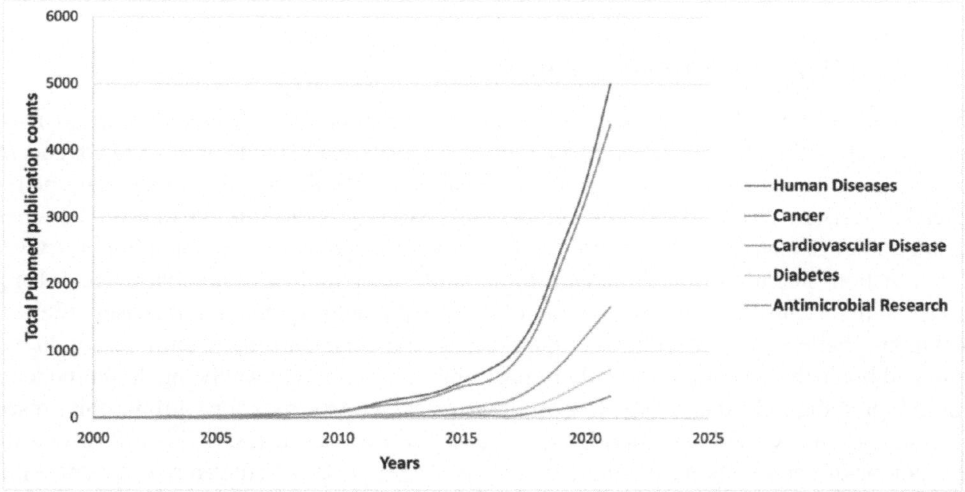

FIGURE 9.3 Number of publications in PubMed using machine learning in different human diseases over time.

Antibiotic resistance can be minimized by prescribing antibiotics appropriately, as improper use of antibiotics can contribute to the growth of treatment-resistant bacteria. An antibiotics susceptibility test also called antibiogram is a broadly used test for antibiotic resistance which is carried out by technicians and analyzed in antibiogram incubator readers, which is very costly and complex in terms of interpretative reading. Recently (Pascucci et al., 2021) a new AI-based mobile application for antibiogram analysis has been developed. The application capture images data which is efficiently processed by the user-friendly graphical interface on a smartphone. The aim of this mobile application is to facilitate the adaptation of the disk diffusion antibiotics susceptibility test in resource limited hospitals and laboratories where the test is not available or poorly interpreted because of lack of technical expertise.

9.7 COVID-19 AND ANTIMICROBIAL RESISTANCE

The COVID-19 pandemic has increased the ongoing AMR due to increase in the use of antibiotics to treat COVID-19 patients. Patients hospitalized with COVID-19 are frequently prescribed with broad-spectrum antibiotics, which act as a fuel for antimicrobial resistance (Chen et al., 2020; Kariyawasam et al., 2022). There are several reasons patients with COVID-19 receive antimicrobials. For example, for a patient with post-COVID-19 structural lung damage, frequent courses of antibiotics and hospital admission may lead to increased risk of colonization and infection with *Pseudomonas aeruginosa* (Knight et al., 2021; Rodrigo-Troyano and Sibila, 2017). According to the International Severe Acute Respiratory and Emerging Infections Consortium report, nearly 62% of patients with COVID-19 received antimicrobial therapy (Rawson et al., 2020). This may favour the emergence of AMR. Apart of that, during hospitalization, COVID-19 patients can also get secondary microbial co-infections which again require antimicrobial treatment. For example, in a study by Zhou et al., in a retrospective cohort analysis of 191 patients from two hospital in Wuhan, China (Zhou et al., 2020), 95% of patients were treated with antibiotics while 21% were treated with antivirals. Another cohort analysis (Huang et al., 2020) of 37 hospitalized patients in three hospitals in Wuhan, China, shows that 97.3% of patient received antiviral treatment, 78.5% received probiotics, 75.7% were treated with traditional Chinese medicine, and 59.50% received antibiotics.

The limited data on COVID-19 and antimicrobial infections has made it difficult to anticipate the potential impact of the pandemic on antimicrobial resistance (Adebisi et al., 2021). The consequence of this pandemic increases the burden of AMR. The pandemic may have an impact on the society through the unintended propagation of AMR (Knight et al., 2021; Rawson et al., 2020).

9.8 DISCUSSION AND CONCLUSION

Antimicrobial resistance has become a major concern in human healthcare. The use of machine learning to address this issue is still in its infancy, but it has already made promising advancements (Macesic et al., 2017). Machine learning can also be used to optimize existing treatments and reduce the spread of resistant strains. For example, machine learning algorithms can analyze data on how different antibiotics interact with bacteria to identify the most effective combination of drugs for a specific infection. The use of machine learning in addressing antimicrobial resistance holds great potential and represents an exciting new frontier in the fight against infectious diseases. While there are still many challenges to be addressed, the future of machine learning in antimicrobial research is bright and holds the potential to greatly improve human health and wellbeing. Machine learning is a rapidly growing field that offers new and innovative technologies for antimicrobial research. This technology can be used to (1) build, predict, and discover new drugs; (2) identify which strains are resistant or susceptible to particular drugs; (3) predict and classify the enzymes responsible for antibiotic resistance; and (4) predict and classify antimicrobial peptides.

In this chapter, we provide a comprehensive overview of the machine learning tools and their applications in antimicrobial research. We also discuss the challenges associated with the use of machine learning in this field and present possible solutions to address these challenges. Finally, we also provide an outlook on future developments and advancements in the field of machine learning for antimicrobial research.

In the future, we expect that machine learning methods have the potential to further improve antimicrobial resistance diagnostics, provide deeper insight into biological mechanisms of resistance, and expand pathways for antibiotic developments. Machine learning will accelerate the discovery of new antimicrobial agents, resulting in more effective treatments for microbial infections. It will continue to play a critical role in the fight against AMR by providing new insights and tools to tackle this growing global challenge. It is important to note that building an effective machine learning model for AMR requires close collaboration among an analytical team of diverse backgrounds, such as clinicians, bioinformaticians, IT engineers, and behavioural scientists. The potential applications of machine learning in antimicrobial research are numerous and exciting. In 2022, Ma et al. (2022) developed a multiple natural language processing neural network model for the identification of AMP from human gut microbiome data. This contribution to emerging research opens the door for the antibiotic discovery field to transition from traditional, time-consuming, trial-and-error methods to a new era where computer technology enables rapid discovery of molecules. By harnessing the power of big data and advanced algorithms, machine learning has the potential to revolutionize the way we diagnose, treat, and prevent antimicrobial resistance. We believe that this technology will be a critical tool in the ongoing fight against this global health challenge.

9.9 ACKNOWLEDGMENTS

All authors are thankful to Bandana Kumari for critically review the manuscript and her suggestions. We are thankful to the reviewers for providing very detailed and constructive suggestions, which helped us in improving the chapter. All authors also acknowledge the time and effort that the reviewers and editor have put in to assess the book chapter.

REFERENCES

Adebisi YA, Alaran AJ, Okereke M, et al. COVID-19 and antimicrobial resistance: A review. Infect Dis (Auckl) 2021;14:11786337211033870.

Alcock BP, Huynh W, Chalil R, et al. CARD 2023: Expanded curation, support for machine learning, and resistome prediction at the comprehensive antibiotic resistance database. Nucleic Acids Res 2023;51(D1):D690–D699.

Arango-Argoty G, Garner E, Pruden A, et al. DeepARG: A deep learning approach for predicting antibiotic resistance genes from metagenomic data. Microbiome 2018;6(1):23.

Aytan-Aktug D, Clausen P, Bortolaia V, et al. Prediction of acquired antimicrobial resistance for multiple bacterial species using neural networks. mSystems 2020;5(1).

Azam MW, Kumar A, Khan AU. ACD: Antimicrobial chemotherapeutics database. PLoS One 2020;15(6):e0235193.

Beltran Lissabet JF, Belen LH, Farias JG. AntiVPP 1.0: A portable tool for prediction of antiviral peptides. Comput Biol Med 2019;107:127–130.

Bonin N, Doster E, Worley H, et al. MEGARes and AMR++, v3.0: An updated comprehensive database of antimicrobial resistance determinants and an improved software pipeline for classification using high-throughput sequencing. Nucleic Acids Res 2023;51(D1):D744–D752.

Boolchandani M, D'Souza AW, Dantas G. Sequencing-based methods and resources to study antimicrobial resistance. Nat Rev Genet 2019;20(6):356–370.

Burnham CD, Leeds J, Nordmann P, et al. Diagnosing antimicrobial resistance. Nat Rev Microbiol 2017;15(11):697–703.

Chandrasekaran S, Cokol-Cakmak M, Sahin N, et al. Chemogenomics and orthology-based design of antibiotic combination therapies. Mol Syst Biol 2016;12(5):872.

Chen H, Engkvist O, Wang Y, et al. The rise of deep learning in drug discovery. Drug Discov Today 2018;23(6):1241–1250.

Chen N, Zhou M, Dong X, et al. Epidemiological and clinical characteristics of 99 cases of 2019 novel coronavirus pneumonia in Wuhan, China: A descriptive study. Lancet 2020;395(10223):507–513.

Das P, Sercu T, Wadhawan K, et al. Accelerated antimicrobial discovery via deep generative models and molecular dynamics simulations. Nat Biomed Eng 2021;5(6):613–623.
de Kraker ME, Stewardson AJ, Harbarth S. Will 10 million people die a year due to antimicrobial resistance by 2050? PLoS Med 2016;13(11):e1002184.
Desselle MR, Neale R, Hansford KA, et al. Institutional profile: Community for open antimicrobial drug discovery – crowdsourcing new antibiotics and antifungals. Future Sci OA 2017;3(2):FSO171.
Durrant JD, Amaro RE. Machine-learning techniques applied to antibacterial drug discovery. Chem Biol Drug Des 2015;85(1):14–21.
Fanelli U, Pappalardo M, Chine V, et al. Role of artificial intelligence in fighting antimicrobial resistance in pediatrics. Antibiotics (Basel) 2020;9(11).
Farrell LJ, Lo R, Wanford JJ, et al. Revitalizing the drug pipeline: AntibioticDB, an open access database to aid antibacterial research and development. J Antimicrob Chemother 2018;73(9):2284–2297.
Feretzakis G, Sakagianni A, Loupelis E, et al. Machine learning for antibiotic resistance prediction: A prototype using off-the-shelf techniques and entry-level data to guide empiric antimicrobial therapy. Healthc Inform Res 2021;27(3):214–221.
Fu H, Cao Z, Li M, et al. ACEP: Improving antimicrobial peptides recognition through automatic feature fusion and amino acid embedding. BMC Genomics 2020;21(1):597.
Gull S, Shamim N, Minhas F. AMAP: Hierarchical multi-label prediction of biologically active and antimicrobial peptides. Comput Biol Med 2019;107:172–181.
He S, Leanse LG, Feng Y. Artificial intelligence and machine learning assisted drug delivery for effective treatment of infectious diseases. Adv Drug Deliv Rev 2021;178:113922.
Huang Q, Deng X, Li Y, et al. Clinical characteristics and drug therapies in patients with the common-type coronavirus disease 2019 in Hunan, China. Int J Clin Pharm 2020;42(3):837–845.
Ivanenkov YA, Zhavoronkov A, Yamidanov RS, et al. Identification of novel antibacterials using machine learning techniques. Front Pharmacol 2019;10:913.
Jamal S, Grover A, Grover S. Machine learning from molecular dynamics trajectories to predict Caspase-8 inhibitors against Alzheimer's disease. Front Pharmacol 2019;10:780.
Jhong JH, Chi YH, Li WC, et al. dbAMP: An integrated resource for exploring antimicrobial peptides with functional activities and physicochemical properties on transcriptome and proteome data. Nucleic Acids Res 2019;47(D1):D285–D297.
Joseph S, Karnik S, Nilawe P, et al. ClassAMP: A prediction tool for classification of antimicrobial peptides. IEEE/ACM Trans Comput Biol Bioinform 2012;9(5):1535–1538.
Kariyawasam RM, Julien DA, Jelinski DC, et al. Antimicrobial resistance (AMR) in COVID-19 patients: A systematic review and meta-analysis (November 2019-June 2021). Antimicrob Resist Infect Control 2022;11(1):45.
Khan ZA, Siddiqui MF, Park S. Current and emerging methods of antibiotic susceptibility testing. Diagnostics (Basel) 2019;9(2).
Kingsford C, Salzberg SL. What are decision trees? Nat Biotechnol 2008;26(9):1011–1013.
Knight GM, Glover RE, McQuaid CF, et al. Antimicrobial resistance and COVID-19: Intersections and implications. Elife 2021;10:e64139.
Kumar R, Srivastava A, Kumari B, et al. Prediction of beta-lactamase and its class by Chou's pseudo-amino acid composition and support vector machine. J Theor Biol 2015;365:96–103.
Lata S, Mishra NK, Raghava GP. AntiBP2: Improved version of antibacterial peptide prediction. BMC Bioinformatics 2010;11 Suppl 1(Suppl 1):S19.
Liu Z, Deng D, Lu H, et al. Evaluation of machine learning models for predicting antimicrobial resistance of actinobacillus pleuropneumoniae from whole genome sequences. Front Microbiol 2020;11:48.
Lv J, Deng S, Zhang L. A review of artificial intelligence applications for antimicrobial resistance. Biosaf Health 2021;3(1):22–31.
Ma Y, Guo Z, Xia B, et al. Identification of antimicrobial peptides from the human gut microbiome using deep learning. Nat Biotechnol 2022;40(6):921–931.
Macesic N, Polubriaginof F, Tatonetti NP. Machine learning: Novel bioinformatics approaches for combating antimicrobial resistance. Curr Opin Infect Dis 2017;30(6):511–517.
Meher PK, Sahu TK, Saini V, et al. Predicting antimicrobial peptides with improved accuracy by incorporating the compositional, physico-chemical and structural features into Chou's general PseAAC. Sci Rep 2017;7:42362.
Mirzaei M, Furxhi I, Murphy F, et al. A machine learning tool to predict the antibacterial capacity of nanoparticles. Nanomaterials (Basel) 2021;11(7).
Mishra M, Kadam P, Doshi B, et al. FEAMR: A database for surveillance of food and environment-associated antimicrobial resistance. Interdiscip Sci 2022;14(4):833–840.

Moradigaravand D, Palm M, Farewell A, et al. Prediction of antibiotic resistance in Escherichia coli from large-scale pan-genome data. PLoS Comput Biol 2018;14(12):e1006258.
Murray CJL, KSL, Fablina S, et al. Global burden of bacterial antimicrobial resistance in 2019: A systematic analysis. Lancet 2022;399(10325):629–655.
Natekin A, Knoll A. Gradient boosting machines, a tutorial. Front Neurorobot 2013;7:21.
Pascucci M, Royer G, Adamek J, et al. AI-based mobile application to fight antibiotic resistance. Nat Commun 2021;12(1):1173.
Pirtskhalava M, Amstrong AA, Grigolava M, et al. DBAASP v3: Database of antimicrobial/cytotoxic activity and structure of peptides as a resource for development of new therapeutics. Nucleic Acids Res 2021;49(D1):D288–D297.
Rahman S, Irfan M, Raza M, et al. Performance analysis of boosting classifiers in recognizing activities of daily living. Int J Environ Res Public Health 2020;17(3).
Rawson TM, Moore LSP, Castro-Sanchez E, et al. COVID-19 and the potential long-term impact on antimicrobial resistance. J Antimicrob Chemother 2020;75(7):1681–1684.
Ren Y, Chakraborty T, Doijad S, et al. Prediction of antimicrobial resistance based on whole-genome sequencing and machine learning. Bioinformatics 2021;38(2):325–334.
Reygaert WC. An overview of the antimicrobial resistance mechanisms of bacteria. AIMS Microbiol 2018;4(3):482–501.
Rodrigo-Troyano A, Sibila O. The respiratory threat posed by multidrug resistant Gram-negative bacteria. Respirology 2017;22(7):1288–1299.
Said Agrebi aAL. Use of artificial intelligence in infectious diseases. Artificial Intelligence in Precision Health 2020:415–438.
Schaduangrat N, Nantasenamat C, Prachayasittikul V, et al. Meta-iAVP: A sequence-based meta-predictor for improving the prediction of antiviral peptides using effective feature representation. Int J Mol Sci 2019;20(22).
Sharma R, Shrivastava S, Kumar Singh S, et al. AniAMPpred: Artificial intelligence guided discovery of novel antimicrobial peptides in animal kingdom. Brief Bioinform 2021;22(6).
Shi G, Kang X, Dong F, et al. DRAMP 3.0: An enhanced comprehensive data repository of antimicrobial peptides. Nucleic Acids Res 2022;50(D1):D488–D496.
Song YY, Lu Y. Decision tree methods: Applications for classification and prediction. Shanghai Arch Psychiatry 2015;27(2):130–135.
Srivastava A, Kumar R, Kumar M. BlaPred: Predicting and classifying beta-lactamase using a 3-tier prediction system via Chou's general PseAAC. J Theor Biol 2018;457:29–36.
Stokes JM, Yang K, Swanson K, et al. A deep learning approach to antibiotic discovery. Cell 2020;180(4):688–702 e613.
Svetnik V, Liaw A, Tong C, et al. Random forest: A classification and regression tool for compound classification and QSAR modeling. J Chem Inf Comput Sci 2003;43(6):1947–1958.
Thakur N, Qureshi A, Kumar M. AVPpred: Collection and prediction of highly effective antiviral peptides. Nucleic Acids Res 2012;40(Web Server issue):W199–W204.
Thomas J, Navre M, Rubio A, et al. Shared platform for antibiotic research and knowledge: A collaborative tool to SPARK antibiotic discovery. ACS Infect Dis 2018;4(11):1536–1539.
Van Camp PJ, Haslam DB, Porollo A. Prediction of antimicrobial resistance in gram-negative bacteria from whole-genome sequencing data. Front Microbiol 2020;11:1013.
Vasala A, Hytonen VP, Laitinen OH. Modern tools for rapid diagnostics of antimicrobial resistance. Front Cell Infect Microbiol 2020;10:308.
Veltri D, Kamath U, Shehu A. Deep learning improves antimicrobial peptide recognition. Bioinformatics 2018;34(16):2740–2747.
Waghu FH, Barai RS, Gurung P, et al. CAMPR3: A database on sequences, structures and signatures of antimicrobial peptides. Nucleic Acids Res 2016a;44(D1):D1094–1097.
Waghu FH, Barai RS, Idicula-Thomas S. Leveraging family-specific signatures for AMP discovery and high-throughput annotation. Sci Rep 2016b;6:24684.
Waghu FH, Idicula-Thomas S. Collection of antimicrobial peptides database and its derivatives: Applications and beyond. Protein Sci 2020;29(1):36–42.
Wang G, Li X, Wang Z. APD3: The antimicrobial peptide database as a tool for research and education. Nucleic Acids Res 2016;44(D1):D1087–1093.
Webb S. Deep learning for biology. Nature 2018;554(7693):555–557.
Weis C, Cuenod A, Rieck B, et al. Direct antimicrobial resistance prediction from clinical MALDI-TOF mass spectra using machine learning. Nat Med 2022;28(1):164–174.

Yang Y, Niehaus KE, Walker TM, et al. Machine learning for classifying tuberculosis drug-resistance from DNA sequencing data. Bioinformatics 2018;34(10):1666–1671.

Ye G, Wu H, Huang J, et al. LAMP2: A major update of the database linking antimicrobial peptides. Database (Oxford) 2020;2020:baaa061.

Zhou F, Yu T, Du R, et al. Clinical course and risk factors for mortality of adult inpatients with COVID-19 in Wuhan, China: A retrospective cohort study. Lancet 2020;395(10229):1054–1062.

10 Tracking Antimicrobial Resistance through Metagenomics

Ketaki Ramani, Manan Desai, and Kruti Dangar

10.1 INTRODUCTION

Antimicrobials are compounds that are used to inhibit bacterial growth (bacteriostatic) or kill them (bactericidal). These can be produced naturally, semisynthetically, or synthetically and are classified based on their efficacy as narrow or broad spectrum. Penicillin was the first antibiotic discovered, and it marked the dawn of the antibiotic age. Prior to the discovery of antibiotics, a minor cut or wound could lead to a fatal condition. From the time of its discovery by Alexander Fleming in 1928 to its large-scale production in 1944, penicillin was deemed as the greatest advance in medicine. However, over time, microorganisms have developed resistance to antibiotics, and this has led to the problem of antimicrobial resistance (AMR). Microorganisms can evolve and adapt to become resistant to the drugs that are used to treat them, leading to standard treatments becoming ineffective. In some cases, no available drugs provide an effective cure, resulting in the spread of disease, increased morbidity, and higher mortality rates. The World Health Organization (2014) had issued a warning of a post-antibiotic era, where minor infections and injuries could once again become fatal. This highlights the importance of controlling the spread of AMR and developing new strategies to tackle microbial infections.

AMR is a natural evolutionary process following the exposure to antibiotics. Microorganisms are the epitome of adaptation given a doubling time of 20 minutes under ideal conditions for *Escherichia coli* whereas humans have an average lifespan of 75 years. Research shows that the increase in use of antibiotics leads to the development of resistance. There is an increasing use and abuse of antibiotics, especially in the agriculture sector. This has a direct impact on human health. Resistance through the use of antibiotics in veterinary science and agriculture has led to outbreaks of infectious disease caused by multidrug-resistant organisms, acquired through food sources, thus creating global health issues (Pruden et al., 2006).

The acquisition, development, and spread of antibiotic resistance genes (ARGs) are crucial components of combating antimicrobial resistance, and the entire microbial population contributes to the creation of the antimicrobial resistome, rather than just one source organism that carries an ARG (Bello-López et al., 2019). As a result, identifying and comprehending horizontal gene transfers (HGTs) among both pathogenic and non-pathogenic species may help uncover the mechanisms behind resistance propagation and distribution. The application of high-throughput sequencing technologies has made it possible and convenient to examine ARG sequences, with metagenomics playing a significant role in the evaluation of antimicrobial resistance in communities.

Resistance to antimicrobials is also developed through ARGs. Clinical pathogens adapt AGRs through diverse mechanisms such as horizontal gene transfer and plasmids, transposons, bacteriophages, conjugation, transformation, transduction-mediated transfer of mobile genetic elements (MGEs), and many more (Rizzo et al., 2013; Christaki et al., 2020) as shown in figure 10.1.

Researchers are utilizing both traditional approaches and omics to create a blueprint for tackling antimicrobial resistance genes as a means of either slowing or overcoming them. This approach has

DOI: 10.1201/9781003269380-10

been documented in studies such as those conducted by McCarthy (2015), Christaki et al. (2020), and De Abreu et al. (2021). Table 10.1 lists some of the known organisms that have developed resistance against standard antimicrobial agents.

In the wake of the Covid-19 pandemic, it became critical to take a multisectoral approach to address the threats posed by AMR and antimicrobial resistant bacteria. Despite the tremendous progress made in the development of antimicrobial, antifungal, antiviral, and other treatments, the reality is that microorganisms are highly adaptable and can still cause infectious diseases in humans and animals. This adaptability also enables them to defend themselves against various antimicrobial agents. However, researchers have made significant progress in understanding the essential factors and mechanisms that contribute to the persistence and expansion of antimicrobial resistance.

One of the challenges in studying microorganisms is that there is a vast array of them that can cause disease, but only a small percentage, typically 1%–3%, can be successfully cultivated in the laboratory, while the remaining majority remain uncultured (Dangar et al., 2017). As a result, the role of uncultivable microbes in the development of antimicrobial resistance is still relatively unexplored, despite their likely significant contribution to the problem.

Studying AMR is a time-consuming process, with several ongoing research efforts aimed at decoding the underlying mechanisms. This can be significantly accelerated using metagenomic approaches, which involves the direct study of genetic material obtained from environmental samples, such as soil,

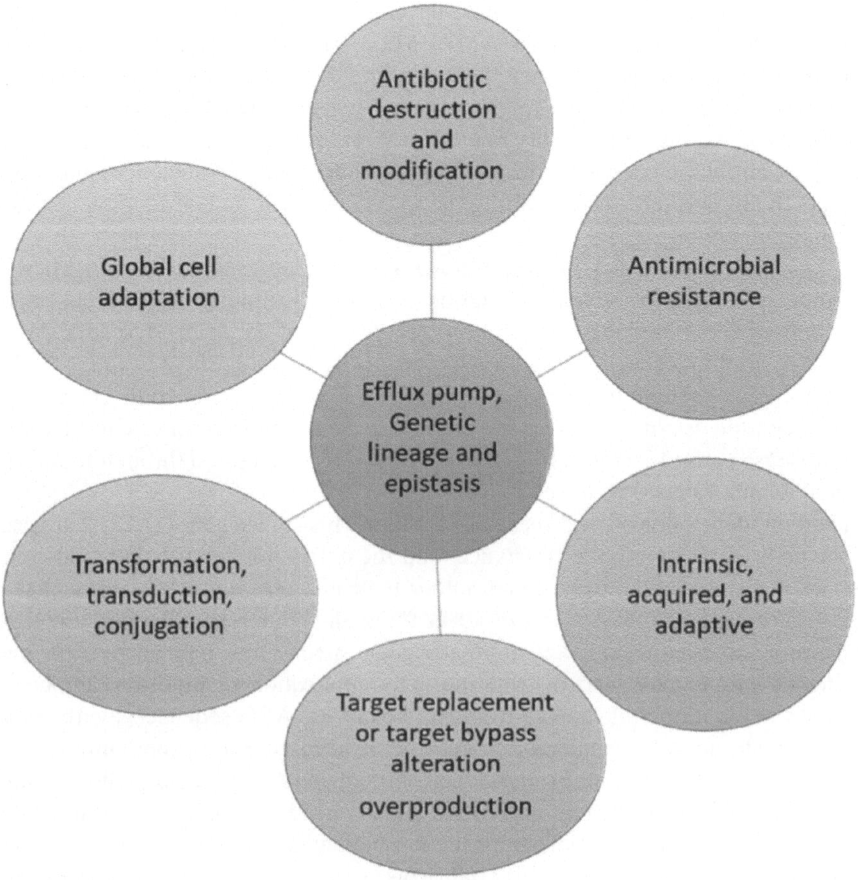

FIGURE 10.1 Various mechanisms to attain antimicrobial resistance.

TABLE 10.1
Antimicrobial Agents Used for the Treatment of Microorganisms

Sr. no.	Antimicrobial agent	Microorganism	Reference
1	Aztreonam	*Pseudomonas aeruginosa*	Jorth et al., 2017
2	Cephamycin	*Plasmidic β-lactamase CMY-2*	Bauernfeind et al., 1996
3	Chloramphenicol	*Typhoid Bacillus*	Anderson and Smith 1972
4	Chloramphenicol	*Salmonella*	Schwarz et al., 2021
5	Clindamycin	*Enterococcus faecium HM1025*	Bozdogan et al., 1999
6	Clindamycin	*Staphylococcus aureus*	Dragana Drinkovic et al., 2001
7	Daptomycin		Hayden et al., 2005
8	Florfenicol	*Escherichia coli*	Schwarz et al., 2021
9	Glycopeptide	*Enterococcus faecium*	Xiaogang Xu et al., 2010
10	Glycopeptides	*Staphylococcus aureus*	Tenover et al., 2001
11	β-Lactamase	*Klebsiella pneumoniae*	Bauernfeind et al., 1996
12	Lincomycin	*Enterococcus faecium HM1025*	Bozdogan et al., 1999
13	Linezolid	*Staphylococcus epidermidis*	Rodrigo et al., 2010
14	Linezolid	*Enterococcus faecium* and *Enterococcus faecalis*	Marshall et al., 2002
15	Macrolide	*Streptococcus pneumoniae*	Schwarz et al., 2021
16	Methicillin	*Staphylococcus aureus*	Hayden et al., 2005
17	Nitrofurantoin	*Escherichia coli*	Sandegren et al., 2008
18	Pirlimycin	*Staphylococcus aureus*	Schwarz et al., 2021
19	Quinupristin–dalfopristin	*Enterococcus faecium*	Sabol et al., 2005
20	Rifampicin	*Mycobacterium tuberculosis*	Zaw et al., 2018
21	Tiamulin	*Actinobacillus pleuropneumoniae*	Schwarz et al., 2021
22	Vancomycin	*Staphylococcus aureus*	Tenover et al., 2001

water, or the human gut, without the need to isolate individual microorganisms. For example, metagenomics analyses of bacterial populations in wastewater has revealed wastewater treatment plants have been found to serve as reservoirs for AMR genes, potentially contributing to the spread of resistance throughout the wider environment. Similarly, studying human gut microbiomes using metagenomic analysis has shown that exposure to antibiotics can lead to significant changes in the abundance and diversity of bacteria, as well as resistance genes, potentially increasing the risk of AMR transmission.

By utilizing metagenomic approaches, researchers can better understand the complex interactions between microorganisms and their environments, including the ways in which exposure to antimicrobial agents can drive the development and spread of resistance. This understanding is critical to developing effective strategies for combating antimicrobial resistance, particularly in the face of emerging threats like the Covid-19 pandemic.

10.2 METAGENOMICS TO TRACK ANTIMICROBIAL RESISTANCE

Handelsman et al. coined the term "metagenomics" in 1998 to describe the analysis of genetic material obtained directly from environmental samples. This approach overcomes the limitations of traditional cultivation-based methods and enables the identification of both cultivable and uncultivable microbial taxa and their activities in their native habitats (Dangar et al., 2017). Metagenomics studies employ various techniques to identify the taxonomic and functional characteristics of microbial communities in a given environment (Prayogo et al., 2020). The isolation of high-quality genomic DNA, followed by amplification, sequencing, and analysis of the sequencing data, is a key aspect of metagenomics research (Raiyani and Singh, 2020). Additionally, metagenomic libraries can be generated by inserting total genomic DNA into suitable vectors, which can then be used to examine the expression of desired genes, proteins, and enzymes (Dhanjal et al., 2020). Apart from this, total genomic DNA was inserted into suitable or biased plasmids/vectors to generate metagenomic libraries. Future metagenomic libraries were examined for the expression of desired/bias genes, proteins, and enzymes of interested clones (Suenaga et al., 2007; Singer et al., 2020). In recent years, metagenomics has been used extensively to investigate AMR in different environments, including the gut, sea water, saline soil, wastewater, and polluted soil, among others. As a result, the database of antimicrobial resistance genes and organisms has become enriched. However, many questions regarding AMR, such as the mechanisms underlying the acquisition of resistance, the factors and genes involved in transmission, and the role of uncultivable microbes in promoting resistance, remain unanswered (Moon et al., 2020; Raza et al., 2021). Table 10.2 shows the metagenomics approach used in several environmental samples.

10.2.1 Sequencing-Based Methods to Track Antimicrobial Resistance

The metagenomic sequence data widely depends upon the source of samples, methods for isolation of genomic DNA, amplification, library preparations, and sequencing lengths, along with

TABLE 10.2
Metagenomics Approach Used in Several Environmental Samples

Collection site	Sample type	DNA extraction method	Sequencing platform	Reported taxa	References
Bodo West community in Gokana Local Government Area of Rivers State, Nigeria	Soil	ZRSoil Microbe DNA Microprep	Illumina MiSeq platform	*Sneathiella, Parvibaculum, Sphingobium, Oceanicaulis*	Chikere et al., 2019
Hypersaline and meromictic Soap Lake, Washington	Water	FastDNA SPIN kit	Roche 454 FLX_ platform	*Firmicutes, Proteobacteria, Tenericutes, Actinobacteria, Bacteroidetes*	Hawley and Hess, 2014
Aloe Vera	Plant	GF-1 bacterial DNA extraction kit	MiSeq instrument	*Proteobacteria, Firmicutes, Actinobacteria, Bacteroidetes*	Akinsanya et al., 2015
Serofluid dish	Food		Illumina HiSeq PE250 platforms	*Ascomycota, Basidiomycota, Zygomycota*	Chen et al., 2016
Rice	Root	Protocol based on Garcias-Bonet et al., 2012	Illumina platform	*Proteobacteria, Firmicutes, Cyanobacteria, Actinobacteria*	Sengupta et al., 2017
Bat guano	Soil	Fast DNA spin kit	Illumina MiSeq	*Chloroflexi, Crenarchaeota, Actinobacteria, Bacteroidetes, Proteobacteria, Planctomycetes*	De Mandal et al., 2015

sequencing platforms and many more. The generated sequences are analyzed by two approaches; the first is based on the read method, and another is the assembly-based method. The major difference in these two approaches is sequences must be assembled into contigs (contiguous fragments) and mapping the reads directly to the databases. Numerous researchers have developed a huge number of tools, databases, websites, algorithms, and software to track antimicrobial resistance.

10.2.2 Read-Based Tools to Track Antimicrobial Resistance

For searching antimicrobial resistance, the assembly reads are directly aligned to different tools or databases. Bowtie 2 is based on hardware-accelerated programming algorithms that make alignment speedily and accurately (Langmead and Salzberg, 2012). SRST2 is a read mapping-based tool to detect antimicrobial genes (Inouye et al., 2014). Search Engine for Antimicrobial Resistance (SEAR) is used for the detection of ARGs from raw sequencing reads obtained from metagenomes as well as sequences of cultivable isolates. It quickly analyses the sequence data with relative abundance valuation. It also provides the identification genes along with sequences (Rowe et al., 2015).

Several metagenomics tools with increased efficiency and larger data processing abilities have been developed lately. One such tool is antiSMASH (antibiotics and secondary metabolite analysis shell), which helps predict gene cluster boundaries, phylogenetic profiling, biosynthetic gene clusters, and many more. It is available as a webserver and tool based on the SANDPUMA and CASSIS algorithm with the latest version antiSMASH 4 (Kai et al., 2017). A graphing Resistance Out Of meTagenomes (GROOT) has the ability to process two gigabytes of data in two minutes by using single central processing units (CPUs) and generates the resistome profiling from metagenomic sequences reads using a scoring scheme (Will and Martyn et al., 2018). The DeepARG is available as a command base version along with Web service at http://bench.cs.vt.edu/deeparg. It is used for identification of genes along with relative abundance of ARGs from the short read sequences and full gene length sequences (Arango-Argoty et al., 2018). The Pathosystems Resource Integration Center (PATRIC, www.patricbrc.org) is a database with a collection of 15,000 genomes and 16,10,744 proteins designed to predict the antimicrobial resistance phenotypes and whole genomes or individual genes responsible for resistance from reads (Dionysios et al., 2019).

10.2.3 Assembly-Based Tools to Track Antimicrobial Resistance

Short reads and long reads can be gained from the various sequencing platforms such as Illumina, Next sequences, PacBio RSII, and Oxford Nanopore. The reads are analyzed for quality control and assembled as contigs. The various programmes, tools, databases, and websites are available for contigs assembly, binning, and annotation. K–SLAM is used to generate contigs based on k-mer by using local sequence alignments (Ainsworth et al., 2017). MEGAHIT is a Next Generation Sequencing tool and is available in CPU and GPU. Its *de novo* assembler was aligned to assemble large and complex metagenomics data in a shorter period of time along with fourfold improvement (Dinghua et al., 2015). metaSPAdes software associations algorithmically confirmed to address various assemblies such as generating strain-contigs, transforming assembly graph into consensus assembly graph, and repeat resolution using strain-paths Nurk et al., 2017). ResistoMap is an online dataset with visualization of the occurrence of resistome enriched with 1,593 gut metagenomes including both healthy and patients from 12 studies covering 15 countries. The displayed features are such as antibiotic resistance genes with relative abundance and levels of mutations and correlates the gut resistome variations (Konstantin et al., 2017). PointFinder Web-based tool used chromosomal gene and chromosomal mutation databases to detect the ARGs in FASTA format and codon mutation conferring antimicrobial resistance (Ea Zankari et al., 2017).

The updated SARG version 2.0 contains sequences from Comprehensive Antibiotic Resistance Database (CARD), Antibiotic Resistance Database (ARDB), and NCBI-NR databases with accurate, moderate, and loose accuracy level. SARG v2.0 has improved coverage of ARGs type subtype-reference sequence and ARGs detection, annotation, and classification of ARGs in metagenomes. ARGs-OAP v2.0 now provides model-based identification of profiles of ARG subtypes from assembled sequences using SARGam, a hidden Markov model (HMM), in built with ARGs-OAP v2.0 (Xiaole Yin et al., 2018). MaxBin 2.0 is based on EM algorithm. It showed enhancement of binning of metagenomic contigs to identify and compare bacterial and community compositions of diverse habitats (Yu-Wei Wu et al., 2016).

Semi-supervised Spectral Normalized Cut for Binning (SolidBin) generated two types of information: must-link constraints for contigs allied to the same group and cannot-link constraints for contigs linked to different groups (Ziye Wang et al., 2019).

The user-friendly tool PathoFact was created to identify virulence factors, toxin genes, antimicrobial resistance genes, and mobile genetic elements containing pathogenicity factors within metagenomic data. It was found that PathoFact has a higher potential for identification of AMR genes along with the resistance difference case and control individuals/groups (de Nies et al., 2021).

10.3 TRACKING OF ANTIMICROBIAL RESISTANCE THROUGH SEQUENCE-BASED METAGENOMICS

In the last three decades, researchers have tried to explore all the aspects of antimicrobial resistance existing in diverse habitats. However, a solid scientific approach to answer all the ambiguities related to antimicrobial resistance is yet to be deciphered. With the advancement of sequencing technologies, the paradigm shifted towards metagenomic analysis from traditional cultivable approaches to hunting antimicrobial resistance (Yadav and Kapley, 2021). Until this date, over 20,000 articles have described the presence of antimicrobial resistance genes, their mechanisms, and many microbes with resistance profiles in metagenomic data from various environmental samples. For example, metagenomic studies of sludge have successfully analyzed diverse ARGs for resistance to tetracycline, sulphonamides, macrolide, and integron-integrase gene (Xu et al., 2018; Tian et al., 2016; Jang et al., 2017; Diehl and Lapara, 2010) by pyrosequencing. The results indicated the presence of beta-lactam, multidrug, aminoglycoside, tetracycline resistance gene (Kristiansson et al., 2011).

The presence of ARGs in various water sources such as drinking water and deep ocean sediment was also observed (Chao et al., 2013; Chen et al., 2013). A total of 159 ARGs within 14 types in metagenome from wastewater samples were collected from the Wastewater Treatment Plant (Nanjing, China).

The extraction of total genomic DNA was performed using the FastDNA@ Spin Kit and sequencing was done by using llumina HiSeq 2500 platforms. To capture the ARGs in metagenome sequences were analyzed using Antibiotic Resistance Genes Database (ARDB), Integron Database (INTEG RALL), Insertion Sequences (IS) Database (IS Finder), and NCBI Reference Sequence Database (NCBI RefSeq) (Hu et al., 2016). Drinking water was studied with two approaches; one is 16S rRNA gene cloning, another is metagenomic analysis. Comparing both the results shows the higher antibiotic resistant bacteria dominance of *Proteobacteria* along with ARGs in the sample (Shi et al., 2013).

Previously, *Proteobacteria* and *Bacteroidetes* were reported with ARGs of β-lactams, tetracycline, and aminoglycosides (Forsberg et al., 2014). Water was collected from the holy river Ganga from Varanasi city, India, and DNA was extracted; the sequencing platform Illumina NextSeq 500 sequencer was used for metagenomic study. Comprehensive Antibiotic Resistance Database version 1.1.9 (McArthur et al., 2013) and BacMet (Pal et al., 2014) version 1.1 using the DIAMOND were used for identification of ARGs and MRGs (metal resistance genes) in metagenome. The result showed the higher abundance phyla and genera *Proteobacteria*, *Actinobacteria*, and *Arcobacter*, *Terrimicrobium*, *Acidibacter*, and *Pseudomonas*.

Antimicrobial resistance genes tuf, parY, ileS, and mfd were present with higher abundance. Further profiling indicated the beta-lactam and elfamycin act as dominant ARGs type (Reddy and Dubey, 2019). A metagenomic approach was used to identify the residing microbes of samples collected from 35 soil samples from industrial parks in China by using the Illumina HiSeq 2500 platform for sequencing. By analysis of 14,383,065 contigs, the most abundant phylum of *Proteobacteria* and ARGs were discovered showing resistance towards rifampisin, fluoroquinolones, and beta-lactams. The present study gave a screenshot of the resistome (Zheng et al., 2021).

Xu et al. (2020) studied the metagenomics of municipal sewage sludge of Changsha, China. They extracted the DNA by using FastDNA SPIN Kit (MP bio, USA). Further, the genomics DNA was sequenced by using shotgun metagenomic sequencing Illumina Platform. The annotation of function genes and ARGs was carried out using the Kyoto Encyclopedia of Genes and Genomes (KEGG) database and the Comprehensive Antibiotic Resistance Database (CARD) respectively. A result showed the presence of *Firmicutes*, *Proteobacteria*, *Chloroflexi*, and many more phyla along with 1,360 ARGs in samples. Results indicated that the dominated six ARGs, macB, tetA(58), bcrA, evgS, mtrA, and oleC, showed resistance towards antibiotics macrolide, tetracycline, peptide, multiple drug classes, penam/macrolide, and macrolide antibiotic respectively in sludge.

10.4 FUNCTIONAL METAGENOMIC TO TRACK ANTIMICROBIAL RESISTANCE

10.4.1 Tracking of Antimicrobial Resistance through Functional-Based Metagenomics

The extracted total community DNA is surrogate into a suitable plasmid or expression vector. Metagenomic libraries were constructed and screened for antimicrobial resistance genes and studied expression and growth of recombinant cells. After the desired/biased expression of functionally selected DNA were subjected to sequencing, annotation was accomplished by using a reference database and capturing the abundance and mechanism of antimicrobial resistance in the environment (Boolchandani et al., 2019; Yadav and Kapley, 2021).

Soil metagenomic DNA was collected through PowerMax soil DNA isolation kit and library was constructed using pUC 18 as vector and *Escherichia coli* as host. After that, the selection and sequencing analysis of clones conferring resistance of trimethoprim and rifampin (McGarvey et al., 2012) were carried out. The plano silt loam soil was collected from West Madison, Wisconsin, USA. Metagenomic four libraries were constructed by using pBeloBAC11, pCF430and pJN105 vectors and surrogate into *E. coli*. The size of total metagenomic libraries was 4.1 gigabases (Gb). A total of 10 clones showed antimicrobial resistance: 9 and 1 clone expressing antibiotic resistance to aminoglycoside and tetracycline respectively (Riesenfeld et al., 2004).

The total genomic DNA was extracted from unpolluted Alaskan soil where there was non-appearance of anthropogenic activity. The metagenomic DNA was inserted into vectors pCF430, pCC1BAC, pCC1FOS, and *E. coli* as hosts to generate the libraries. The researcher captured the genes conferring the resistance to ß-lactam was found in 13 clones (Allen et al., 2009). Metagenomic libraries were built using metagenomic DNA extracted from the human oral cavity. DNA was incorporated into the TOPO-XL vector using *E. coli* as host. The recombinant clones showed the presence of tetracycline resistance gene (tet37) (Diaz-Torres et al., 2003). A metagenomic library consisting of 96,000 clones of activated sludge was explored for identification of resistance gene. The result showed bleomycin resistance genes along with two novel genes (Mori et al., 2008). Environmental DNA was collected from apple orchard soil and ligated into the pCC1FOS fosmid vector, hosted into the EPI300 strain of *E. coli* generated metagenomic library. Out of 446,000 clones, 9, 1, and 3 clones were conferring resistance to β-lactams, tetracycline, and aminoglycoside respectively (Donato et al., 2010). Functional metagenomic screening of AMR of cultivated microbes of gypsy moth larval midguts, the research identified genes in metagenomic libraries code for resistance of penicillin-type β-lactams, erythromycin, erythromycin, carbenicillin, and chloramphenicol (Allen et al., 2009). Ten artificial chromosome clones were constructed through the metagenomic functional approach of DNA collected from organic pig gut as carrying the known genes conferring resistance to tetracyclines with the presence of genes tet(C), tet(W), and tet(40) along resistance to the minocycline and doxycycline (Kazimierczak et al., 2009).

Four metagenomic libraries were constructed to investigate the antibiotic resistance potential of the native human oral microbiota and harness their antibiotic resistance capabilities. The genes code for tetracycline resistance (tet(M), tet(O), tet(Q), tet(W), tet37, and tet(A)) and β-lactamase was investigated (Diaz-Torres et al., 2006). A novel gene (pexA) code for florfenicol resistance and chloramphenicol resistance was categorized with functional metagenomic aspect in Alaskan soil (Lang et al., 2010). Metagenomic libraries with 3.6 Gb of soil genomic DNA were assessed for identification of antimicrobial resistance genes. Total 11 (3 + 2 + 2 + 4) clones possess antimicrobial resistance genes conferring resistance to ampicillin, gentamicin, chloramphenicol, and trimethoprim respectively (Torres-Cortés et al., 2011).

The PowerSoil DNA extraction kit was utilized to extract DNA from river sediments in India, and metagenomic libraries were built using *E. coli* as the host and the pZE21-MCS vector. The selection of resistance conferring clones was carried out by MuellerHinton agar plates containing antibiotics. Further clones were sequenced using Sanger and PacBio RSII platforms and annotated by using reference databases. By comparing the results obtained from shotgun metagenomics sequencing and functional genomics was similar. The presence of six beta-lactamases and one amikacin resistance gene were identified in functional metagenomics (Marathe et al., 2018). Soil samples from Yunnan and Tibet were collected to extract metagenomic DNA, and metagenomic libraries were constructed using pWEB-TNC as vector and transfected into *E. coli*. Metagenomics DNA from soil samples of Sichuan was incorporated into the pJTU2554 vector in the *E. coli* (JTU007) to build the metagenomic libraries. Efflux transporter genes, efflux pump genes, tetracycline resistance genes, and one multidrug and toxic compound extrusion (MATE) transporter gene were detected in the metagenomic library through the functional metagenomics (Wang et al., 2017).

From the saliva samples, human oral metagenomic DNA was extracted and introduced into the pSMART BAC vector with *Escherichia coli* as the host. The library was screened for various antimicrobial resistance genes, revealing the presence of genes such as msr, mef(A), tetAB(46), fabK, recA, and accB. The gene conferring the resistance for chlorhexidine, triclosan, erythromycin, tetracycline, and sodium hypochlorite was also detected (Wigand et al., 2021).

10.5 FUTURE SCOPE

Knowledge gaps regarding AMR in agriculture, wastewater, and soil and the roles of these sites as sources and sinks of AMR needs to be identified and worked upon in order to curb instances of AMRs. Stringent policies and regulations are required to govern the manufacture, distribution, dispensing, and prescription of antibiotics. To aid developing nations in implementing effective surveillance of antibiotic use and resistance, international collaboration with relevant organizations is necessary. As the public health issue of antibiotic resistance grows, the adoption of policies like voluntary agreements, taxes, or permits is being encouraged to reduce antimicrobial usage in animals used for food production. Several countries have established that restricting the use of antimicrobial agents in food animals can be followed by a decrease in antimicrobial resistance in humans without compromising animal health or significantly increasing the cost of production. Appropriate use of antimicrobial agents in humans and food animals is an important factor in maintaining their effectiveness. Effective surveillance programs of soil, plants, animal, and water analysis could help develop effective data to generate risk analyses of different ecological compartments.

10.6 CONCLUSION

While several tools have been developed to study antimicrobial resistance and the genes responsible thereof, one of the most effective strategies to reduce antimicrobial resistance is to monitor the dose of antimicrobials consumed, thereby limiting the scope of resistance development. Metagenomics approaches help ease the identification of antimicrobial resistant genes in the sample and thereby aid in developing effective strategies to culminate the development of related health concerns. Metagenomics analysis is an effective tool to study ARGs in samples. However, as with any new technology, it is important to use metagenomics in conjunction with other methods and approaches to ensure a comprehensive and accurate understanding of the problem of AMR. Overall, metagenomics has proven to be a powerful tool for investigating microbial communities and their activities in their natural habitats, and it has the potential to contribute significantly to our understanding of AMR and other important biological processes.

The most important considerations in a metagenomic resistome study are understanding the nature of the dataset being analyzed and the support that is available for its analysis. If one takes into account the large quantity of data and the complexity of the biological mechanisms involved in antibiotic resistance, it may be preferable to adopt reductionist approaches to decrease bias and increase the objectivity of analysis. Drawing attention to the decreasing expenses associated with algorithms, computers, and analytical tools is crucial. This has led to the increasing popularity of *in silico* predictions based on machine learning, which have the potential to predict resistance beyond existing databases. This advancement will enable the development of high-throughput data analysis methods and assist in addressing more complex questions related to antimicrobial resistance.

REFERENCES

Ainsworth D, Sternberg MJE, Raczy C, et al. (2017) k-SLAM: Accurate and ultra-fast taxonomic classification and gene identification for large metagenomic data sets. Nucleic Acids Res 45: 1649–1656. https://doi.org/10.1093/nar/gkw1248

Akinsanya MA, Goh JK, Lim SP, Ting ASY (2015). Metagenomics study of endophytic bacteria in Aloe vera using next-generation technology. Genomics Data 6: 59–163. https://doi.org/10.1016/j.gdata.2015.09.004

Allen HK, Cloud-Hansen KA, Wolinski JM et al. (2009) Resident microbiota of the gypsy moth midgut harbors antibiotic resistance determinants. DNA Cell Biol 28: 109–117. https://doi.org/10.1089/dna.2008.0812

Allen HK, Moe L, Rodbumrer J, et al. (2009) Functional metagenomics reveals diverse β-lactamases in a remote Alaskan soil. ISME J 3: 243–251. https://doi.org/10.1038/ismej.2008.86

Anderson ES, Smith HR (1972) Chloramphenicol resistance in the typhoid bacillus. *Br Med J* 3(5822): 329–331. https://doi.org/10.1136/bmj.3.5822.329

Arango-Argoty G, Garner E, Pruden A et al. (2018) DeepARG: A deep learning approach for predicting antibiotic resistance genes from metagenomic data. Microbiome 6: 23. https://doi.org/10.1186/s40168-018-0401-z

Bauernfeind A, Alexander Project Collaborative Group (1996) Antibiotic susceptibility patterns of respiratory isolates of Klebsiella pneumoniae in Europe and the USA in 1992 and 1993. J Antimicrob Chem 38(suppl A): 107–115. https://doi.org/10.1093/jac/38.suppl_a.107

Bauernfeind A, Stemplinger I, Jungwirth R, Giamarellou H (1996) Characterization of the plasmidic beta-lactamase CMY-2, which is responsible for cephamycin resistance. Antimicrob Agents Chemother 40(1): 221–224. https://doi.org/10.1128/AAC.40.1.221

Bello-López JM, Cabrero-Martínez OA, Ibáñez-Cervantes G, Hernández-Cortez C, Pelcastre-Rodríguez LI, Gonzalez-Avila LU, Castro-Escarpulli G. Horizontal gene transfer and its association with antibiotic resistance in the genus Aeromonas spp. *Microorganisms*. 2019 Sep 18;7(9):363. https://doi.org/10.3390/microorganisms7090363

Boolchandani M, D'Souza AW, Dantas G (2019) Sequencing-based methods and resources to study antimicrobial resistance. Nat Rev Gen 20(6): 356–370. https://doi.org/10.1038/s41576-019-0108-4.

Bozdogan B, Berrezouga L, Kuo MS, et al. (1999) A new resistance gene, linB, conferring resistance to lincosamides by nucleotidylation in Enterococcus faecium HM1025. Antimicrob Agents Chemother 43(4): 925–929. https://doi.org/10.1128/AAC.43.4.925

Chao Y, Ma L, Yang Y et al. (2013) Metagenomic analysis reveals significant changes of microbial compositions and protective functions during drinking water treatment. Sci Rep 3:3550. https://doi.org/10.1038/srep03550

Chen B, Yang Y, Liang X, et al. (2013) Metagenomic profiles of Antibiotic Resistance Genes (ARGs) between human impacted estuary and deep ocean sediments. Environ Sci Tech 47(22): 12753–12760. https://doi.org/10.1021/es403818e

Chen P, Zhao Y, Wu Z, et al. (2016) Metagenomic data of fungal internal transcribed spacer from serofluid dish, a traditional Chinese fermented food. Genomics Data 7: 134–136. https://doi.org/10.1016/j.gdata.2015.12.028

Chikere CB, Mordi IJ, Chikere BO, et al. (2019) Comparative metagenomics and functional profiling of crude oil-polluted soils in Bodo West Community, Ogoni, with other sites of varying pollution history. Ann Microbiol 69(5): 495–513. https://doi.org/10.1007/s13213-019-1438-3

Christaki E, Marcou M, Tofarides A (2020) Antimicrobial resistance in bacteria: Mechanisms, evolution, and persistence. J Mol Evol 88(1): 26–40. https://doi.org/10.1007/s00239-019-09914-3

Dangar KG, Raiyani NM, Pandya RD, Singh SP. (2017) Uncultivated lineages and host–microbe interaction in saline environment. *Understanding Host-Microbiome Interactions-An Omics Approach: Omics of Host-Microbiome Association.* 2017:13–28.

de Abreu VAC, Perdigão J, Almeida S (2021) Metagenomic approaches to analyze antimicrobial resistance: An overview. Front Genet 11: 575592. https://doi.org/10.3389/fgene.2020.575592

De Mandal S, Panda AK, Bisht SS, Kumar NS. (2015 Jun 1) First report of bacterial community from a bat guano using Illumina next-generation sequencing. *Genomics Data.* 4:99–101. https://doi.org/10.1016/j.gdata.2015.04.001

De Nies L, Lopes S, Busi SB et al. (2021) PathoFact: A pipeline for the prediction of virulence factors and antimicrobial resistance genes in metagenomic data. Microbiome 9: 49. https://doi.org/10.1186/s40168-020-00993-9

Dhanjal DS, Chopra C, Chopra RS (2020) Metagenomic DNA Sequencing: Technological Advances and Applications. In: Chopra, R.S., Chopra, C., Sharma, N.R. (eds) *Metagenomics: Techniques, Applications, Challenges and Opportunities.* Springer, Singapore. https://doi.org/10.1007/978-981-15-6529-8_3

Diaz-Torres ML, McNab R, Spratt DA, Villedieu A, et al., (2003) Novel tetracycline resistance determinant from the oral metagenome. *Antimicrobial agents and chemotherapy.* Apr; 47(4):1430–1432. https://doi.org/10.1128/aac.47.4.1430-1432.2003

Diaz-Torres ML, Villedieu A, Hunt N, et al. (2006) Determining the antibiotic resistance potential of the indigenous oral microbiota of humans using a metagenomic approach. FEMS Microbiol Lett 258(2): 257–262. https://doi.org/10.1111/j.1574-6968.2006.00221.x

Diehl DL, Lapara TM (2010) Effect of temperature on the fate of genes encoding tetracycline resistance and the integrase of Class 1 integrons within anaerobic and aerobic digesters treating municipal wastewater solids. Environ Sci Tech 44(23): 9128–9133. https://doi.org/10.1021/es102765a

Dinghua Li, Chi-Man L, Ruibang L, et al. (2015) MEGAHIT: An ultra-fast single-node solution for large and complex metagenomics assembly via succinct de Bruijn graph. Bioinformatics 31(10): 1674–1676. https://doi.org/10.1093/bioinformatics/btv033

Dionysios A Antonopoulos, Rida Assaf, et al., (2019) PATRIC as a unique resource for studying antimicrobial resistance, *Briefings in Bioinformatics,* 20(4), 1094-1102..https://doi.org/10.1093/bib/bbx083

Donato JJ, Moe LA, Converse BJ, et al. (2010) Metagenomic analysis of apple orchard soil reveals antibiotic resistance genes encoding predicted bifunctional proteins. App Environ Microbiol 76(13): 4396–4401. https://doi.org/10.1128/AEM.01763-09

Dragana Drinković, Arthur J. Morris, Sudha Pottumarthy, Donald MacCulloch, Teena West. (November 2003) Bacteriological outcome of combination versus single-agent treatment for staphylococcal endocarditis, *Journal of Antimicrobial Chemotherapy,* 52(5), 820–825, https://doi.org/10.1093/jac/dkg440

Forsberg KJ, Patel S, Gibson MK, et al. (2014) Bacterial phylogeny structures soil resistomes across habitats. Nature 509(7502): 612–616. https://doi.org/10.1038/nature13377

Garcias-Bonet N, Arrieta JM, de Santana CN, Duarte CM, Marbà N (2012) Endophytic bacterial community of a Mediterranean marine angiosperm (Posidonia oceanica) *Front. Microbiol.* 3:342. https://doi.org/10.3389/fmicb.2012.00342

Handelsman J, Rondon MR, Brady SF et al (1998) Molecular biological access to the chemistry of unknown soil microbes: a new frontier for natural products. *Chem Biol* 5(10):R245–R249 http://biomednet.com/elecref/10745521005R0245

Hawley ER, Hess M (2014) Metagenome sequencing of the prokaryotic microbiota of the hypersaline and meromictic Soap Lake, Washington. Genome Ann 2(1): e01212–e01213. https://doi.org/10.1128/genomeA.01212-13

Hayden MK, Rezai K, Hayes RA, et al. (2005) Development of daptomycin resistance in vivo in methicillin-resistant Staphylococcus aureus. J Clin Microbio 43(10): 5285–5287. https://doi.org/10.1128/JCM.43.10.5285-5287.2005

Hu Q, Zhang X-X, Jia S, et al. (2016) Metagenomic insights into ultraviolet disinfection effects on antibiotic resistome in biologically treated wastewater. Water Res 101: 309–317. https://doi.org/10.1016/j.watres.2016.05.092

Inouye M, Dashnow H, Raven LA, et al. (2014) SRST2: Rapid genomic surveillance for public health and hospital microbiology labs. Genome Med 6: 90. https://doi.org/10.1186/s13073-014-0090-6

Jang HM, Shin J, Choi S, et al. (2017) Fate of antibiotic resistance genes in mesophilic and thermophilic anaerobicdigestion of chemically enhanced primary treatment (CEPT) sludge. Biores Tech 244 (Pt 1): 433–444. https://doi.org/10.1016/j.biortech.2017.07.153

Jorth P, McLean K, Ratjen A, et al. (2017) Evolved aztreonam resistance is multifactorial and can produce hypervirulence in Pseudomonas aeruginosa. mBio 8(5): e00517–17. https://doi.org/10.1128/mBio.00517-17

Kai Blin, Thomas Wolf, et al. (2017) antiSMASH 4.0—improvements in chemistry prediction and gene cluster boundary identification, *Nucleic Acids Research,* 45(W1):W36–W41. https://doi.org/10.1093/nar/gkx319

Kazimierczak KA, Scott KP, Kelly D, Aminov RI (2009) Tetracycline resistome of the organic pig gut. Appl Environ Microbiol 75: 1717–1722. https://doi.org/10.1128/AEM.02206-08

Konstantin SY, Boris AK (2017) ResistoMap—online visualization of human gut microbiota antibiotic resistome. *Bioinformatics,* 33(14):2205–2206. https://doi.org/10.1093/bioinformatics/btx134

Kristiansson E, Fick J, Janzon A, et al. (2011) Pyrosequencing of antibiotic-contaminated river sediments reveals high levels of resistance and gene transfer elements. PLoS One 6(2): e17038. https://doi.org/10.1371/journal.pone.0017038

Lang KS, Anderson JM, Schwarz S, et al. (2010) Novel florfenicol and chloramphenicol resistance gene discovered in Alaskan soil by using functional metagenomics. Appl Environ Microbiol 76 (15): 5321–5326. https://doi.org/10.1128/AEM.00323-10

Langmead B, Salzberg S (2012) Fast gapped-read alignment with Bowtie 2. Nat Methods 9: 357–359. https://doi.org/10.1038/nmeth.1923

Marathe NP, Janzon A, Kotsakis SD, et al. (2018) Functional metagenomics reveals a novel carbapenem-hydrolyzing mobile beta-lactamase from Indian river sediments contaminated with antibiotic production waste. Environ Int 112: 279–286. https://doi.org/10.1016/j.envint.2017.12.036

Marshall SH, Donskey CJ, Hutton-Thomas R (2002) Gene dosage and linezolid resistance in Enterococcus faecium and Enterococcus faecalis. Antimicrob Agents Chemother 46(10): 3334–3336. https://doi.org/10.1128/AAC.46.10.3334-3336.2002

McArthur AG, Waglechner N, Nizam F, et al. (2013) The comprehensive antibiotic resistance database. Antimicrob Agents Chemother 57(7): 3348–3357. https://doi.org/10.1128/AAC.00419-13

McCarthy M (2015) White House proposes doubling spending on combating antibiotic resistance to $1.6bn in 2016. BMJ 350: h534. https://doi.org/10.1136/bmj.h534

McGarvey KM, Queitsch K, Fields S (2012) Wide variation in antibiotic resistance proteins identified by functional metagenomic screening of a soil DNA library. Appl Environ Microbiol 78(6): 1708–1714. https://doi.org/10.1128/AEM.06759-11

Moon K, Jeon JH, Kang I, et al. (2020) Freshwater viral metagenome reveals novel and functional phage-borne antibiotic resistance genes. Microbiome 8: 75. https://doi.org/10.1186/s40168-020-00863-4

Mori T, Mizuta S, Suenaga H, Miyazaki K (2008) Metagenomic screening for bleomycin resistance genes. Appl Environ Microbiol 74(21): 6803–6805. https://doi.org/10.1128/AEM.00873-08

Nurk S, Meleshko D, Korobeynikov A, Pevzner PA (2017) metaSPAdes: A new versatile metagenomic assembler. Genome Res (5): 824–834. https://doi.org/10.1101/gr.213959.116

Pal C, Bengtsson-Palme J, Rensing C, et al. (2014) BacMet: Antibacterial biocide and metal resistance genes database. Nucleic Acids Res 42: D737–D743. https://doi.org/10.1093/nar/gkt1252

Prayogo FA, Budiharjo A, Kusumaningrum HP, Wijanarka W, Suprihadi A, Nurhayati N (2020). Metagenomic applications in exploration and development of novel enzymes from nature: a review. *J Genet Eng Biotechnol.* Dec;18(1):1-0. https://doi.org/10.1186/s43141-020-00043-9

Pruden A, Pei R, Storteboom H, Carlson KH (2006). Antibiotic resistance genes as emerging antibiotic resistance genes as emerging contaminants: Studies in northern Colorado. Environ Sci Tech 40(23): 7445–7450. https://doi.org/10.1021/es060413 l

Raiyani NM, Singh SP (2020). Taxonomic and functional profiling of the microbial communities of Arabian Sea: A metagenomics approach, Genomics, 112(6): 4361–4369. https://doi.org/10.1016/j.ygeno.2020.07.024

Raza S, Jo H, Kim J et al. (2021) Metagenomic exploration of antibiotic resistome in treated wastewater effluents and their receiving water. Sci Total Environ 765: 142755. https://doi.org/10.1016/j.scitotenv.2020.142755

Reddy B, Dubey SK (2019) River Ganges water as reservoir of microbes with antibiotic and metal ion resistance genes: High throughput metagenomic approach. Environ Pollut 246: 443–451. https://doi.org/10.1016/j.envpol.2018.12.022

Riesenfeld CS, Goodman RM, Handelsman J (2004) Uncultured soil bacteria are a reservoir of new antibiotic resistance genes. Environ Microbiol 6(9): 981–989. https://doi.org/10.1111/j.1462-2920.2004.00664.x

Rizzo L, Manaia C, Merlin C, et al. (2013) Urban wastewater treatment plants as hotspots for antibiotic resistant bacteria and genes spread into the environment: A review. Sci Total Environ 447: 345–360. https://doi.org/10.1016/j.scitotenv.2013.01.032

Rodrigo EM, Lalitagauri D, et al. (2010) First Report of Staphylococcal Clinical Isolates in Mexico with Linezolid Resistance Caused by cfr: Evidence of In Vivo cfr Mobilization. *Journal of Clinical Microbiology* 48(8). https://doi.org/10.1128/jcm.00880-10

Rowe W, Baker KS, Verner-Jeffreys D, et al. (2015) Search engine for antimicrobial resistance: A cloud compatible pipeline and web interface for rapidly detecting antimicrobial resistance genes directly from sequence data. PLoS One 10(7): e0133492. https://doi.org/10.1371/journal.pone.0133492

Sabol K, Patterson JE, et al. (2005) Emergence of Daptomycin Resistance in Enterococcus faecium during Daptomycin Therapy. *Antimicrobial Agents and Chemotherapy* 49(4).

Sandegren L, Lindqvist A, Kahlmeter G, Andersson DI (2008) Nitrofurantoin resistance mechanism and fitness cost in Escherichia coli. J Antimicrob Chemother 62(3): 495–503. https://doi.org/10.1093/jac/dkn222

Schwarz S, Zhang W, et al. (2021) Mobile Oxazolidinone Resistance Genes in Gram-Positive and Gram-Negative Bacteria ASM Journals, *Clinical Microbiology Reviews* 34(3). https://doi.org/10.1128/cmr.00188-20

Shi P, Jia S, Zhang XX, et al. (2013) Metagenomic insights into chlorination effects on microbial antibiotic resistance in drinking water. Water Res 47(1): 111–120. https://doi.org/10.1016/j.watres.2012.09.046

Sengupta S, Ganguli S, Singh PK. (2017 Jun 1) Metagenome analysis of the root endophytic microbial community of Indian rice (O. sativa L.). *Genomics data.* 12:41–43. https://doi.org/10.1016/j.gdata.2017.02.010

Singer GA, Shekarriz S, McCarthy A, Fahner N, Hajibabaei M. (2020 Mar) The utility of a metagenomics approach for marine biomonitoring. *BioRxiv*. 18:2020-03. https://doi.org/10.1101/2020.03.16.993667

Suenaga H, Ohnuki T, Miyazaki K (2007 Sep) Functional screening of a metagenomic library for genes involved in microbial degradation of aromatic compounds. *Environmental microbiology*. 9(9):2289-97. https://doi.org/10.1111/j.1462-2920.2007.01342.x

Tenover FC, Biddle JW, Lancaster MV (2001) Increasing resistance to vancomycin and other glycopeptides in Staphylococcus aureus. Emerg Infect Dis 7(2): 327–332. https://doi.org/10.3201/eid0702.010237

Tian Z, Zhang Y, Yu B, Yang M (2016) Changes of resistome, mobilome and potential hosts of antibiotic resistance genes during the transformation of anaerobic digestion from mesophilic to thermophilic. Water Res 98: 261–269. https://doi.org/10.1016/j.watres.2016.04.031

Torres-Cortés G, Millán V, Ramírez-Saad HC, et al. (2011) Characterization of novel antibiotic resistance genes identified by functional metagenomics on soil samples. Environ Microbiol 13(4): 1101–1114. https://doi.org/10.1111/j.1462-2920.2010.02422.x

Wang S, Gao X, Gao Y, et al. (2017) Tetracycline resistance genes identified from distinct soil environments in China by functional metagenomics. Front Microbiol 8:1406. https://doi.org/10.3389/fmicb.2017.01406

Wang Z, Wang Z, Lu YY, et al. (2019) SolidBin: Improving metagenome binning with semi-supervised normalized cut. Bioinformatics 35(21): 4229–4238. https://doi.org/10.1093/bioinformatics/btz253

Wigand J, Tansirichaiya S, Winje E, Al-Haroni M (2021) Functional screening of a human saliva metagenomic DNA reveal novel resistance genes against sodium hypochlorite and chlorhexidine. BMC Oral Health (1): 1–3. https://doi.org/10.1186/s12903-021-02000-5

Will P M Rowe, Martyn D Winn (2018) Indexed variation graphs for efficient and accurate resistome profiling, *Bioinformatics*, 34(21):3601–3608. https://doi.org/10.1093/bioinformatics/bty387

World Health Organization (2014) report. https://www.who.int/southeastasia/news/detail/30-04-2014-who-s-first-global-report-on-antibiotic-resistance-reveals-serious-worldwide-threat-to-public-health

Wu, YW, Simmons, BA, Singer, SW (2016) MaxBin 2.0: An automated binning algorithm to recover genomes from multiple metagenomic datasets. Bioinformatics 32(4): 605–607. https://doi.org/10.1093/bioinformatics/btv638

Xu R, Yang Z, Wang Q, et al. (2018) Rapid startup of thermophilic anaerobic digester to remove tetracycline and sulfonamides resistance genes from sewage sludge. Sci Total Environ 612(112): 788–798. https://doi.org/10.1016/j.scitotenv.2017.08.295

Xu R, Zhang Y, Xiong W, et al. (2020) Metagenomic approach reveals the fate of antibiotic resistance genes in a temperature-raising anaerobic digester treating municipal sewage sludge. J Cleaner Prod 20(277): 123504. https://doi.org/10.1016/j.jclepro.2020.123504

Xu X, Lin D, Yan G (2010) vanM, a new glycopeptide resistance gene cluster found in Enterococcus faecium. Antimicrob Agents Chemother 54(11): 4643–4647. https://doi.org/10.1128/AAC.01710-09

Yadav S, Kapley A (2021) Antibiotic resistance: Global health crisis and metagenomics. Biotechnology Reports 29: e00604. https://doi.org/10.1016/j.btre.2021.e00604

Yin X, Jiang X-T, Chai B, et al. (2018) ARGs-OAP v2.0 with an expanded SARG database and Hidden Markov Models for enhancement characterization and quantification of antibiotic resistance genes in environmental metagenomes. Bioinformatics 34(13): 2263–2270. https://doi.org/10.1093/bioinformatics/bty053

Zankari E, Allesøe R, Joensen KG, et al. (2017) PointFinder: A novel web tool for WGS-based detection of antimicrobial resistance associated with chromosomal point mutations in bacterial pathogens. J Antimicrob Chemo 72(10): 2764–2768. https://doi.org/10.1093/jac/dkx217

Zaw MT, Emran, NA, Lin, Z (2018) Mutations inside rifampicin-resistance determining region of rpoB gene associated with rifampicin-resistance in Mycobacterium tuberculosis. J Infect Public Health 11(5): 605–610. https://doi.org/10.1016/j.jiph.2018.04.005

Zheng B, Liu W, Xu H, Li J, Jiang X (2021) Occurrence and distribution of antimicrobial resistance genes in the soil of an industrial park in China: A metagenomics survey. Environ Pollut 15(273): 116467. https://doi.org/10.1016/j.envpol.2021.116467

Index

Note: Page numbers in *italics* indicate a figure on the corresponding page.